Bayesian Models of Perception and Action

Bayesian Models of Perception and Action

An Introduction

Wei Ji Ma, Konrad Paul Kording, and Daniel Goldreich

The MIT Press
Cambridge, Massachusetts
London, England

The MIT Press would like to thank the anonymous peer reviewers who provided comments on drafts of this book. The generous work of academic experts is essential for establishing the authority and quality of our publications. We acknowledge with gratitude the contributions of these otherwise uncredited readers.

This book was set in Times New Roman by Westchester Publishing Services. Printed and bound in the United States of America.

Library of Congress Cataloging-in-Publication Data

Names: Ma, Wei Ji, author. | Kording, Konrad, author. | Goldreich, Daniel, author.
Title: Bayesian models of perception and action : an introduction / Wei Ji Ma, Konrad Paul Kording, Daniel Goldreich.
Description: [Cambridge, MA] : MIT Press, [2023] | Includes bibliographical references and index.
Identifiers: LCCN 2022033539 (print) | LCCN 2022033540 (ebook) | ISBN 9780262047593 (hardcover) | ISBN 9780262372824 (epub) | ISBN 9780262372831 (pdf)
Subjects: LCSH: Bayesian statistical decision theory.
Classification: LCC BF39.2.B39 M38 2023 (print) | LCC BF39.2.B39 (ebook) | DDC 150.1/519542—dc23/eng/20230209
LC record available at https://lccn.loc.gov/2022033539
LC ebook record available at https://lccn.loc.gov/2022033540

1 0 9 8 7 6 5 4 3 2 1

We dedicate this book to the memory of David Knill (1961–2014). All three of us have learned a good part of what we know about Bayesian modeling of perception and action from him. As a caring and patient mentor and as an excellent teacher, he also made studying this topic a lot more enjoyable for all of us. The field of Bayesian modeling of perception and action would not be where it is without him and this book would probably never have been written.

Contents

Acknowledgments

This book has been a long time in the making, and we are indebted to many people. We first came up with the idea in June 2009, when—together with Alan Stocker and Jonathan Pillow—we taught a computational neuroscience course at the Instituto Gulbenkian de Ciência in Oeiras, Portugal. At the time, in an impressive display of unbridled optimism, Konrad predicted that we would be done by December 2009. A short fourteen years later, we have the book in hand. The delay has come with benefits, though: over the years, we have used chapter drafts and the book's ideas to teach Bayesian modeling to hundreds of undergraduate students, graduate students, and postdocs in our courses at McMaster University, Baylor College of Medicine, Northwestern University, New York University, the University of Pennsylvania, and in tutorials at conferences and summer schools. Many of these students, as well as our teaching assistants—notably Ronald van den Berg, Anna Kutschireiter, Lucy Lai, Jennifer Laura Lee, Julie Lee, Jorge Menendez, Sashank Pisupati, Anne-Lene Sax, Shan Shen, Bei Xiao, and Hörmet Yiltiz—and lab members (too many to list) contributed numerous corrections, comments, problem solutions, and problem suggestions. We thank Nuwan de Silva for testing all problems in an earlier version of the book. We thank readers of various drafts, in particular Luigi Acerbi, Robert Jacobs, Michael Landy, Zili Liu, and Javier Traver, for providing deep and useful feedback on the content and exposition; we also thank Robert and Zili for being two of our most steadfast supporters over the years. We are deeply grateful to Brennan Klein, a postdoc at Northeastern University, for professionalizing and redesigning our figures throughout the book, as well as for many fun drawings and for making us more dutiful as authors. This book would never have been finished without the help of 1,3,7-trimethylxanthine and we are thankful for its existence. We are grateful to Robert Prior from MIT Press, by name destined to be our editor, who repeatedly set us firm deadlines, which he patiently consented to extend when we invariably missed them, and who made a free online version possible. Finally, we would like to thank our families, who have been unreasonably supportive all these years.

The Four Steps of Bayesian Modeling

Example from chapters 3–4

Step 1: Generative model

(a) **Draw a diagram** where each node is a variable and each arrow a dependency. Observations/measurements are at the bottom.

(b) For each variable, **write an equation for its probability distribution.** For each observation, assume a noise model. For others, get the distribution from the experimental design or from natural statistics. If there are incoming arrows, the distribution is a conditional one.

Stimulus

s $\quad p(s) = \mathcal{N}(s; \mu, \sigma_s^2)$

x $\quad p(x|s) = \mathcal{N}(x; s, \sigma^2)$

Measurement

Step 2: Bayesian inference (decision rule)

(a) **Compute the posterior over the world state of interest given observaions.** The optimal observer does this using the distributions in the generative model. Alternatively, the observer might assume different distributions (e.g., wrong beliefs). Marginalize (average) over variables other than the observations and the world state of interest.

(b) **Specify the readout of the posterior.** Assume a utility function, then maximize expected utility under the posterior. (Alternative: sample from the posterior.) Result: decision rule (mapping from observations to decision). When utility is accuracy, the readout is to maximize the posterior (maximum in posteriori decision rule).

$$\mathcal{L}(s;x) = p(x|s)$$

$$p(s|x) \propto \mathcal{L}(s;x)p(s)$$

$$p(s|x) = \mathcal{N}\left(s; \frac{J_s\mu + Jx}{J_s + J}, \frac{1}{J_s + J}\right)$$

$$\hat{s} = \frac{J_s\mu + Jx}{J_s + J}$$

$$J_s \equiv \frac{1}{\sigma_s^2} \text{ and } J \equiv \frac{1}{\sigma^2}$$

Step 3: Response probabilities

For every unique trial in the experiment, **compute the probability of the observer making each possible decision given the stimuli on that trial.** To do so, use the distribution of the observations given the stimuli (from step 1) and the decision rule (from step 2).

- Good method: sample observations according to step 1; for each, apply decision rule; tabulate responses.
- Better: integrate numerically over observations.
- Best (when possible): integrate analytically over observations.

Optional: add response noise or lapses.

$$p(\hat{s}|s) = \int p(\hat{s}|x)p(x|s)dx$$

$$= \mathcal{N}\left(\hat{s}; \frac{J_s\mu + Js}{J_s + J}, \frac{J}{(J_s + J)^2}\right)$$

Step 4: Model fitting and model comparison

(a) **Compute the parameter log likelihood**, the log probability of the subject's responses across all trials for a hypothesized parameter combination.

(b) **Maximize the parameter log likelihood.** Result: parameter estimates and maximum log likelihood. Test for parameter recovery and summary statistics recovery using synthetic data. Use more than one algorithm.

(c) **Obtain fits to summary statistics** by rerunning the fitted model.

(d) **Formulate alternative models** (e.g., vary step 2). **Compare maximum Log likelihood across models.** Correct for number of parameters (e.g., using Akaike Information Criterion). Test for model recovery using synthetic data.

(e) **Check model comparison results** using summary statistics.

(f) Optional: **Evaluate absolute goodness of fit.**

$$\log \mathcal{L}(\sigma; \text{data}) = \sum_{i=1}^{n_{\text{trials}}} \log p(\hat{s}_i | s_i, \sigma)$$

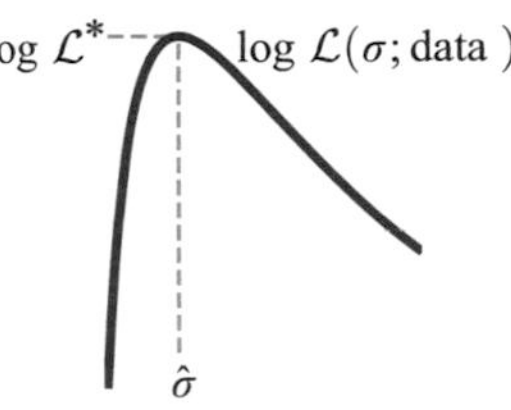

List of Acronyms

AIC	Akaike Information Criterion
ANN	artificial neural network
BIC	Bayesian Information Criterion
CCSD	class-conditioned stimulus distribution
CDF	cumulative distribution function
HMM	hidden Markov model
LLR	log likelihood ratio
LPR	log posterior ratio
LR	likelihood ratio
MAP	maximum a posteriori [estimate]
MLE	maximum-likelihood estimate
MSE	mean squared error
PDF	probability density function
PME	posterior mean estimate
PPV	positive predictive value
PSE	point of subjective equality
RMSE	root-mean-square error
ROC	receiver operating characteristic

Introduction

This book provides an introduction to Bayesian models of perceptual decision making and action. In such models, the human mind behaves like a competent data scientist (or diagnosing physician, or crime scene investigator—pick your metaphor) when extracting meaning from noisy and ambiguous sensory observations. In recent decades, the Bayesian approach to perception and action has become increasingly popular and the resulting models have been widely tested. The approach is exciting to us because it is principled and has successfully accounted for many experimental findings.

We decided to write this book because there was no text that taught from scratch how to build Bayesian models and how to think about them. This is not to suggest that excellent materials on Bayesian modeling are unavailable. However, primary research papers tend to be narrow and to leave out necessary background, while review papers are generally too qualitative and too focused on recent results to be practical to an aspiring modeler. In our own research, we have extensively perused edited volumes, such as [45, 94, 181], but those are not meant to be didactically coherent. Finally, [171] is an excellent introduction to Bayes' Rule but it does not focus on the construction of models for use in research practice.

This book does not require any previous knowledge of probability theory, but it does necessarily involve mathematics. Readers with a basic understanding of calculus will find the book accessible; others will still be able to understand the majority of the content. If this is your first foray into mathematical modeling: congratulations! Using the language of mathematics avoids ambiguity, and mathematical models yield quantitative predictions. We understand very well that math is anxiety-inducing for some, but we believe that with sufficient effort, the material in this book is within reach for any university student in the sciences.

We recommend that readers take the time to work through the exercises within the chapters, which are usually conceptual or involve short calculations, and the problems at the end of each chapter, which are a mix of concepts, math, and simulations. To build understanding, however, we believe that there is no substitute for a certain amount of struggling with mathematics and the concepts linking to reality.

Perception as Inference

The fundamental challenge of perception is that the state of the world is for the most part not directly accessible. The eyes, ears, skin, and other sensory organs register physical signals and convert them into electrical impulses that travel toward the brain, a sort of neural Morse code. The eyes register patterns of light but do not identify the visual scene. The skin senses pressure and vibration but does not itself identify the external object causing these stimuli. The ears detect sound waves but do not indicate their meaning. There is thus a fundamental problem of how to interpret sensory observations.

The brain's job is to figure out the underlying causes of sensory observation in order to guide behavior. This job is hard because the sensory observation typically allows for multiple possible interpretations. The sensory observation might be of low quality; for example, objects might be poorly lit, far away, or moving fast. But even high-quality sensory information does not guarantee a single possible interpretation. A two-dimensional retinal image is compatible with infinitely many three-dimensional objects. A well-camouflaged animal might be indistinguishable from the background. The words "red" and "read" usually produce the same auditory output. Tension in the muscles does not unambiguously signal the configuration of the body. Multiple interpretations of the same sensory observations are possible.

Drawing conclusions about the state of the world based on low-quality or incomplete observations is a form of *inference*. Specifically, *probabilistic inference* or *Bayesian inference* is inference performed by assigning probabilities to possible states of the world. The theory expounded in this book is that perceptual decision making is a form of Bayesian inference. Through Bayesian inference, the brain is able to assign probabilities to different world states, a crucial step in deciding which actions to take.

In Bayesian inference, probabilities are assigned based on a statistical description of how observations are generated—a *generative model*. This description includes prior beliefs as well as beliefs about statistical dependencies between variables. While the nature of the sensory information and the structure of the generative model are task-dependent, the Bayesian computation itself always follows the same rules of probability calculus. The Bayesian approach thus unifies an enormous range of otherwise apparently disparate behavior within one coherent framework.

The generative model embodies the assumptions made by the observer and can therefore take many forms. Of special importance is the case in which the generative model is a *correct* statistical description of the observations. In that case (and not necessarily otherwise), Bayesian inference is *optimal*, in the sense that the resulting probabilities assigned to world states are a component of the performance-maximizing strategy—whether performance is measured in terms of accuracy or otherwise. Stated otherwise, among all strategies an observer could have for a perceptual task, Bayesian inference based on the correct generative model is the best possible one.

Optimal does not mean error-free. It means drawing the best conclusion possible given the information available to the observer. The performance of the optimal strategy will depend on the informativeness of the sensory observation. Perceptual illusions often involve erroneous percepts, yet illusions can often be explained in terms of Bayesian inference, as we will see in several examples.

Positioning Bayesian Models within the Space of Models

Scientists and philosophers have long sought ways of characterizing the logical structure of models. One particularly useful scheme is to categorize models as *descriptive models*, *process models*, or *normative models*.

A *descriptive model* (sometimes called a *What model*) is a mathematical description of behavioral variables (such as accuracy or reaction time) in terms of observed variables (such as intensity of a stimulus or a quantified personality trait). These descriptions may take the form of a regression, a generalized linear model, or a machine-learning model. The kind of statement supported by fitting descriptive models to data is simply that a model exists that can fit data with a certain amount of error.

A *process model* (sometimes called a *How model*) is more ambitious: it tries to dissect the mapping from observation to output into generalizable and psychologically or physiologically meaningful component processes, specifying how the observer/agent makes a decision based on the information available. Examples of such component processes would be "Gaussian measurement noise is added to a stimulus variable" or "the observer maps the decision variable to a decision by applying a criterion." The goal of a process model is not just to fit the data but to commit to internal representations and to a particular flow of information. The kinds of statements supported by fitting process models to data are that if the process is as hypothesized then the measured behavior would result.

A *normative model* (sometimes called a *Why model*) is ambitious in a different way. In a normative model, we ask why behavior is the way it is. More specifically, we ask why, in a particular ecological niche, certain behaviors are beneficial. For example, a normative model may (explicitly or implicitly) assume that maximizing accuracy is important in our lives. Such a model might then make assumptions about the world that are viewed as immutable (e.g., that there is a fixed amount of noise in vision). The model could then derive the optimal solution to the problem, which can be compared with actual behavior. The kinds of statements supported by fitting normative models to data are that if the world or ecological niche is as hypothesized, then the measured behavior would be beneficial.

Bayesian models are typically used as normative models: We specify the objective an agent wants to optimize, the assumptions the agent has about its world, and the types of observations the agent has available. The Bayesian formalism then allows us to discover the behavior that optimizes the agent's objective. For a Bayesian model with defined observations, prior knowledge, and goals, there is one well-defined optimal action, and this book explains how to calculate it. In this sense, a Bayesian model is a prototypical example of a normative model.

Bayesian models are also often used as process models. In one way, Bayesian models are obviously process models, namely, they divide the mapping from stimulus to response into two stages by taking the point of view of the observer. In the "encoding" or "representation" stage, they specify the nature of the data that the observer works with, which can differ from the stimulus—we do not directly sense world states but rather obtain sensory observations from the world. The second stage is a decision stage, in which the data available to the observer are converted into actions. This two-stage sequence constitutes a process model. In another way, however, the process nature of Bayesian models is less clear. Namely, the

Bayesian computation in the decision stage is performed by calculating likelihood functions and a posterior distribution and producing an action based on the latter. Opinions can differ as to whether these steps and their associated constructs are psychologically meaningful or simply a way to derive a normative decision rule. Depending on one's stance, the decision stage of a Bayesian model may or may not be considered a process model. We discuss this further in chapter 15.

Bayesian models can also be positioned on the behavioral-neural axis. As optimality considerations pertain to the functioning of the whole organism in the world, Bayesian models of the brain are primarily behavioral models. If a Bayesian model successfully describes behavior in a particular task, we may then hope to constrain our understanding of the underlying neural processes and to develop neural or implementation-level models. We touch on this in chapter 14. This approach contrasts with a bottom-up one, in which one might start by modeling circuits of neurons in biological detail, and then attempt to build up the models by combining multiple circuits.

A final relevant axis is the perception-cognition axis. While this book focuses on perception and action, Bayesian models are widely useful in other realms of cognitive science and psychology. In particular, there is a rich history on Bayesian models in higher-level cognition [34], dating back at least to the work of Jonathan Evans [48] and John Anderson [11], with great contributions to the understanding of cognitive development (e.g. [68, 210]). Higher-level cognition makes occasional appearances in this book, especially in chapters 6 ("Learning as Inference"), 12 ("Inference in a Changing World"), and 13 ("Combining Inference with Utility"). In chapter 15 ("Bayesian Models in Context"), we comment on differences between perception and cognition.

What This Book Is Not

Bayesian models of perception and action model observers who infer the state of the world from sensory observations. This contrasts with Bayesian statistical data analysis, in which an experimenter infers the value of a model parameter from collected data. The mathematical formalism is the same, but in this book, we focus on how the brain perceives or decides, not on how data are analyzed. That being said, Bayesian models, like all models, have parameters whose values need to be inferred. For this reason, we include Appendix C on model fitting and model comparison; this appendix, however, still does not cover Bayesian methods for data analysis. For those, we recommend [59, 103, 164].

A Disclaimer on Citations

In this book, we cite the works of many excellent scientists. Without detracting from their achievements, we wish to recognize that academic science has for long—whether deliberately or inadvertently—erected barriers to the participation of individuals from many groups, including genders other than men, people of color, people with disabilities, immigrants, and people of lower socioeconomic status. Consequently, members of these groups have had, and continue to have, fewer chances to come up with the same ideas or findings, and even when they do they often face greater challenges in receiving recognition for their work. We hope that the reader will keep this in mind when reflecting on citations and will contribute to eliminating barriers to participation.

In Conclusion

We hope you enjoy the book, and we welcome your feedback on the book's website, www.bayesianmodeling.com. Extra material, including solutions to problems, interactive demonstrations, and further reading, will in future be available there.

1

Uncertainty and Inference

How do we transform our sensory observations into beliefs about the state of the world?

Whenever we perceive something, make a prediction, or deliberate over a decision, we are reasoning with probabilities, even if we do not realize it. We are using the information we have at hand to infer or estimate something else that interests us. The information we have is usually incomplete or noisy, so our inference is not certain. For instance, if we observe a shiny floor (available information), this *suggests* that it may be wet (the focus of our interest). Using the available sensory information and any relevant knowledge we may have, we must determine the probability of each interpretation (wet or dry). How can we make sound judgments in such situations?

Plan of the Chapter

We outline the perceptual inference process, emphasizing the uncertainty that is inherent in perception. Using simple examples, we introduce the probabilities involved in perceptual inference, the likelihood, the prior, and the posterior, focusing on the underlying intuitions. We then illustrate the ubiquity of perceptual inference in daily life with a series of examples involving visual and auditory perception. We do not use mathematics in this chapter but instead explore each example qualitatively and graphically. Our goal is to provide an intuitive understanding of the perceptual inference process, which will serve as a foundation for the more rigorous mathematical treatments in the following chapters.

1.1 The Goal of Perception

Nothing can determine save the reason of the mind.
Eyes can't grasp the true nature of things. So do not claim
The fault's with them, when really it's the mind that is to blame.
—Lucretius, *De rerum natura* [111]

Humans and other animals are endowed with a collection of exquisite sensory organs through which they detect properties of the environment. Sensory organs respond to physical properties as diverse as light (eyes), sound (ears), temperature (skin), material texture (skin), chemical composition (nose, tongue), and body position (joint and muscle receptors, vestibular organs). Our sensory organs form an integral part of ourselves, so much so that we usually take their presence for granted. To appreciate the role that our senses play, try to imagine life without vision, hearing, touch, smell, or taste.

As intricate as the sensory organs are, their activation by physical stimuli is only the first step in perception. We do not primarily care about the pattern of light wavelengths (colors) and intensities (brightness) entering our eyes, or about the pattern of acoustic energy, varying in amplitude and time, entering our ears. Rather than these sensory activation patterns per se, which we call the sensory *inputs* or *observations*, we care about their interpretation. In fact, our quality of life—and often our life itself—depends on our ability to come up with correct interpretations. Does that pattern of light reflect the face of a friend? Is that acoustic waveform the sound of the wind, the howl of a dog, or the voice of our companion? In short, our interest lies not in sensory input per se, but in the information the input provides about the relevant states of the world.[1]

To make the interpretative transition from sensation (the activation of the sensory organs) to perception (a conclusion regarding the state of the world) is a challenging task. Broadly speaking, this book is about how the brain can optimally accomplish this task. A large and rapidly growing body of experimental and theoretical work shows that perception is, at least implicitly, a probabilistic inference process, in which the organism attempts to infer the most probable state of the world, using sensory inputs and all relevant knowledge at its disposal. As Lucretius's quote illustrates, perception sometimes goes wrong, but the mistakes can often be understood as byproducts of a sensible inference strategy.

1.2 Hypotheses and Their Probabilities

The transition from sensation to perception requires *conditional probabilities*. A conditional probability is a probability of one event given another: for example, the probability that you are in a good mood given that it is raining outside. We denote conditional probabilities as $p(B \mid A)$, read "the probability of B given A." Whether people are aware of it or not, we make conditional probability judgments very frequently in daily life (figure 1.1).

Importantly, conditional probabilities are not symmetric. In general $p(A \mid B) \neq p(B \mid A)$. For instance, most professional basketball players are tall, but most tall people are not professional basketball players. If A is "being a basketball pro" and B is "being tall," then this case example is an example of $p(B \mid A) > p(A \mid B)$.

Exercise 1.1 In each case below, which of the two conditional probabilities is larger and why?

- p(rain | clouds) or p(clouds | rain)
- p(speaks French | born and raised in Paris) or
  p(born and raised in Paris | speaks French)
- p(unmarried | college student) or p(college student | unmarried)
- p(you understand Bayes' rule | you read this book) or
  p(you read this book | you understand Bayes' rule)

One way to visualize probabilities is to use areas of rectangles (figure 1.2). The area of rectangle A is proportional to the probability of event A, denoted $p(A)$ and the area of

1. Some perception researchers refer to the world state as the *distal stimulus* and the observation as the *proximal stimulus*.

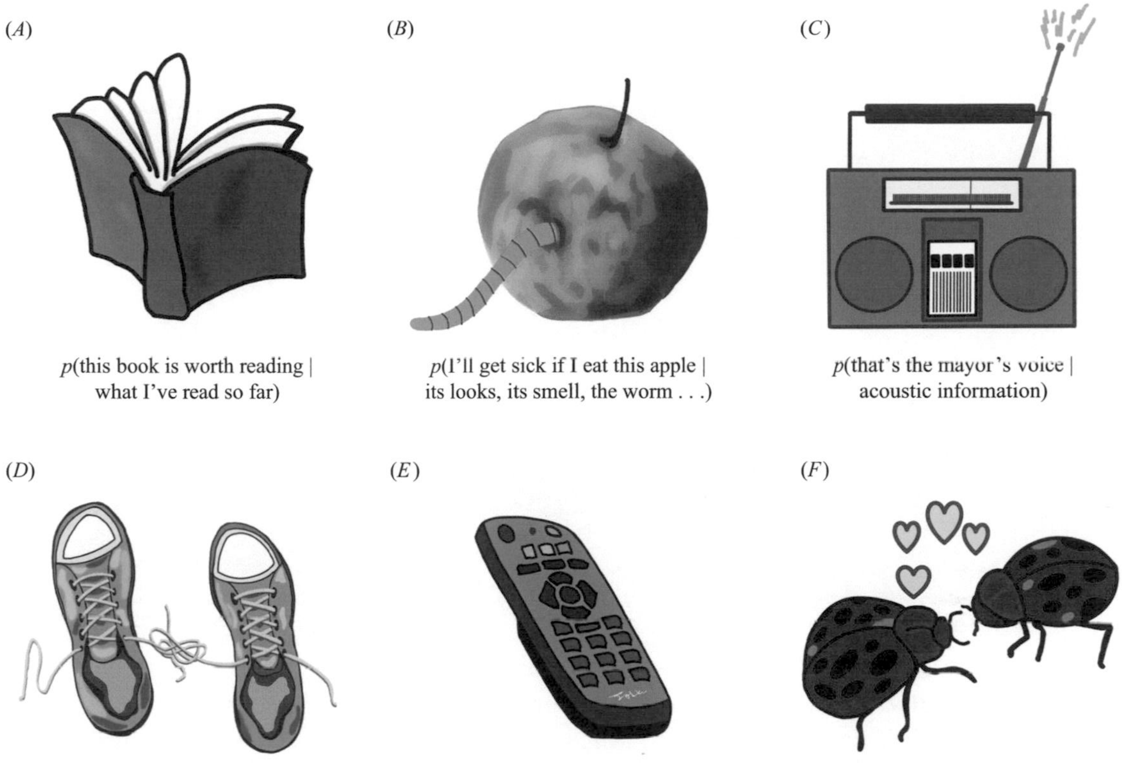

(*A*) p(this book is worth reading | what I've read so far)

(*B*) p(I'll get sick if I eat this apple | its looks, its smell, the worm . . .)

(*C*) p(that's the mayor's voice | acoustic information)

(*D*) p(I'm going to trip and fall | my shoelaces are tied in a knot)

(*E*) p(switching channels on TV | the remote has no batteries)

(*F*) p(they're the one | personality, behavior, appearance)

Figure 1.1
Different scenarios with probability judgments. The notation $p(B \mid A)$ is read "the probability of event B given event A."

p(tall | professional basketball player) > p(professional basketball player | tall)

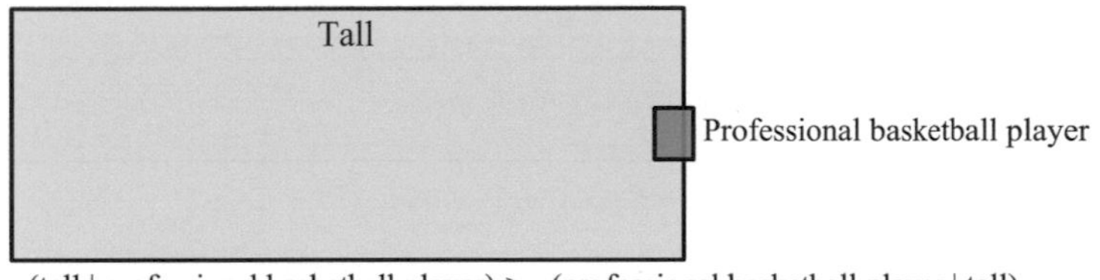

Figure 1.2
$p(A \mid B)$ does not in general equal $p(B \mid A)$. The area of each rectangle represents the probability of the event in the overall relevant set (here, say, all human beings). The overlap of the two rectangles represents $p(A, B)$. The sizes of the rectangles and their overlap here are conceptual and not calibrated against actual basketball participation data.

rectangle B is proportional to the probability of event B, denoted $p(B)$. Returning to the basketball example, there are vastly more tall people than professional basketball players, so the area of the lavender rectangle is much greater than the area of the teal rectangle. The overlap area occupies nearly all of the teal rectangle area, showing that the probability of being tall, given that one is a professional basketball player, is nearly 100 percent. However, the overlap area is much less than the lavender rectangle area, showing that

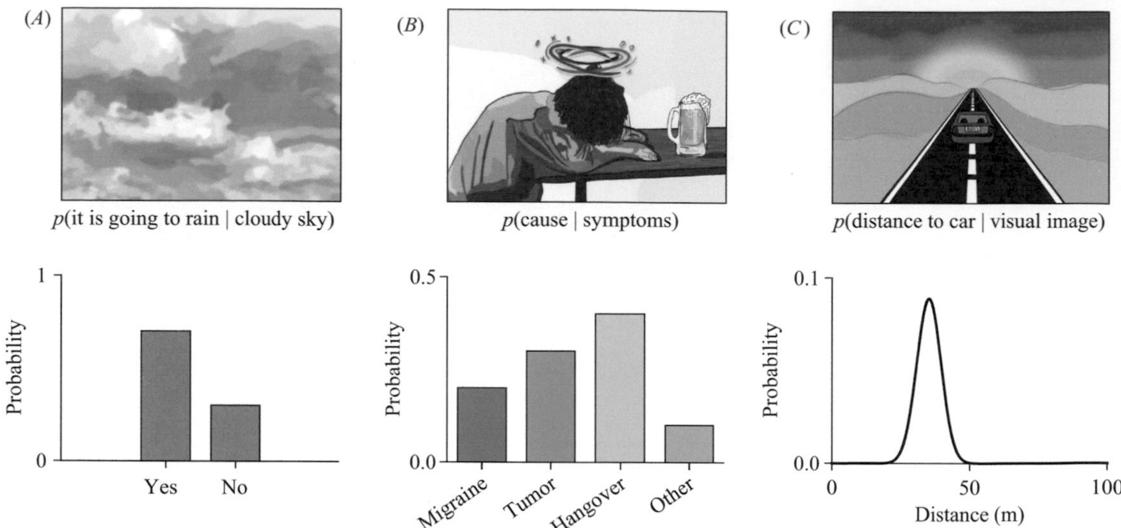

Figure 1.3
Different kinds of probability distributions. (*A*) Two hypotheses. (*B*) Multiple hypotheses. (*C*) Continuous (infinitely many) hypotheses.

the probability of being a professional basketball player, given that one is tall, is very small. The distinction between $p(A \mid B)$ and $p(B \mid A)$ is apparent in a multitude of real-world examples.

In perception, what is given are the sensory inputs or observations that are directly available to the observer, for example the activation pattern of photoreceptors in the retina. Given these observations (A), the observer would like to infer the current state of the world (B). Since the observer does not know the true state of the world, B is a hypothesis that the observer is entertaining, and we refer to B as the hypothesized world state. The observer's goal is to evaluate the probability that the world is in one possible state or another. For example, the observer might want to know how probable it is that a floor is wet (hypothesized world state B), given that the floor is shiny (observation A).

The conditional probability of interest to the observer is $p(B \mid A)$, the probability of a hypothesized world state given the sensory observations.

Depending on the situation, the observer may be concerned with evaluating the conditional probabilities of just two hypothesized world states (the floor is wet or dry), multiple distinct world states (the animal on the path ahead is a dog, a cat, a rabbit, a raccoon, or a skunk), or even a continuum (infinite number) of world states. Ultimately, we would like to express the results of our inference by calculating the probability of each world state, given the observation (figure 1.3). This would allow us to make an informed decision about the world.

Is That My Friend?

Suppose you see a person in the distance who appears to be walking toward you, and you wonder whether they are your friend (figure 1.4). Whatever conclusion you reach, you will have some degree of confidence, and your degree of confidence may change over time as

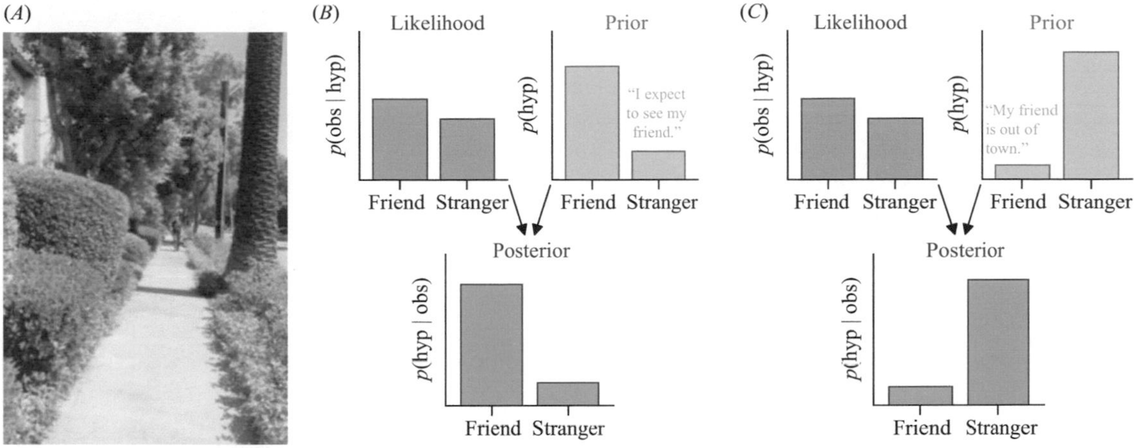

Figure 1.4
Recognizing a friend. (*A*) A visual scene offers a low-resolution view of a person in the distance who resembles your friend. (*B*) You consider the probability that the visual image would result from your friend to be greater than the probability that it would result from a stranger (likelihood function). You expected to meet your friend at this time and place (prior distribution). Therefore, you believe the person in question is probably your friend (posterior distribution). (*C*) In this *alternate* scenario, you thought your friend was out of town, so your prior distribution sharply favors the *stranger* hypothesis. Given the same observation (likelihood function), you conclude that the person in question is probably not your friend.

you continue to view the scene. We perceive visual scenes with little conscious effort, but scene recognition is in fact a computationally challenging endeavor. Like all forms of perception, scene recognition is challenging because the sensory input captured by the nervous system (the visual image in this case) is typically compatible with multiple interpretations. The visual image could be that of your friend or of another person. The image may provide sufficient information to recognize that the object is, in fact, a person, and it may provide information regarding the approximate shape (height, girth, etc.) of the person. Over time, the moving image may additionally provide information about the person's gait. Nevertheless, the person is far away, and your visual observations are compatible with many possible individuals.

The *likelihood* of a hypothesis is the probability of the observation given the hypothesized world state, p(observation | hypothesized world state). A plot of the likelihoods of all the hypotheses summarizes how well a visual image allows you to distinguish one possible world state from another (e.g., friend or stranger). This plot is known as the *likelihood function*.

In general, many factors affect the likelihood function. In our current example, the person's height (your friend is tall), hair color (brown), and way of holding their head (tilted) would each affect the likelihoods of the hypotheses. If you were to continue to watch as the person approaches, your likelihood function would sharpen over time as you obtained clearer and more detailed observations. We will leave aside at present how we might arrive at the exact form of the likelihood function. For now, it is sufficient to understand that the likelihood function represents the full information content of the image relevant to the question at hand (is that my friend?). Specifically, it represents the probability that your friend would give rise to the visual image, compared to the probability that another person would give rise to the same visual image.

Although the likelihood function is a crucial component of our inference process, it is not sufficient to solve the problems we want to solve. The likelihood function plots the probability of the observation given each hypothesized world state: p(observation | world state). What we want to know is the posterior distribution: the probability of each possible world state, given the observation: p(world state | observation). To determine the posterior distribution, we combine the likelihood function with a *prior distribution* that plots the *prior probability* of each world state. The prior probability of a hypothesized world state, denoted p(world state), is the probability of the world state based on all knowledge that you have apart from the observation—for instance, your belief that your friend would be present, before you even look up the street.

Let's consider two different scenarios that would cause you to have different prior distributions.

- Scenario 1: You had arranged to meet your friend on the street shown, and at the time shown, when you see the person walking toward you who looks like your friend.
- Scenario 2: When you see the person walking toward you who looks like your friend, you are surprised, because you thought your friend was still away on vacation and not planning to return to town until the following week.

The sensory input is identical in the two scenarios (figure 1.4A), but your perceptual inference would differ dramatically. Under scenario 1, p(friend) was high, and you would conclude that the person walking toward you probably is your friend; under scenario 2, p(friend) is low, and you would conclude that the person probably is not your friend. Clearly, your prior probabilities play a crucial role in your perceptual inference process.

Bayes' rule, a fundamental theorem in probability theory, shows how to optimally combine expectation, represented by the prior distribution, with the observation, represented by the likelihood function, in order to calculate a posterior distribution (figure 1.4B–C). The posterior probability of each world state is your belief in that world state based on all relevant information at your disposal. Bayes' rule states that the posterior probability is proportional to the product of the prior and the likelihood:

$$\text{Posterior} = \text{constant} \cdot \text{Prior} \cdot \text{Likelihood} \tag{1.1}$$

The posterior is based on all knowledge we have (i.e., our current sensory observation and relevant prior knowledge). When we start using actual numbers in chapter 2, we will discuss the constant in this equation, but for now, this equation captures the relevant intuitions. When both the prior and the likelihood are higher for hypothesis A than for hypothesis B, then the posterior will also be higher for hypothesis A. However, if the prior favors hypothesis A and the likelihood hypothesis B, then the posterior could go either way, depending on the exact numbers. For example, if one quantity (the prior or likelihood) only mildly favors hypothesis A but the other quantity strongly favors hypothesis B, then the posterior will tilt in favor of hypothesis B.

1.3 Sensory Noise and Perceptual Ambiguity

In any perceptual system, the flatter (i.e., more level) the posterior distribution, the more ambiguous—that is, open to multiple interpretations—the observations are. As we've seen, the shape of the posterior distribution results from the shapes of the likelihood function and

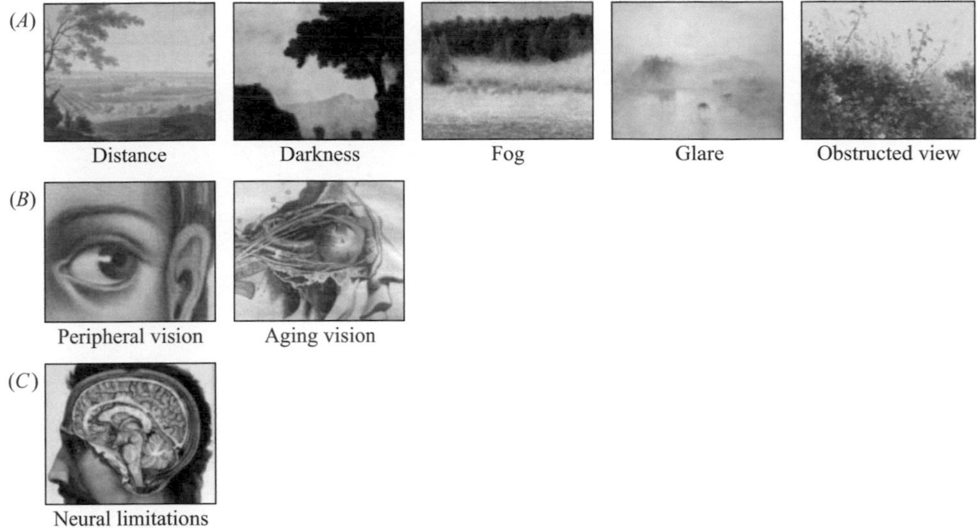

Figure 1.5
Sources of sensory degradation that reduce the quality of visual inputs, causing likelihood functions to flatten. (*A*) Physical features of the environment. (*B*) Limitations of the observer's sensory organs. Most of the factors shown in A and B have analogues in the other senses. For example, in the case of audition, distance, soft speech, ambient noise, and ageing ears all result in low-quality inputs. (*C*) The observer's nervous system. In every sensory system, neural limitations such as faulty background knowledge and neural noise also pose a challenge to perception. (1) Jean Chaufourier (1679–1757), *Landscape with Town or Palace in the Distance*. (2) Unknown British artist, *Landscape with a Dark Tree*, Late 18th C. (3) Camille Pissarro (1830–1903), *Setting Sun and Fog, Eragny* (1891). (4) J. M. W. Turner (1775–1851), *Norham Castle, Sunrise* (ca. 1845). (5) Myles Birket Foster (1825–1899), *A Roadside with Hedge-Roses*. (6) Peripheral vision: Wellcome Library, Eye surgery to correct strabismus. (7) Structure of the Eye and Optic Nerves, illustration from Peter Degravers, *A Complete Physico-Medical and Chirurgical Treatise on the Human Eye and a Demonstration of Natural Vision* (London: B. Law, 1780). (8) Jean-Baptiste Bourgery (1797–1849), Brain, from *Traité complet de l'anatomie de l'homme* (1831–1854).

prior distribution. Perceptual ambiguity often results from a wide likelihood function in the absence of a countervailing sharp prior.

Many factors can reduce the quality of sensory data and thereby widen the likelihood function. These include physical features of the environment, limitations of the observer's sensory organs, and limitations in the nervous system (figure 1.5). One ubiquitous likelihood function widening factor is sensory noise. By sensory noise we mean stochastic variability inherent in a physical process that generates a sensory observation. Because of noise, a stimulus that is repeated identically over multiple trials typically produces a somewhat different sensory observation each time. The scattering of light, random variability in ambient sounds, or biophysical variability in the firing rates of sensory neurons are all examples of sensory noise. As these examples indicate, sensory noise can occur both in the external world and within the observer.

While sensory noise is ubiquitous, it is not the only cause of wide likelihood functions. Even if all sensory input could somehow be made noiseless, many visual likelihood functions would be wide for purely geometric and optical reasons. (figure 1.6). For instance, information is necessarily lost when the three-dimensional visual world maps onto a two-dimensional retinal image (or when the three-dimensional auditory world maps onto the two ears). This collapsing across a dimension gives rise to many instances of ambiguity, including size-distance ambiguity. Another common geometric source of wide likelihood

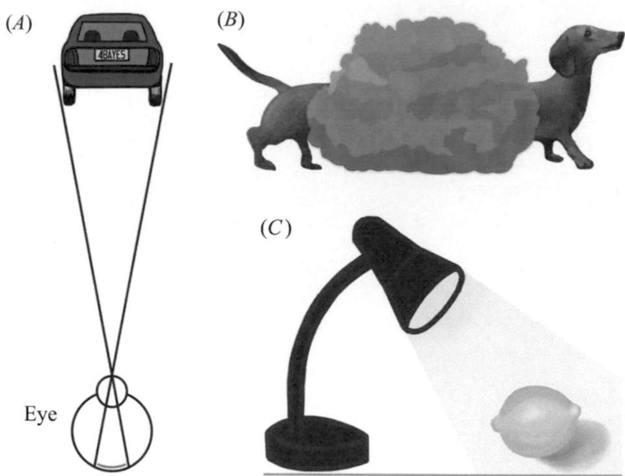

Figure 1.6
Likelihood functions widening caused by factors other than sensory noise. (*A*) The same retinal image size can result from a small object closer to the observer or a larger object farther away. (*B*) A bush prevents the observer from knowing whether this scene contains two (or even more) dogs or just one longer dog. (*C*) The observer can't be sure of the shade of a surface without also knowing the intensity of the illuminating light.

functions is occlusion, in which an object partially obstructs the observer's view, such that the scene is compatible with a variety of alternative configurations. As a final example, consider the apparently simple task of perceiving the shade of a gray surface from the intensity of light that reflects off the surface and enters your eyes. A particular light intensity entering your eyes is consistent with multiple combinations of the true shade of the surface and the intensity of the illuminating light source. For example, the same intensity of light entering the eyes can be produced by dark paper in sunlight or white paper in dim light. Therefore, unless the intensity of the illuminating light is known, the likelihood function in this scenario is wide. We consider these and other examples mathematically in later chapters.

The perceptual ambiguity resulting from noise or other likelihood-widening factors can be lessened or prevented altogether if the observer has a sharp prior distribution. Prior probabilities are based on background knowledge and therefore can evolve over time as an observer acquires new knowledge. Priors also may differ from one observer to another. In general, those with greater relevant knowledge have more realistic priors, facilitating accurate perception. Consider figure 1.6B. What prior knowledge about dogs might help one observer experience less ambiguity in this situation than another observer who lacks that knowledge?

1.4 Bayesian Inference in Visual Perception

As perceptual inference is so important, we want to illustrate it with additional examples drawn from everyday life. Our goal is to develop an intuitive understanding of likelihoods, priors, and posteriors, and an appreciation for the remarkable explanatory power of Bayesian inference as a model of perception. We will see that each example has unique features, yet each is based on the joining of a likelihood function and a prior distribution via Bayes' rule, in order to generate a posterior perceptual inference. We hope that these examples begin to

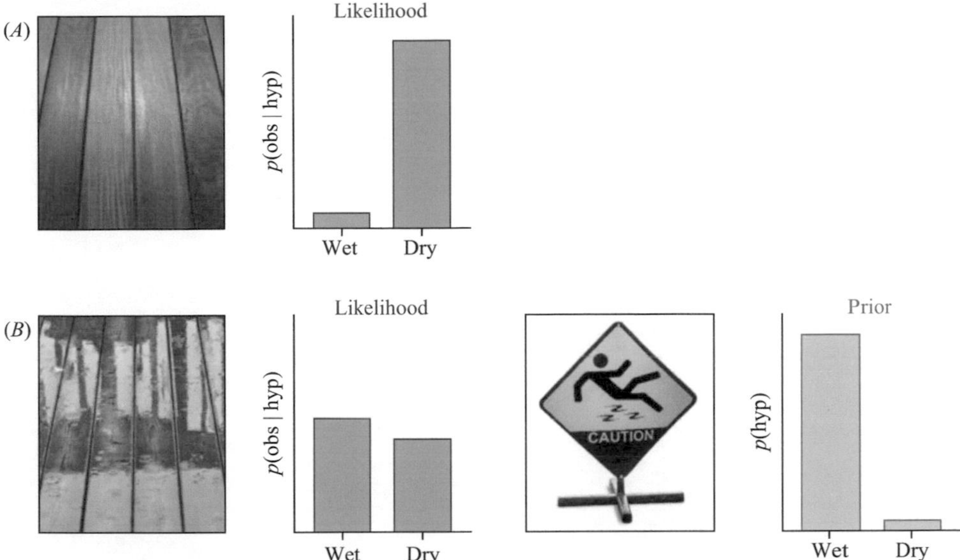

Figure 1.7
Perception of wetness. (*A*). The likelihood function resulting from the visual image of this wood surface favors the "dry" world state: $p(\text{observation} \mid \text{dry}) \gg p(\text{observation} \mid \text{wet})$. (*B*). The shiny surface results in a likelihood function that favors the "wet" world state: $p(\text{observation} \mid \text{wet}) > p(\text{observation} \mid \text{dry})$. A caution sign would result in a sharper prior in favor of the "wet" world state.

reveal both the richness of perceptual inference and the wide applicability of the Bayesian perceptual framework.

Perception of Wetness

As humans move through the world, we rely on our senses to avoid hazards. In the modern world, hazards come in many forms, for instance an object in our path, a rapidly approaching car, or a downward step such as a curb. Another hazard of modern life is a wet walking surface. Is the surface wet (figure 1.7)? If it is—or might be—caution is warranted, and we may want to take small, careful steps. If it is not, we can safely proceed with long, purposeful strides. How can perception distinguish the two possibilities?

The probability of a surface being shiny if it is wet is greater than the probability of it being shiny if it is dry. The observation of a shiny surface thus results in a likelihood function that favors wet. It is important to understand that not only our priors but also our likelihoods depend on background knowledge. In general, to determine likelihoods, the observer needs to have an (implicit) understanding of the process by which different world states generate sensory data. In this case, the observer needs an intuitive understanding of optics, that is, that a wet surface tends to reflect light to a greater degree.

As explained above, the brain combines likelihoods, $p(\text{observation} \mid \text{hypothesized world state})$, with prior probabilities, $p(\text{hypothesized world state})$, to generate the probabilities it most wants to know: $p(\text{hypothesized world state} \mid \text{observation})$. These latter probabilities are called *posterior probabilities* to indicate that, unlike prior probabilities, they are formed *after* the observation. The *posterior probability distribution* represents the brain's belief in each possible world state, based on all relevant information (i.e., observation and

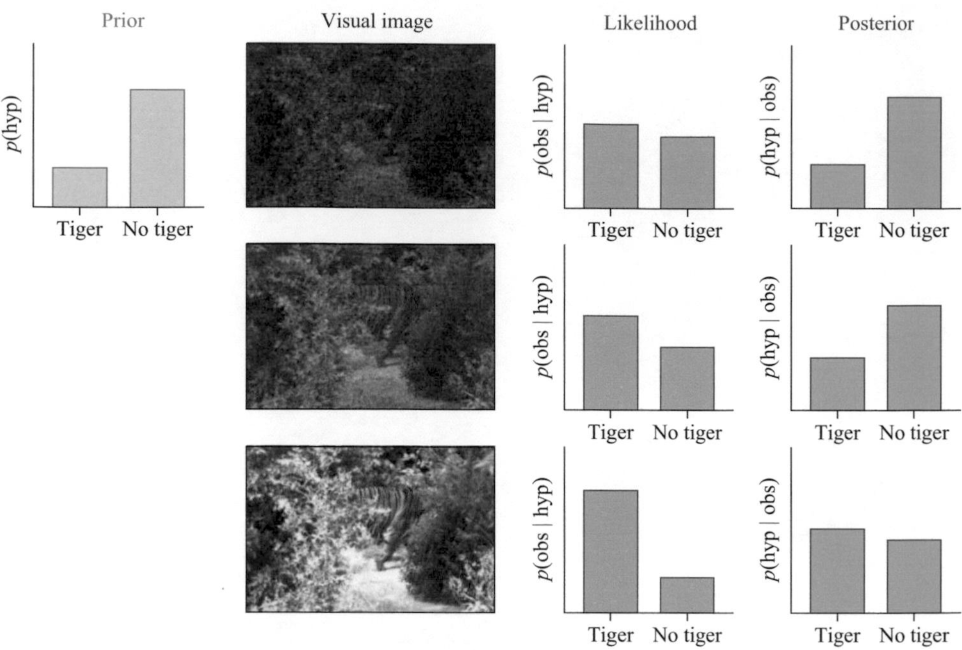

Figure 1.8
Effect of visual acuity on the posterior distribution. Three prey who hold identical (20 percent) prior expectation for the presence of a tiger (upper left) differ in visual acuity, and therefore experience different likelihood functions when confronted with the same visual scene. The flatter the likelihood function, the more the posterior distribution resembles the prior distribution. (*top*) To this animal with poor visual acuity, the visual scene evokes a nearly flat likelihood function. The animal's posterior distribution is therefore similar to its prior distribution; it has learned little from the visual observation. (*middle*) An animal with intermediate visual acuity has a likelihood function that is not flat. This animal's posterior distribution differs slightly from its prior distribution. (*bottom*) For this animal with excellent visual acuity, the scene results in a sharp likelihood function in favor of the tiger's presence. The animal's posterior distribution indicates slightly greater than 50 percent probability that a tiger is present.

expectation). We need to calculate the posterior probability of each hypothesized world state, $p(\text{world state} \mid \text{observation})$. This calculation involves multiplying priors and likelihoods. Recall that the prior probability, $p(\text{wet})$, reflects the observer's expectation regarding the wetness of the surface, independently of the visual observation. For instance, before even entering the area, what probability would the observer assign to the hypothesis that the surface will be wet? How does the observer acquire such priors?

The background knowledge that informs priors may have been acquired over a lifetime of previous experience or very recently. An observer may have a prior that favors dry, because similar locations in the observer's experience are dry the majority of the time. However, if the observer sees a caution sign that indicates a slippery surface, their prior may change to favor the wet hypothesis. To recognize the dependence of the prior on the observer's background knowledge, we sometimes write the prior as $p(\text{world state} \mid B)$, where B again signifies information obtained through previous experience.

Since both prior and likelihood depend on background knowledge, the posterior, too, depends on background knowledge. To recognize this dependency, we sometimes write the posterior probability of each world state as $p(\text{world state} \mid \text{observation}, B)$.

Figure 1.9
Likelihood function-flattening features in the animal kingdom. (*A*) The peppered moth caterpillar (*Biston betularia*) changes its color to blend in with the background (willow or birch tree). (*B*) A flying lizard (*Draco dussumieri*) blends in impressively with the tree bark background. (*C*) A flounder rests on the sea floor. (*D*) A well-camouflaged leopard in South Africa (*Panthera pardus*).

Camouflage

In the animal kingdom, survival often depends on seeing but not being seen. As noted previously, a sharp likelihood function indicates that an observation is highly informative, whereas a flatter likelihood function provides little information. In general, then, it is to an animal's benefit to have keen senses that produce sharp likelihood functions when the animal views the world (figure 1.8), but at the same time for it to engage in behaviors or have physical features that produce relatively level likelihood functions in other species.

Indeed, many species have evolved traits and behaviors that serve to disguise their presence or identity (figure 1.9). These diverse examples of camouflage and mimicry in the animal kingdom can be understood as evolved strategies aimed at leveling the likelihood functions of onlookers. Consider, for instance, the peppered moth caterpillar (figure 1.9A). Remarkably, individuals of this species assume the color of the tree bark on which they live. By blending in with the background, these caterpillars protect themselves from predatory birds. The visual image observed by a bird provides scant indication of the caterpillar's presence. Predators, too, benefit from camouflage. Consider the image of a leopard lying in wait for prey (figure 1.9D). In the high golden grass, whose coloration closely resembles its own, the leopard is nearly invisible to an unsuspecting onlooker. Although they can run fast, leopards and other large cats lack stamina for long chases. Their success in hunting depends on their ability to approach prey unnoticed. Examples of camouflaged predators and prey abound in the animal kingdom.

As long as an observer's likelihood function is not perfectly flat, the observer will learn something from the sensory input. However, when a well-camouflaged animal is viewed, the observer's likelihood function is nearly uninformative. Importantly, the shape of the likelihood function depends not exclusively on the visual scene but also on the sensory acuity and acumen of the observer. An animal that to one observer is nearly perfectly camouflaged may be noticed by another observer who has finer vision or better understanding. To an observer who knows from experience that peppered moth caterpillars tend to be slightly wider than the twigs of the tree they inhabit, or that leopards' spots differ slightly in appearance from the surrounding vegetation, the same visual scene will result in a sharper likelihood function than it does for an observer lacking this background knowledge.

Along with camouflage, evolution has given rise to sophisticated sensory systems—and cognitive abilities—that function to reduce uncertainty about the presence and locations of other animals. In an arms race of sorts, animals have evolved progressively keener sensory systems to detect their progressively better-hidden opponents. The evolution of mammalian visual, auditory, and olfactory systems are cases in point, as is the evolution of highly specialized detection systems such as the ultrasonic echolocation used by insect-eating bat species. In general, animals benefit if they perceive others sharply while others have difficulty perceiving them. Animals have therefore evolved impressive perceptual systems to achieve sharper likelihood functions for themselves, while at the same time evolving camouflage to force wider likelihood functions on others.

1.5 Bayesian Inference in Auditory Perception

So far, we have considered visual examples. However, perceptual inference occurs in every sensory modality. For instance, humans live in an acoustically rich environment: birds chirp, the wind howls, dogs bark, car horns blare, music plays, and, perhaps most importantly, we talk to one another. Whenever we attempt to identify the source of a sound (is that a dog barking?), perceive its location (where is that barking dog?), or interpret its meaning (what was that word you just said?), we are undertaking perceptual inference. In order to optimize auditory perception, a Bayesian observer would combine prior probabilities with acoustic likelihoods to obtain the most accurate perceptual inference possible.

Birds on a Wire

Humans rely at least in part on our sense of hearing to locate objects. We and other mammals localize sounds sources by using unconscious calculations, including comparing the intensity and time of arrival of sounds at the two ears.

Suppose that you are walking outside on a beautiful sunny morning, when you notice the silhouettes of five birds perched on a wire (figure 1.10A). Suddenly, one of the birds (you cannot see which) bursts into melodious song. Which bird sang? Your auditory system rapidly processes the acoustic observation, yielding a wide likelihood function. This likelihood function is a continuous function over location; that is, the sound you heard is compatible with a source along a continuum of locations. Nevertheless, certain locations are associated with higher likelihoods than others. Interestingly, the location of highest

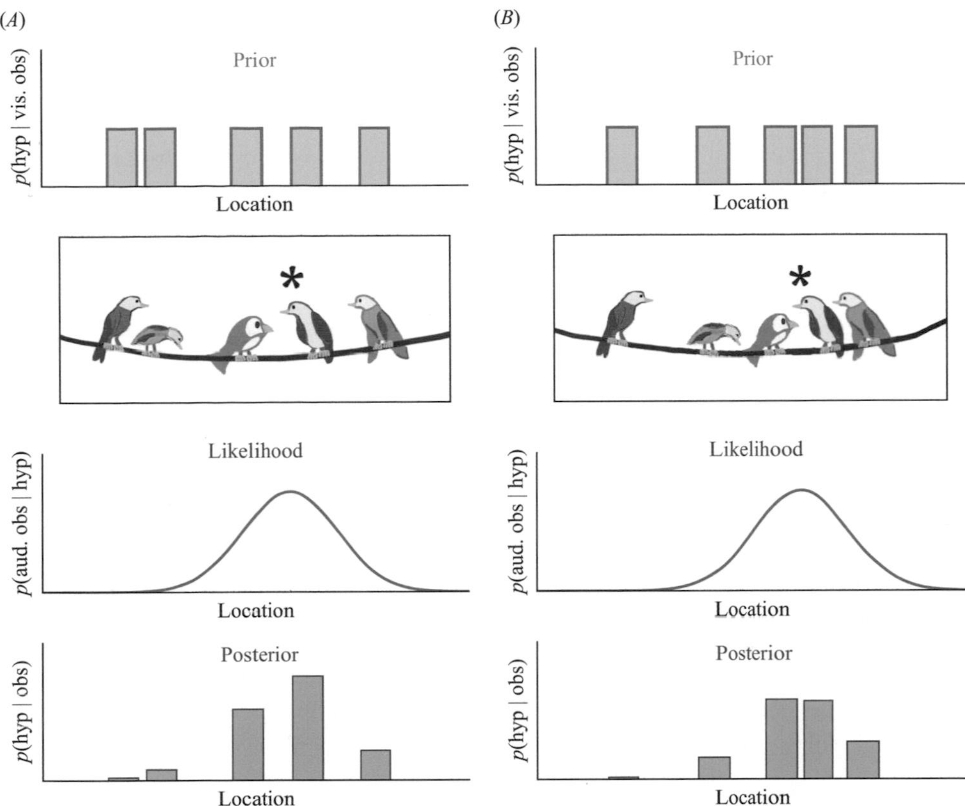

Figure 1.10
Sound source localization. (*A*) The visual image of the birds provides the basis for a prior distribution over sound source location. The wide likelihood function reflects the imprecision of the acoustic observation. The posterior distribution favors the hypothesis that the fourth bird from the left sang (*). (*B*) Perceptual uncertainty increases markedly if the birds crowd closer together.

likelihood may not coincide with the exact location of any bird. This situation is common in acoustic perception and can be caused by many factors. For instance, if the bird that chirped was not facing you directly, then the sound it produced may have deflected off nearby objects before reaching your ears. Even if sound waves were able to reach your ears without deflection, stochastic variability in the response of your nervous system can cause the likelihood function to peak at a location that is slightly offset from the source location.

Unlike the likelihood function, the visual information in the present example is not continuous, but rather discrete. You see five individual birds. Your visual observation, which occurred before the bird chirped, provides you with a prior distribution. Thus, the prior distribution is nonzero at five discrete locations (we are assuming that your visual perception is highly accurate for this high-contrast scene). Note that the prior probabilities are taken to be equal across the five birds, and that the prior probability that the sound source would occupy an empty location on the wire is zero. This simply means that, prior to hearing the song, you considered it equally probable that any one of the birds would sing. Using Bayes' rule, we can now calculate the posterior distribution for the location of the sound source. For each of the five hypothesized locations, we multiply the likelihood by the prior. The resulting posterior distribution indicates the fourth bird from the left as the most likely source of the pleasing melody.

Intuition suggests that if the birds had been closer together on the wire, our inference would be less certain. This result indeed emerges from the Bayesian inference, as shown in figure 1.10B. Here, we show the singing bird at the same location, but with three of the other birds closer to it than they were before. Our prior distribution reflects the new positions of the birds, but the acoustic observation and therefore the likelihood function are the same as before. The posterior distribution is now wider and lower, indicating that, although the same bird is the most probable singer, our uncertainty has markedly increased. Indeed, in our judgment the singing bird could nearly equally probably be the third or the fourth bird from the left.

Before leaving this example, we would like to draw your attention to two alternative approaches to solving the problem that would have led to the same answer. In one alternative approach, we could have started, before looking at the wire, with a wide, uniform prior over hypothesized bird locations, reflecting the fact that, before looking, we had no idea where any birds would be perched. We could then have incorporated the subsequent visual observation into a likelihood function and combined this with our flat prior distribution to produce a posterior distribution over the birds' locations. Indeed, it was this original *posterior* distribution from the visual input that we used here as a *prior* distribution for our analysis of the auditory observation. This illustrates an important general feature of Bayesian inference: it can be done iteratively, the posterior distribution from one inference being used as the prior distribution for the next.

We will learn about a second alternative approach to this problem in chapter 5, which again would reach the same answer: Starting with a flat prior over position, we could incorporate simultaneously both the visual and the acoustic observations as likelihood functions, in a procedure known as *cue combination*. In this approach, we would not use the visual information to generate a prior distribution for the subsequent auditory observation but would instead combine a visual likelihood function that has five discrete peaks with a continuous acoustic likelihood function. In essence, when we have two or more independent sources of information, we can choose whether to incorporate the different sources sequentially, with the posterior from each observation being used as the prior for the next, or all at once, with all the observations entering through likelihood functions. Thus, there is often a blurring of boundaries between likelihood functions and priors, with the choice of how to incorporate the information left up to the Bayesian modeler. This flexibility is not a problem but an asset of the Bayesian approach. The internal consistency of the rules of Bayesian inference ensures that, as long as all the information is incorporated, the resulting posterior distribution will be the same regardless of the route taken. Within the Bayesian framework, there are often multiple ways of arriving at the same solution.

Mondegreens

Although our brains do it automatically and apparently effortlessly, speech perception requires sophisticated inference on a variety of levels. Most obviously, we must correctly perceive the spoken word. It is easy to misinterpret even a single word spoken in isolation, particularly when one is in the presence of ambient noise (the drone of a car engine, street sounds, chatter from nearby people speaking, running water in a sink, building ventilation noises, and so on). Under such conditions, akin to low-contrast vision, likelihood functions are wide, and one word may be misperceived for another that sounds similar. In fact, misperceived speech is such a common occurrence that humans often pass by these moments without a second thought. You may want to try to keep a list of such occurrences yourself. The results are both educational and amusing. For instance, in conversations with others, we have misheard *Mongolia* as *magnolia*, *fumaroles* as *funerals*, *hogs* as *hawks*, *census* as *senses*, *a moth* as *I'm off*, *maple leaf* as *make believe*, *this lime* as *the slime*, *below-knee* as *baloney*, and *peaches and strawberries too* as *peaches and strawberries stew*. As these examples illustrate, in addition to the similar sounds of different words, a challenge to speech perception arises because the pauses between spoken words are often no longer than the pauses between syllables within a single word. Consequently, it is by no means a trivial task to infer where one word ends and the next begins. This parsing difficulty can lead to errors in which syllables from different words combine improperly in our perception.

As a child, the author Sylvia Wright enjoyed listening to the popular seventeenth-century Scottish ballad "The Bonny Earl o'Moray," spoken to her frequently by her mother. She was particularly fond of the sad but beautiful lines describing the murders of the earl and the love of his life, Lady Mondegreen:

Ye Highlands and ye Lowlands,
Oh, where hae ye been?
They hae slain the Earl Amurray,
And Lady Mondegreen (208: 48–51).

As impactful as they were, the words heard by the young Sylvia Wright were not those that her mother spoke. In fact, the ballad makes no mention whatsoever of a Lady Mondegreen. The unfortunate dead earl was placed on the grass, alone; they "laid him on the green." Sylvia Wright's creative but mistaken interpretation of the spoken ballad reflects a perceptual parsing error. She interpreted the sounds "laid hi-" as "lady," and "-m on the green" as "Mondegreen." Sylvia Wright later coined the term "mondegreen" to refer to a misheard word or phrase [208]. Given the inherent phonetic ambiguity of spoken language, examples of mondegreens abound. When Queensland, Australia, was inundated by tropical cyclone Tasha, the *Morning Bulletin* of Rockhampton (January 6, 2011) reported the tragic news that, as a result of the flooding, "More than 30,000 pigs have been floating down the Dawson River since last weekend" [28]. This startling story, based on an interview between the reporter and the owner of a local piggery, was staggeringly incorrect. The owner had spoken, not of "30,000 pigs," but of "30 sows and pigs" swept downstream! The *Morning Bulletin* published a correction the following day.

Books and many websites are devoted to listing peoples' favorite mondegreens, particularly those resulting from misheard song lyrics, which we can all enjoy. It is instructive to visit websites on which listeners post their particular misheard versions of the same songs. The many different misheard versions of a line such as "Lucy in the sky with diamonds" presumably reflect both the phonetic ambiguity (wide likelihood function) and improbable content (low prior probability) of the original lyrics. With respect to prior probabilities, "There's a bathroom on the right" is surely a more common sentence than "There's a bad moon on the rise," and "submarine" is arguably more plausible than "summer breeze" as a mode of transport (figure 1.11).

Lucy in the sky with diamonds
- The Beatles

"Lucy and this guy with diamonds"
"Lucy in disguise with diamonds"
"Lucy in the sky with Simon"

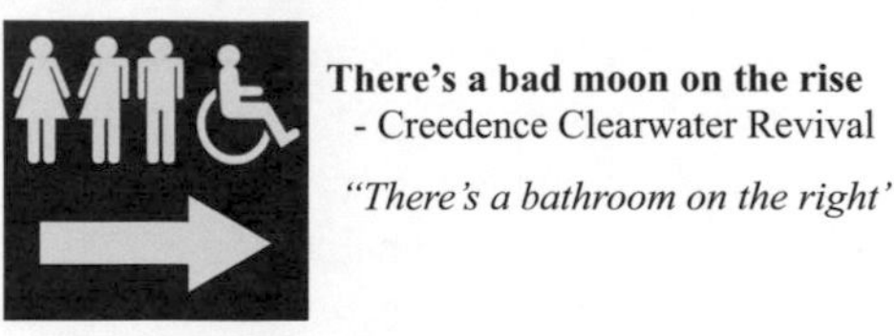

The Death of Lady Mondegreen
- Harper's Magazine (Nov. 1954)

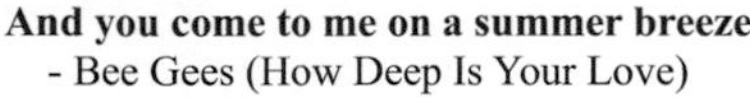

"And you come to me on a submarine"

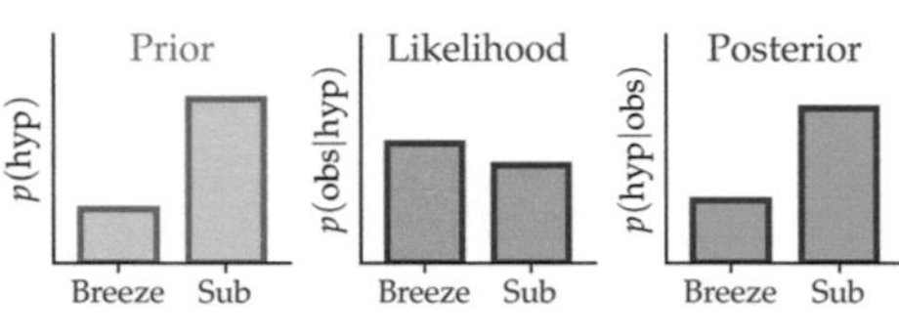

Figure 1.11
Mondegreens result from phonetic ambiguity (wide likelihood functions) coupled with low expectation for the actual phrase that is sung or spoken (prior distribution in favor of the "wrong" hypothesis).

The occurrence of mondegreens suggests strongly that speech perception, like visual perception, results from the combination of likelihood functions and prior distributions. Humans generally perceive speech accurately, but of course, occasional mistakes are inevitable. Indeed, "The more unintelligible the original lyrics, the more likely it is that listeners will hear what they want to hear" [139]. Rephrased in terms of Bayesian inference, the flatter the likelihood function, the greater will be the influence of the prior distribution on the resulting posterior distribution. Thus, the incorporation into perceptual inference of prior expectation, which in most circumstances improves perceptual accuracy, can backfire to create mondegreens on occasions when we are faced with an unexpected word (low prior) that sounds like (wide likelihood) another, more expected (high prior) word.

Keeping this in mind, it is rather easy to evoke mondegreens in others. Simply select two different words or phrases that sound alike, ensure that your listener has a prior distribution in favor of one of the words or phrases, then speak the other. For instance, you could tell a friend that "You know, humans are very good at speech recognition; in fact, we can understand speech much better than even the best computer programs can. We really know how to wreck a nice beach. Now, what did I just say? We really know how to . . .?" If you spoke the words "wreck a nice beach" naturally and without taking pains to enunciate clearly, your friend will probably have perceived "recognize speech," rather than the words you actually spoke. In Bayesian terms, the wide likelihood function experienced by your friend will combine with a sharp prior distribution (given the previous content of your discourse) to favor the "recognize speech" hypothesis. For more information about this particular misheard phrase, see [177].

Even when listeners perceive every word correctly, they face a final challenge: to identify the intended meaning of the string of words. Once again, this often requires evaluating

Figure 1.12
Semantic ambiguity in language.

multiple hypotheses. Suppose someone were to tell you that a particular bridge "is being held up by red tape." Statements such as this one, even when perfectly heard or read on a printed page, are nevertheless consistent with two or more interpretations [39]; that is, such statements evoke a wide likelihood function, due not to phonetic ambiguity but rather to ambiguity regarding the meaning of the words: semantic ambiguity (figure 1.12). In other cases, the sentence structure itself is ambiguous, a situation known as syntactic ambiguity. When we hear or read an ambiguous sentence, we naturally combine the likelihood functions with a prior distribution, and usually reach the correct perception. We are, however, sometimes bemused—and amused—momentarily, as both interpretations cross our minds. This occurred to one of the authors when a person told him of a wilderness trip he took with his parents. "There was wildlife everywhere," he exclaimed, "In fact, I saw a bear, walking with my mother" (figure 1.13). Although this book will not focus on such situations, we point them out here to illustrate that uncertainty and inference play a role at multiple levels of perceptual and cognitive processing.

1.6 Historical Background: Perception as Unconscious Inference

Bayes' rule is named after the English minister and mathematician Thomas Bayes (1702–1761), who was interested in problems of inverse probability, essentially how to calculate $p(B \mid A)$ when $p(A)$ and $p(A \mid B)$ are known. Bayes' *An Essay towards Solving a Problem in the Doctrine of Chances*, published posthumously in 1763, introduced the foundation for the conditional probability calculus, a field of statistical reasoning now called Bayesian inference. Bayes' rule was later derived independently by the French mathematician and physicist Pierre-Simon Marquis de Laplace (1749–1827). Laplace applied the formula with great effect to problems in a wide range of disciplines. Importantly, Laplace also recognized the pervasiveness of probability, stating that "the most important questions of life . . . are indeed, for the most part, only problems in probability. One may even say, strictly speaking, that almost all our knowledge is only probable" [106]. Indeed, today, Bayesian inference is playing a rapidly growing role in an extraordinarily diverse set of disciplines covering nearly

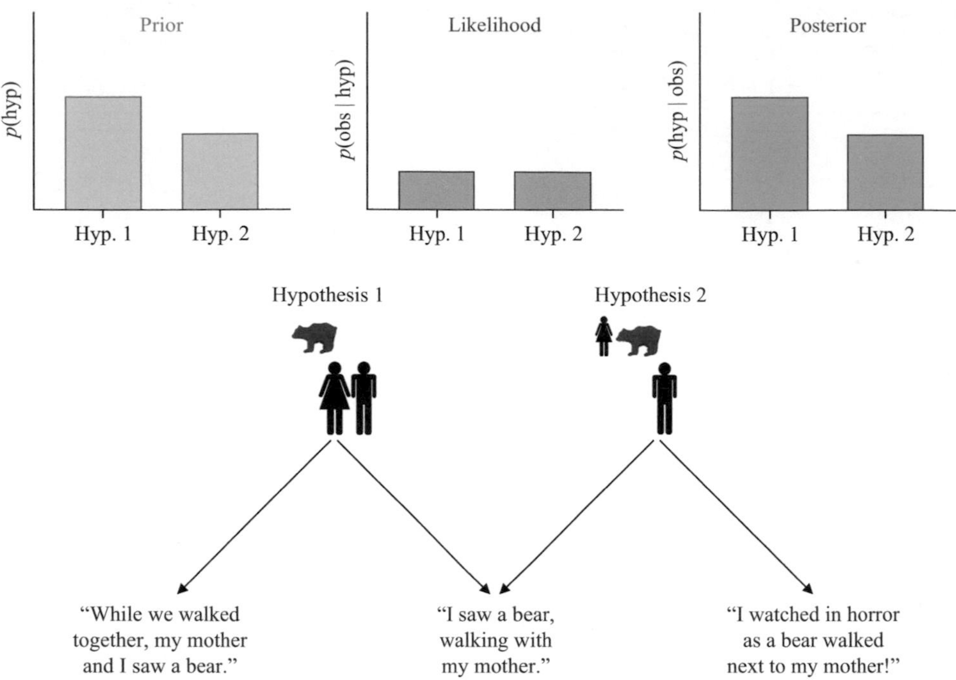

Figure 1.13
Perceptual inference under syntactic ambiguity. Each world state (hypothesis) could be described in many different ways, three of which are shown. The speaker happened to choose an expression that could describe both world states: "I saw a bear, walking with my mother." Background knowledge suggests that bears are less likely to walk alongside people than to be seen by them at a distance, so the prior distribution favors hypothesis 1. The likelihood function shows that the spoken sentence has about equal probability under the two hypotheses. The posterior distribution therefore favors hypothesis 1.

all fields of science and engineering: neuroscience, psychology, evolutionary and molecular biology, geology, astronomy, economics, robotics, and computer science, to name but a few.

The idea that perception is a form of unconscious inference, however, arose independently of Bayes and Laplace. Several scientists contributed to this notion. The early Arab physicist and polymath Ibn al-Haytham (965–ca. 1040), recognized presciently that "not everything that is perceived by sight is perceived through brute sensation; instead, many visible characteristics will be perceived through judgment . . . in conjunction with the sensation of the form that is seen." Thus, "familiar visible objects are perceived by sight through defining features and through previous knowledge" [10]. Much later, the German physician and physicist Hermann von Helmholtz (1821–1894) again expressed the idea that perception is a form of unconscious inference, stating eloquently that "previous experiences act in conjunction with present sensations to produce a perceptual image" [190]. The ideas of al-Haytham and Helmholtz fit beautifully with the view that perception is a form of Bayesian inference (figure 1.14). However, it is difficult to establish whether a particular form of perceptual inference is conscious or unconscious, and we will not comment on that issue in this book.

Thomas Bayes, 1702–1761

Ibn al-Haytham, 965–ca. 1040
"Familiar visible objects are perceived by sight through defining features and through previous knowledge." —*De aspectibus*

Pierre-Simon Laplace, 1749–1827
"The most important questions in life . . . are indeed, for the most part, only problems in probability. One may even say, strictly speaking, that almost all of our knowledge is only probable." —*Philosophical Essay on Probabilities*

Hermann Ludwig von Helmholtz, 1821–1894
"Previous experiences act in conjunction with present sensations to produce a perceptual image." —*Treatise on Physiological Optics*

$$\underbrace{p(\text{hyp} \mid \text{obs})}_{\text{Posterior}} \propto \underbrace{p(\text{obs} \mid \text{hyp})}_{\text{Likelihood}} \cdot \underbrace{p(\text{hyp})}_{\text{Prior}}$$

Figure 1.14
Luminaries in the development of Bayesian inference and the view that perception is unconscious inference. Note that any recognition of individuals in the history of science must happen with the understanding that many groups, especially women and people of color, were systemically excluded from the same opportunities.

1.7 Summary

In this chapter, we have introduced the concept that perception is inherently probabilistic, and as such optimally characterized as a process of Bayesian inference. Regarding Bayesian inference, we have learned the following:

- Conditional probabilities such as $p(A \mid B)$ represent the probability of A given B. In Bayesian perceptual inference, A and B typically represent a world state and an observation.
- The likelihood function, $p(\text{observation} \mid \text{world state})$, captures the information content of the sensory observation, relevant to distinguishing one world state from another.
- The flatter the likelihood function, the less we learn from our senses. If the likelihood function is perfectly flat, then the observer has learned nothing from the observation.
- In some cases, such as the example of "Is that my friend?," the likelihood changes over time.

- The prior distribution over world states, p(world state), summarizes the information content of our past observations, the background knowledge we have about the world. Perception is not based entirely on sensory observation but also on expectation grounded in previous experience.
- Flatter prior distributions mean we know less about the potential states of the world.
- Bayes' rule calculates the posterior probability of each hypothesized world state, p(world state | observation), from the likelihoods and prior probabilities of the world states.
- The procedures of Bayesian inference apply equally to situations in which the hypothesized world states are discrete or in which they are continuous.
- Perceptual situations, whether in vision, audition, or other senses, are subject to various levels of uncertainty.
- Speech perception is fraught with phonetic and syntactic ambiguity, frequently giving rise to flat likelihood functions. The combination of priors and likelihoods can cause misinterpretations such as mondegreens.

1.8 Suggested Readings

- Alhacen. *Alhacen's Theory of Visual Perception: A Critical Edition, with English Translation and Commentary, of the First Three Books of Alhacen's "De Aspectibus," the Medieval Latin Version of Ibn Al-Haytham's "Kitāb Al-Manāz. ir."* Edited by A. Mark Smith. Vol. 1. Philadelphia, PA: American Philosophical Society, 2001.
- Thomas Bayes. "An Essay towards Solving a Problem in the Doctrine of Chances." *Biometrika* 45, no. 3–4 (1958): 296–315.
- Peter Brugger and Susanne Brugger. "The Easter Bunny in October: Is It Disguised as a Duck?" *Perceptual and Motor Skills* 76, no. 2 (1993): 577–578.
- Daniel Burdon. "Pigs Float down the Dawson." *Morning Bulletin*, February 9, 2011. https://realtegan.blogspot.com/2011/02/best-correction-ever.html.
- Gloria Cooper. *Red Tape Holds up New Bridge, and More Flubs from the Nation's Press*. New York: TarcherPerigee, 1987.
- Wilson S. Geisler and Randy L. Diehl. "A Bayesian Approach to the Evolution of Perceptual and Cognitive Systems." *Cognitive Science* 27, no. 3 (2003): 379–402.
- Gary Hatfield. "Perception as Unconscious Inference." In *Perception and the Physical World: Psychological and Philosophical Issues in perception*. Chichester: John Wiley, 2002.
- Hermann von Helmholtz. *Treatise on Physiological Optics*. Edited by James P. C. Southall. Vol. 3, *The Perceptions of Vision*. New York: Optical Society of America, 1925.
- Pierre-Simon Laplace. *A Philosophical Essay on Probabilities*. Edited and translated by Andrew I. Dale. New York: Springer Science and Business Media, 2012.
- Mohamed A. F. Noor, Robin S. Parnell, and Bruce S. Grant. "A Reversible Color Polyphenism in American Peppered Moth (*Biston betularia cognataria*) Caterpillars." *PloS One 3*, no. 9 (2008): e3142.

- Pamela Licalzi O'Connell. "Sweet Slips of the Ear: Mondegreens." *New York Times*, April 9, 1998.
- Russell Smith. "Milk Drinkers Turn to Powder and Other Pun-ishing Headlines." *Globe and Mail*, (September 23,).
- Stephen M. Stigler. "Who Discovered Bayes' Theorem?" *American Statistician* 37, no. 4a (1983): 290–296.
- J. V. Stone, I. S. Kerrigan, and J. Porrill. "Where Is the Light? Bayesian Perceptual Priors for Lighting Direction." *Proceedings of the Royal Society B: Biological Sciences* 276, no. 1663 (2009): 1797–1804.
- Dave Tompkins. *How to Wreck a Nice Beach: The Vocoder from World War II to Hip-Hop, the Machine Speaks*. New York: Melville House, 2011.
- Michel Treisman. "Motion Sickness: An Evolutionary Hypothesis." *Science* 197, no. 4302 (1977): 493–495.
- Sylvia Wright. "The Death of Lady Mondegreen." *Harper's Magazine* 209, no. 1254 (1954): 48–51.

1.9 Problems

Problem 1.1 Rephrase in terms of Bayesian perceptual inference the following statement written by Ibn al-Haytham approximately 1,000 years ago: "When sight perceives a rose-red color among the flowers in some garden, it will immediately perceive that the things in which that color inheres are roses because that color is specific to roses . . . . But this does not happen when sight perceives a myrtle-green color in the garden. For when sight perceives only the myrtle-green in the garden, it will not perceive the myrtle-green to be myrtle simply from the perception of the green, because several plants are green, and, in addition, several plants resemble myrtle in greenness and shape." (*Alhacen, De aspectibus*, book 2, Alhacen's *Theory of Visual Perception*, trans Smith).

Problem 1.2 Why is it that we identify ourselves at the very beginning of a phone conversation, even to people we already know, but we do not do this when we meet in person? Express your answer within the framework of Bayesian perceptual inference.

Problem 1.3 When a conversation partner speaks softly, or when a conversation occurs in the presence of significant ambient noise, we sometimes cup our ears and/or look carefully at the speaker's lips. Why, in Bayesian perceptual terms, do we do this?

Problem 1.4 To explore how a noisy environment engenders uncertainty, consider the word "lunch." Suppose that you see this word written (or hear it spoken), with the letter "l" blocked out (making it unknown): _unch (e.g., by ambient auditory noise). List all source words that are compatible with what you see. Now consider the case in which both the l and the n are blocked: _u_ch. In terms of conditional probabilities relevant to perception, what is the effect of blocking out the l, n, and both?

Problem 1.5 The NATO phonetic alphabet, used by many military, maritime, and other organizations during radio communications, represents each letter with a word: A (Alpha), B (Bravo), C (Charlie), D (Delta), E (Echo), F (Foxtrot), and so on. What purpose does this serve in radio communications? Explain with respect to conditional probabilities. In particular, consider a radio communication under conditions of considerable background noise, in which the sender wishes to spell the word "FACE." Compare p(auditory signal heard by the receiver | FACE spelled by the sender) vs p(auditory signal heard by the receiver | another word, such as FADE, spelled by the sender), when the sender uses the regular alphabet, and again when the sender uses the NATO phonetic alphabet.

Problem 1.6 English speakers sometimes incorrectly perceive English words when they listen to songs sung in a foreign language with which they are unfamiliar, and listeners also mistakenly perceive words in music that is played backward. Provide a Bayesian explanation for these phenomena.

Problem 1.7 Sometimes, when you press the button to call an elevator, you notice that the elevator car starts moving immediately afterward (as shown by a display showing the car's current floor). Argue that in those situations, the likelihood of the hypothesis that there is nobody inside when it arrives is much higher than the likelihood of the hypothesis that there are people inside. Priors do not play a role in this problem.

Problem 1.8 Suppose you see someone you do not know, getting only a brief look at them from a distance of about 10 meters. If you are interested in estimating this person's age, how would you proceed? What factors, including and in addition to the person's appearance, would affect your estimation? Provide a Bayesian description of your reasoning. As part of your answer, draw examples of your likelihood function, prior distribution, and resulting posterior distribution.

Problem 1.9 A research article entitled "The Easter Bunny in October: Is It Disguised as a Duck?" explained that "Very little is known about the looks of the Easter bunny on his non-working days" [26]. To investigate, the authors showed an "ambiguous drawing of a duck/rabbit (unattributed from the October 23 1892 issue of Fliegende Blatter) . . . to . . . 265 subjects on Easter Sunday and to 276 different subjects on a Sunday in October of the same year." The authors report: "Whereas on Easter the drawing was significantly more often recognized as a bunny, in October it was considered a bird by most subjects." The drawing shown by the authors in their study was similar to the following:

Rabbit or duck, Fliegende Blätter, *October 23, 1892 issue of Fliegende Blatter.*

Provide a Bayesian perceptual explanation for the authors' results.

Problem 1.10 The images below show a hollow face mask, rotated progressively from a side view (left) to reveal the hollow back of the mask (right):

The right image looks like a normal, convex face, when in reality it is the hollow (concave) side of the mask. This is called the *hollow-face illusion* or the *hollow-mask illusion*. Provide a Bayesian explanation for this illusion.

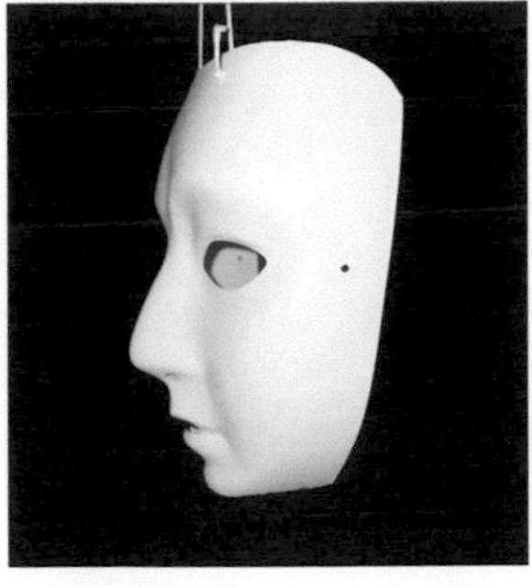

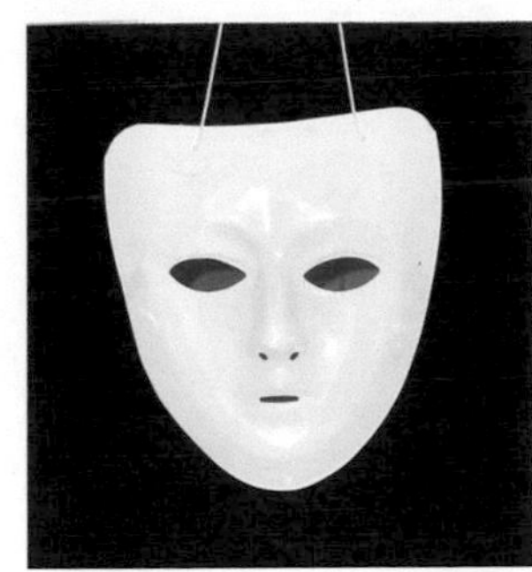

Problem 1.11 Give three daily-life examples (perceptual or cognitive) in which you tried to infer a world state from incomplete or imperfect information. For each example, specify the observation(s), the world state of interest, and the source(s) of uncertainty.

Problem 1.12 Michel Treisman has tried to explain motion sickness in the context of evolution ([179], *Motion Sickness*, 493–495). During the millions of years over which the human brain evolved, accidentally eating toxic food was a real possibility, and that could cause hallucinations. Perhaps our modern brain still uses prior probabilities genetically passed on from those days; those would not be based on our personal experience, but on those of our ancestors! Here, we do not delve into the merits of this theory but try to cast it in Bayesian form. Suppose you are in the windowless room on a ship at sea. Your brain has two sets of sensory observations: visual observations and vestibular observations. Assume that the brain considers three hypotheses for what caused these observations:

Hypothesis 1: The room is not moving and your motion in the room causes both sets of observations.

Hypothesis 2: Your motion in the room causes your visual observations whereas your motion in the room and the room's motion in the world together cause the vestibular observations.

Hypothesis 3: You are hallucinating: your motion in the room and ingested toxins together cause both sets of observations.

(a) In prehistory, surroundings would almost never move. Once in a while, a person might accidentally ingest toxins. Assuming that your innate prior probabilities are based on these prehistoric frequencies of events, draw a bar diagram to represent your prior probabilities of the three hypotheses above. No numbers are needed.

(b) In the windowless room on the ship, there is a big discrepancy between your visual and vestibular observations. Draw a bar diagram that illustrates the likelihoods of the three hypotheses in that situation (i.e., how probable these particular sensory observations are under each hypothesis). No numbers are needed.

(c) Draw a bar diagram that illustrates the posterior probabilities of the three hypotheses. No numbers are needed.

(d) Explain using the posterior probabilities why you might vomit in this situation.

2

Using Bayes' Rule

How do we make quantitative inferences?

In chapter 1, we explained that Bayes' rule is relevant for understanding perception. In this chapter, we describe how to actually do calculations with Bayes' rule. We show with examples how to compute posterior distributions from prior distributions and likelihood functions.

Plan of the Chapter

We describe the steps of Bayesian modeling. We present intuitive derivations of Bayes' rule using mathematical and areal representations. We introduce simple perceptual scenarios with categorical (in particular: binary) variables in order to illustrate how observers perform Bayesian inference.

2.1 Steps of Bayesian Modeling

Every Bayesian model consists of a series of steps that must be followed in order. The first step is to formulate the generative model, which represents the statistical structure of the world and the observations. The second step is inference: given a particular observation, how should the observer's beliefs about the state of the world be updated? The second step concludes by specifying how the observer reaches a decision. We will encounter a third step in the next chapter. As a "cheat sheet," we have listed the steps, with the leading example of this chapter, inside the book's cover.

Step 1: The Generative Model

Let us return to the example from section 1.4, that of estimating whether a surface is wet based on visual information (shiny or not). Here we have two possible world states (wet or dry). We have two potential observations (shiny or not). Every Bayesian model starts with a specification of the process by which we believe the observed data are generated; this is called a *generative model*. The generative model is a full statistical description of what is happening in a task. It always includes the world state of interest and the observer's sensory observations, but it could also include other variables. Mathematically, the generative model specifies the probability distributions of all variables in the problem. Researchers

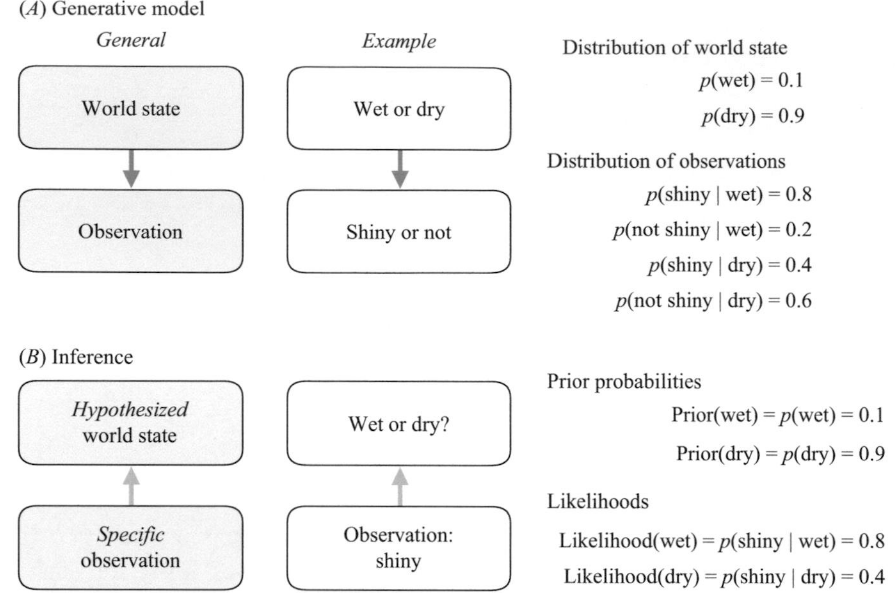

Figure 2.1
(*A*) A generative model can be represented by a diagram with nodes and arrows. Each node is a variable, with the observation(s) always at the bottom. Each node is associated with a probability distribution. Each node with an arrow pointing to it is associated with a conditional probability distribution. Numbers are given for our shiny floor example. (*B*) In inference, the observer has a specific observation (in the example: the floor is shiny) and tries to calculate the probabilities of the hypothesized world states. The same numbers are now used as priors and likelihoods. Inference "inverts" the generative model.

often visualize the generative model in a diagram called a *graphical model* (figure 2.1A). In a graphical model, each node represents a variable and each arrow a statistical influence of one variable on another. The sensory observations should always be at the bottom of the diagram.

The generative model specifies probabilities of world states, for example, that most floors are dry. Suppose that, in our experience, 10 percent of all floors are wet. Then:

$$p(\text{wet}) = 0.1; \tag{2.1}$$

$$p(\text{dry}) = 0.9. \tag{2.2}$$

Probabilities of occurrence in the absence of any observations would be called *base rates* in statistics or *prevalences* in medicine.

The generative model also specifies the probabilities of observations conditioned on world states. Suppose that, in our experience, 80 percent of wet floors are shiny, but only 40 percent of dry floors are shiny. These probabilities are *conditional probabilities* and take the form $p(\text{observation} \mid \text{world state})$. These probabilities must be specified for each combination of observation and world state. In the example, we assume that

$$p(\text{shiny}|\text{wet}) = 0.8; \tag{2.3}$$

$$p(\text{not shiny}|\text{wet}) = 0.2; \tag{2.4}$$

$$p(\text{shiny}|\text{dry}) = 0.4; \tag{2.5}$$

$$p(\text{not shiny}|\text{dry}) = 0.6. \tag{2.6}$$

Step 2a: Inference Using Bayes' Rule

If we make the observation that a floor is shiny, we will want to draw a conclusion about whether or not it is wet. Inference can be done using Bayes' rule, a central rule of probability calculus:

$$p(A|B) = \frac{p(B|A)p(A)}{p(B)}, \tag{2.7}$$

where A and B are any two random variables.

Exercise 2.1 Prove Bayes' rule. Hint: according to the chain rule, the probability for having both A and B is $p(A, B) = p(A)p(B|A)$. For more background, you may want to read appendix section B.11.

In the context of inference, A is a *hypothesized* rather than an actual state of the world, and we will often denote it by H. B is an observation or set of observations, which we will denote by "obs." In the shiny floor example, the hypotheses are that the floor is wet and that it is dry.

Bayes' rule. Bayes' rule in the context of inference is therefore

$$p(H|\text{obs}) = \frac{p(\text{obs}|H)p(H)}{p(\text{obs})}. \tag{2.8}$$

Here, $p(H)$ is called the prior probability of H, $p(\text{obs}|H)$ is called the likelihood of H, and $p(H|\text{obs})$ is called the posterior of H. This equation reflects the intuitive nature of Bayes' rule as it applies to the evaluation of hypotheses: if a hypothesis is more often true in the world, then before we even take the observations into consideration, that hypothesis is more probable (prior), and if the observations are more expected under a hypothesis (likelihood), then that hypothesis is also more probable.

In this chapter and most of the book, we assume that the observer's understanding of the generative model is correct; that is, it is an accurate description of the true statistics of world states and observations. This means that the observer's priors are numerically identical to the frequencies of the corresponding world states in the generative model and that the observer's likelihoods correctly represent the probabilities of the observation given the world states. However, it is possible that a mismatch could happen between the observer's assumed generative model (sometimes called the observer's *internal model*) and the actual generative model. For example, for some reason you could believe that 50 percent of all floors are wet, even though the true proportion is 10 percent. We will comment on such *model mismatch* in several places later in the book.

Prior. The prior distribution reflects the believed frequencies of occurrence of the values of the world state of interest, here wetness. These values are obtained from the frequencies of these world states in the generative model (figure 2.1).

$$\text{Prior(wet)} = p(\text{wet}) = 0.1; \tag{2.9}$$

$$\text{Prior(dry)} = p(\text{dry}) = 0.9. \tag{2.10}$$

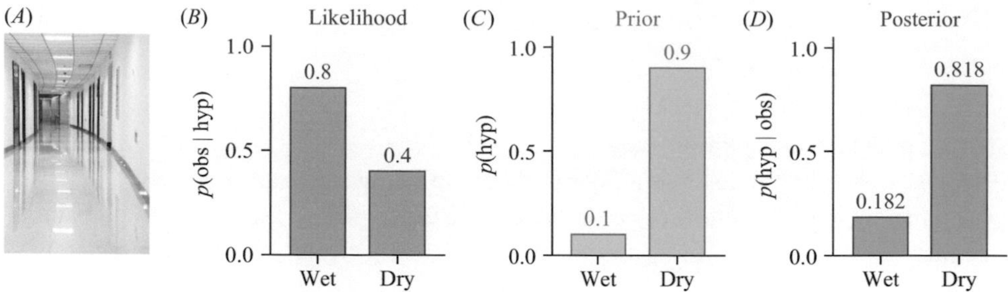

Figure 2.2
A shining example of perceptual inference. (*A*) Is this floor wet? (*B*) A wet floor is more likely than a dry floor to be shiny. (*C*) Most floors are dry. (*D*) The posterior distribution favors the hypothesis that the floor is dry.

Likelihoods Next, we consider the likelihoods. Suppose you make the observation that the floor is shiny (see figure 2.2). The likelihood of "wet" is then the probability that the floor is shiny if it is wet, whereas the likelihood of "dry" is the probability that the floor is shiny if it is dry. These values are obtained from the generative model, but we need only use those probabilities that pertain to the actual observation, "shiny." The likelihoods are

$$\text{Likelihood(wet)} = p(\text{shiny}|\text{wet}) = 0.8; \tag{2.11}$$

$$\text{Likelihood(dry)} = p(\text{shiny}|\text{dry}) = 0.4. \tag{2.12}$$

It is important to understand that likelihoods *do not need to sum to 1*. The likelihood function is not a probability distribution over the world state of interest.

Note on terminology: Never say "the likelihood of the observation." The likelihood is always a function of the hypothesized state of the world. It is equal to the *probability* of a given observation under the hypothesized state of the world.

Protoposterior The numerator in Bayes' rule, the product of likelihood and prior, does not have an official name, but we will call it the *protoposterior* of a hypothesis H:

$$\text{Protoposterior}(H) = p(\text{obs}|H)p(H) = p(H, \text{obs}). \tag{2.13}$$

In our example, the protoposteriors of the two hypotheses are

$$\text{Protoposterior(wet)} = \text{Prior(wet)} \cdot \text{Likelihood(wet)} = 0.1 \cdot 0.8 = 0.08; \tag{2.14}$$

$$\text{Protoposterior(dry)} = \text{Prior(dry)} \cdot \text{Likelihood(dry)} = 0.9 \cdot 0.4 = 0.36. \tag{2.15}$$

The protoposterior is the probability of the hypothesis and the observation together. In our example, 8 percent of floors are wet and shiny, and 36 percent of floors are dry and shiny. Like likelihoods, protoposteriors will in general not sum to 1.

Normalization Bayes' rule, equation (2.8), tells us to divide the protoposteriors by $p(\text{obs})$, the probability of the observations regardless of what world state they were produced by. As it turns out, $p(\text{obs})$ is equal to the sum of the protoposteriors over all hypotheses:

$$p(\text{obs}) = \sum_H \text{Protoposterior}(H) \tag{2.16}$$

$$= \sum_H p(\text{obs}|H)p(H). \tag{2.17}$$

Exercise 2.2 Prove this. Hint: think of all the ways in which a given observation can be reached. Again, for more background, consult appendix section B.11.

In our example,

$$p(\text{shiny}) = \text{Protoposterior(wet)} + \text{Protoposterior(dry)} \tag{2.18}$$

$$= 0.08 + 0.36 = 0.44. \tag{2.19}$$

This calculation indicates that 44 percent of all floors are shiny.

Posteriors In order to obtain posterior probabilities that sum to 1, we normalize the protoposteriors, that is, divide each by their sum, $p(\text{obs})$. In this sense, $p(\text{obs})$ is a *normalization factor* in Bayes' rule. In our example,

$$\text{Posterior(wet)} = p(\text{wet}|\text{shiny}) = \frac{\text{Protoposterior(wet)}}{p(\text{shiny})} = \frac{0.08}{0.44} = 0.182; \tag{2.20}$$

$$\text{Posterior(dry)} = p(\text{dry}|\text{shiny}) = \frac{\text{Protoposterior(dry)}}{p(\text{shiny})} = \frac{0.36}{0.44} = 0.818. \tag{2.21}$$

Thus, $p(\text{obs})$ guarantees that, unlike likelihoods or protoposteriors, posterior probabilities always sum to 1, that is, are *normalized*. As a result of our observation that the floor is shiny, we have updated our beliefs in the hypotheses. Our belief that the floor is dry has reduced from our prior, 90 percent, to our posterior, 82 percent. The likelihood function favors the wet hypothesis (i.e., wet floors tend more often than dry ones to be shiny), so we have lost some confidence in the hypothesis that the floor is dry. Nevertheless, we still favor this hypothesis, because the strength of the evidence in favor of wet, as indicated by the ratio of likelihoods for the hypotheses, $\frac{\text{Likelihood(wet)}}{\text{Likelihood(dry)}} = 2$, is not sufficiently large to offset the ratio of priors, which favors the hypothesis that the floor is dry: $\frac{\text{Prior(wet)}}{\text{Prior(dry)}} = \frac{1}{9}$.

Exercise 2.3 Suppose the floor is not shiny. What is the posterior probability that it is dry?

2.2 Alternative Form of Bayes' Rule

We could combine equations (2.8) and (2.17) to obtain an alternative form of Bayes' rule:

$$p(H|\text{obs}) = \frac{p(\text{obs}|H)p(H)}{\sum_H p(\text{obs}|H)p(H)}. \tag{2.22}$$

This form makes clear that the denominator is a sum of terms such as those in the numerator. The numerator is only for one specific H (the one on the left-hand side of the equation), and the denominator is a sum over all H.

2.3 Areal Representation

In order to deepen our understanding of Bayes' rule, let's recast the shiny floor problem by representing the relevant probabilities as areas of rectangles (figure 2.3). The large outer rectangle, of area 1, represents the universe of possibilities, namely that the floor is either wet or dry. The prior probability that the floor is wet, $p(\text{wet})$, is represented by the area of the left rectangle; the prior probability that the floor is dry, $p(\text{dry})$, is represented by the area

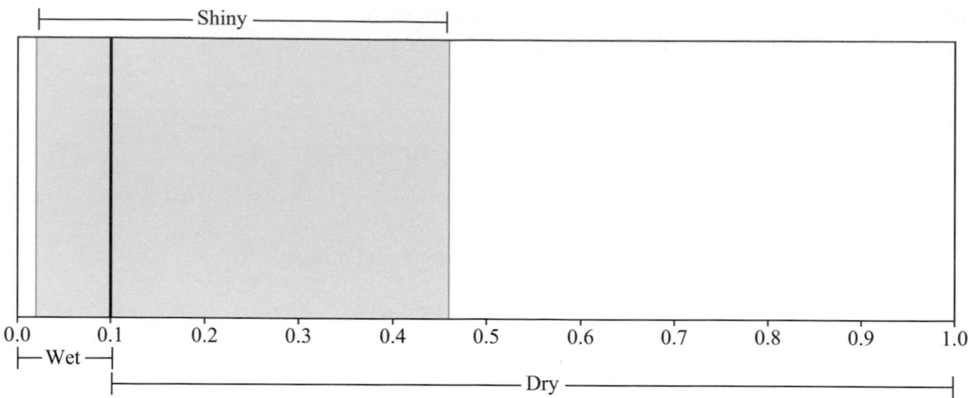

Figure 2.3
Areal representation of Bayesian inference. The posterior probability that the floor is wet, given that it is shiny, is the blue area within the left rectangle divided by the total blue area. The posterior probability that the floor is dry, given that it is shiny, is the blue area within the right rectangle divided by the total blue area.

of the right rectangle. The probability of a shiny floor, p(shiny), is represented by the shaded blue area. Blue fills 80 percent of the left rectangle and 40 percent of the right rectangle; these probabilities represent $p(\text{shiny} \mid \text{wet})$ and $p(\text{shiny} \mid \text{dry})$, respectively. The probability that the floor is both wet and shiny, $p(\text{wet, shiny})$, is the blue area within the left rectangle. The probability that the floor is both dry and shiny, $p(\text{dry, shiny})$, is the blue area within the right rectangle.

Now, let's suppose that you throw a dart that has uniform probability of landing anywhere within the large outer rectangle of area 1. Suppose your dart happens to land somewhere in the blue region (i.e., you have sampled a shiny floor). What is the probability that it is also within the left (wet) rectangle? On reflection, it should be clear that this probability is equal to the proportion of blue area that is also "wet," that is:

$$p(\text{wet} \mid \text{shiny}) = \frac{p(\text{wet, shiny})}{p(\text{shiny})}. \tag{2.23}$$

Similar reasoning reveals that

$$p(\text{dry} \mid \text{shiny}) = \frac{p(\text{dry, shiny})}{p(\text{shiny})} \tag{2.24}$$

To drive this point home, imagine throwing 100 darts that land uniformly randomly within the large outer rectangle of area 1. Whenever a dart happens to land in the blue region, you record whether it is within the Wet or the Dry rectangle. On average, you would find that 44 of your 100 darts land in the blue, and that, of those 44 darts, 8 land in the Wet rectangle and 36 in the Dry rectangle. Therefore, the probability of wet, given shiny, is $\frac{8}{44}$, and the probability of dry, given shiny, is $\frac{36}{44}$, as we found above.

Step 2b: Read out of the Posterior

The output of Bayes' rule is a posterior distribution over the hypothesized state of the world, $p(H|\text{obs})$. However, when we obtain the posterior distribution, we are not yet done, because usually the goal of a model is to predict what an observer would *decide* or *perceive*. A decision is an answer to a question such as "which hypothesis is true?" A percept is what the observer perceives. It could be the answer to the question "What do you see?" (or hear, smell, feel, etc.), but it is not necessarily tied to a query—it could be an unprompted experience. One could argue that all percepts are decisions, but that is mostly a topic for philosophical discussion.

Note on terminology: Do not confuse the terms "percept" and "observation," even though they may appear interchangeable. An observation is the input to the inference process, a percept is the output from it.

A *readout* of the posterior is a mapping from the posterior distribution to a decision or percept. The most obvious readout is to pick the maximum, the hypothesis with the highest posterior probability. This readout is also called *maximum a posteriori* (MAP) estimation. In our example, the MAP percept would be that the floor is dry. Moreover, the observer would be moderately confident about that conclusion, since the corresponding posterior probability is 81.8 percent.

One difference between decision and percept is that the former can incorporate external rewards and costs, whereas the latter does not. For example, it can be much more costly to mistake a wet floor for dry than the other way round. As a consequence, even with a posterior probability of 81.8 percent that the floor is dry, it might still be a good decision to walk more carefully. We will say more about combining inference with utility in chapter 13.

We will later put two caveats on the universality of MAP estimation: one, for a continuous variable, picking the maximum is not always the best readout; two, decision noise might be present and cause deviations from MAP estimation.

2.4 The Prosecutor's Fallacy

We now consider several interesting examples that will serve to consolidate our understanding of the steps outlined above and that also hint at the vast range of scenarios to which Bayesian inference applies. The first of these is the prosecutor's fallacy.

As we have seen, conditional probabilities are not symmetric. In general, $p(A|B) \neq p(B|A)$. For instance, in the shiny floor example, we found that $p(\text{shiny}|\text{wet}) = 0.8$, whereas $p(\text{wet}|\text{shiny}) = 0.182$. This asymmetry is apparent in the areal diagram (figure 2.3), where 80 percent of the Wet rectangle is filled with blue, but only 18.2 percent of the blue region falls within the Wet rectangle. Unfortunately, people who are untrained in probabilistic reasoning sometimes make the mistake of equating $p(A|B)$ and $p(B|A)$. This fallacy is called the *prosecutor's fallacy* or the *conditional probability fallacy*. The prosecutor's fallacy takes its name from the false argument, sometimes put forth in courts of law, that $p(\text{defendant is innocent} \mid \text{evidence}) = p(\text{evidence} \mid \text{defendant is innocent})$. For example, suppose that a partial, smudged fingerprint is found on a weapon left at a crime scene. A fingerprint database search reveals that a person who lives in the same city has a fingerprint that matches the one left on the weapon. A forensic expert testifies that only 1 in 1,000 randomly selected people would provide such a match. The prosecutor argues that, based on the forensic expert's testimony, the probability that the defendant is innocent is only 1 in 1,000. The prosecutor is confusing $p(\text{observation} \mid \text{innocent})$—the testimony of the forensic expert—with $p(\text{innocent} \mid \text{observation})$.

Bayes' rule permits the correct calculation of $p(A|B)$ from $p(B|A)$ and other relevant probabilities and has been used for this purpose in some courts [50]. Let us suppose that the city has 1,000,001 (1 million plus 1) adult inhabitants. Given only that the defendant lives in the city, their prior probabilities of being innocent (H_1) or guilty (H_2) are therefore:

$$\text{Prior}(H_1) = \frac{1000000}{1000001}; \tag{2.25}$$

$$\text{Prior}(H_2) = \frac{1}{1000001}. \tag{2.26}$$

The observation that the defendant's fingerprint matches that at the crime scene results in the likelihoods:

$$\text{Likelihood}(H_1) = \frac{1}{1000}; \tag{2.27}$$

$$\text{Likelihood}(H_2) = 1. \tag{2.28}$$

The protoposteriors are:

$$\text{Protoposterior}(H_1) = \text{Prior}(H_1) \cdot \text{Likelihood}(H_1) = \frac{1000000}{1000001} \cdot \frac{1}{1000} = \frac{1000}{1000001}; \tag{2.29}$$

$$\text{Protoposterior}(H_2) = \text{Prior}(H_2) \cdot \text{Likelihood}(H_2) = \frac{1}{1000001} \cdot 1 = \frac{1}{1000001}. \tag{2.30}$$

The normalization is

$$\text{Normalization} = \text{Protoposterior}(H_1) + \text{Protoposterior}(H_2); \tag{2.31}$$

$$= \frac{1000}{1000001} + \frac{1}{1000001} = \frac{1001}{1000001}. \tag{2.32}$$

Thus,

$$\text{Posterior}(H_1) = \frac{\text{Protoposterior}(H_1)}{\text{Normalization}} = \frac{1000}{1001}; \tag{2.33}$$

$$\text{Posterior}(H_2) = \frac{\text{Protoposterior}(H_2)}{\text{Normalization}} = \frac{1}{1001}. \tag{2.34}$$

The defendant is almost surely innocent, despite the prosecutor's argument! Another way to explain this: The city contains 1,000,001 people, 1,000,000 of whom are innocent and 1 of whom is guilty. Consequently, if we had the fingerprints of everyone in the city, we would expect 1001 matches, only 1 of which is from the guilty citizen. The probability of guilt given a fingerprint match is therefore $\frac{1}{1001}$.

2.5 A Changing Prior: Luggage Carousel Example

We now move to a slightly more complicated example, in which relevant probabilities change over time. Many air travelers have waited expectantly in an airport baggage claim area, watching for their bags to drop down the chute into the circulating luggage carousel

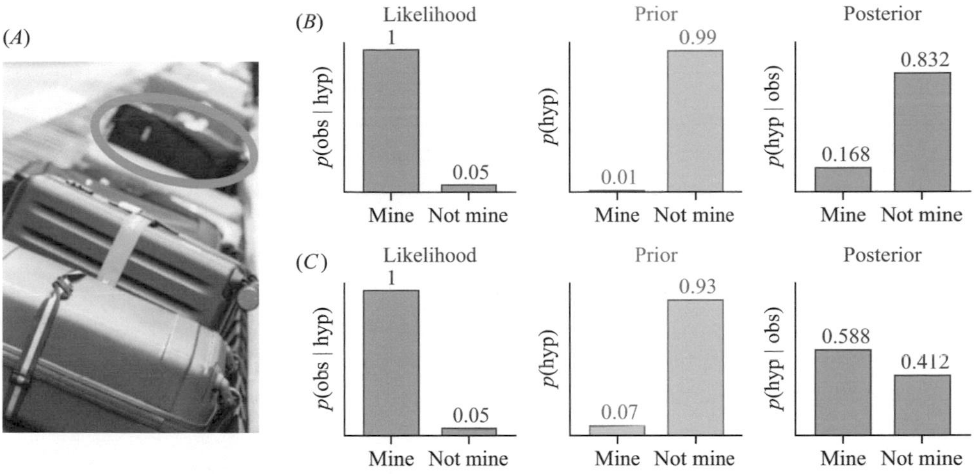

Figure 2.4
Expectation influences perception. (*A*) The first bag and the 86th bag both match yours in shape, size, and color. (*B*) Likelihood function, prior probability distribution, and posterior probability distribution on viewing the first bag enter the luggage carousel. Your posterior distribution indicates that the bag is probably not yours. (*C*) Likelihood function, prior distribution, and posterior distribution on viewing the 86th bag enter the luggage carousel. The same likelihood as in (*A*) combined with a different prior expectation produces a posterior distribution that favors the hypothesis the bag is yours.

(figure 2.4A). This situation presents opportunities for probabilistic reasoning [155] as well as perceptual inference. Let us suppose that you are engaged in this ritual of modern-day air travel along with 99 other passengers from your flight, each of whom, like you, checked one item of luggage. A recording piped through the speakers reminds you that "Many bags look alike. Please check your bag carefully before exiting the terminal." Indeed, your bag is one of the most popular models on the market, a medium-sized black rectangular case used by 5 percent of all travelers. Given your distance from the luggage chute, you can discern only the shape, size and color of the bags that emerge, not any detailed markings on the bags. Now let us suppose that the first bag from your flight to enter the luggage carousel indeed has the same shape, size, and color as your bag. Is it yours?

This question cannot be answered with a definite "yes" or "no." Rather, the question demands a probabilistic judgment. You may consider it more or less likely that the bag is yours, but cannot yet be sure. In lieu of certainty, perception is most often characterized by varying degrees of confidence, which can be expressed as probabilities ranging from impossible to certain, occupying some particular place along the stretch of numbers between 0 and 1 (0 to 100 percent). As you view the bag in the luggage carousel, you will have an intuitive sense of the probability that it is your bag, p(this bag is mine | shape, size, color). But how could you arrive at this probability estimate?

Likelihoods At the root of perceptual uncertainty is the fact that different world states can generate the same sensory observation. Not only do "many bags look alike," but many objects, people, and events produce nearly identical observations of one kind or another (sights, sounds, etc.). Thus, the information provided by the senses is typically imprecise, open to multiple interpretations.

What information is contained in your observation? If the bag you are viewing is in fact your own, it will have the same shape, size, and color. Accordingly, p(observed shape, size, color | my bag) = 1. But even if the bag you are viewing is not your own, it has some chance of matching the shape, size, and color of your bag. Since your bag is the model used by 5 percent of travelers, p(observed shape, size, color | not my bag) = 0.05. These two conditional probabilities are the *likelihoods*. As a reminder, the likelihood of a hypothesis is the probability of the sensory observations if the hypothesis were true or, in other words, how expected the observations are if the hypothesis were true. As before, a plot of the likelihood of every possible world state, known as the *likelihood function*, summarizes the degree to which the observation is compatible with each world state interpretation (figure 2.4). The less informative the observation, the "broader" or "flatter" will be the likelihood function; the more informative the observation, the "narrower" or "sharper" will be the likelihood function.

If you look at your bag close-up, you will notice identifying markings—a name tag, a piece of string you have attached to the handle, and so on—that allow you to unambiguously identify your bag. In other words, as the bag approaches you on the carousel, your likelihood function will generally sharpen considerably. This is how, on a relatively short timescale, you can become convinced that the bag you have been observing is in fact not your own.

Prior Let us now consider how your perceptual inference changes over a longer timescale, as you wait patiently at the carousel, watching bag after bag drop from the chute. When the very first bag from your flight emerges from the chute, and you notice the resemblance to your own bag, you will be hopeful but at the same time probably somewhat doubtful that the bag in question is your own. Your skepticism is justified, because not only do 5 percent of bags look like yours, but the probability that your bag would emerge as the first off the flight is just 1 in 100. After all, your flight carried 100 passengers, each of whom checked one bag. Now let us suppose that you have waited expectantly at the carousel, viewing each bag that emerges and checking more closely those that resembled your own, only to find yourself, 10 minutes and 85 bags later, still without having encountered your bag. At this point, let us suppose that the 86th bag emerges, and it again resembles your own. This time, you will be more confident than before that the bag is yours, despite the fact that the observation, and therefore the likelihood function, is identical for the first and the 86th bags. This illustrates that your perception, p(world state | observation), is not the same as the likelihood, p(observation | world state).

In short, perceptual inference is based not just on the observation (as reflected in the likelihood function), but also on expectation. As explained previously, we represent expectation by *prior probability*. The prior probability of a world state is based on all relevant information except the current observation. In the present example, your experience of waiting patiently as 85 bags emerged onto the carousel, together with your background knowledge that 100 bags were present on your flight, has informed you that the prior probability that your bag will emerge next is 1 in 15 (i.e., 6.7 percent), which is greater than the 1 percent that it was for the first bag. Although prior probabilities are conditioned on experience and background knowledge, in the interest of brevity we usually omit the conditioning symbol (|) and write prior probabilities simply as p(hypothesized world state), for example, p(the bag is mine) and p(the bag is not mine). We plot the prior probability of each hypothesized world state as a *prior probability distribution* (figure 2.4B–C).

Posterior Let us calculate the posterior probability that the first bag that you see emerge onto the luggage chute is your own. We first enumerate the possible world states or hypotheses: H_1 (the bag is mine) and H_2 (the bag is not mine). Next, we write down the prior probabilities of each hypothesis, given our knowledge that this is the first bag to appear:

$$p(H_1) = 0.01; \tag{2.35}$$

$$p(H_2) = 0.99. \tag{2.36}$$

We then write the likelihoods that express the probability of the sensory observation (shape, size, and color of the luggage seen) given each hypothesis

$$p(\text{observation}|H_1) = 1; \tag{2.37}$$

$$p(\text{observation}|H_2) = 0.05. \tag{2.38}$$

Since the prior probability is 1 percent that the first bag is yours, it is 99 percent that the first bag is not yours. Note that, since the visual image shows a bag that matched yours in shape, size, and color, we set the likelihood to 1 for H_1. This is logical, since if it were your bag, the visual image will surely match the shape, size, and color of your bag. Finally, we enter the prior probabilities and likelihoods into Bayes' rule, to calculate the posterior probabilities of the hypotheses:

$$p(H_1|\text{observation}) = \frac{1 \cdot 0.01}{1 \cdot 0.01 + 0.05 \cdot 0.99} = 0.168; \tag{2.39}$$

$$p(H_2|\text{observation}) = \frac{0.05 \cdot 0.99}{1 \cdot 0.01 + 0.05 \cdot 0.99} = 0.832. \tag{2.40}$$

There are several considerations to appreciate at this point:

1. First and foremost, it is important to realize that we have learned from the observation, updating our prior probability for H_1 (0.01) to a posterior probability that is much greater (0.168). Our posterior probably for H_1 has increased because the observation was more consistent with H_1 than with H_2. In general, the more strongly the observation favors one hypothesis over the other, the more we will learn.
2. Nevertheless, we are still more confident that the bag is not ours (83.2 percent) than that it is ours (16.8 percent). Despite the favorable observation, we believe that the bag is most probably not ours, because we started with such a low prior probability for H_1. In essence, the observation of a bag that looks like ours does not sufficiently favor H_1 to overcome our well-justified prior bias against H_1.
3. Importantly that the posterior probability, $p(H_1|\text{observation}) = 16.8$ percent, does not equal the likelihood, $p(\text{observation}|H_1) = 100$ percent. As explained above, in general, $p(A|B) \neq p(B|A)$.
4. Note that in this example the hypothesis with the maximum likelihood (known as the maximum-likelihood estimate, or MLE)—H_1—is not the hypothesis with the maximum posterior probability (the MAP estimate)—H_2. This situation is not uncommon in perceptual inference. Sometimes, the MLE and the MAP estimate are the same, but often they are not.

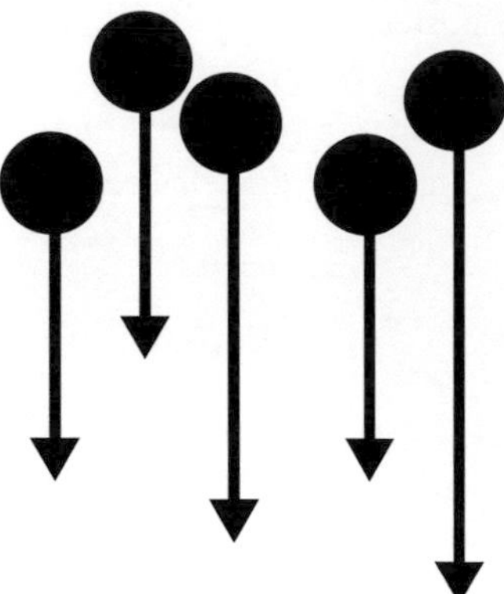

Figure 2.5
Five dots that all move downward.

Now suppose that we continue to wait for our bag to appear, failing to see it among the first 85 bags to enter the carousel. To calculate the posterior probability that the 86th bag, which also matches ours in shape, size, and color, is our own, we follow the same procedure, but with new prior probabilities of $\frac{1}{15}$ for H_1 and $\frac{14}{15}$ for H_2 (figure 2.4C).

Exercise 2.4 Verify that the posterior probabilities when evaluating the 86th bag is roughly $p(H_1|\text{observation}) = 0.588$ and $p(H_2|\text{observation}) = 0.412$.

Thus, the probability that the bag we are viewing is our own has now increased dramatically, from 16.8 percent (first bag seen) to 58.8 percent (86th bag seen), despite the fact that in the two cases the observation, and therefore the likelihood functions, are the same. This makes very clear that the posterior distribution depends not only on the sensory data but also on the prior distribution.

2.6 A Flat Prior: Gestalt Perception Example

After these more cognitive examples, we return to perception. A common misunderstanding about Bayesian models of perception is that unequal priors must always play a role. In fact, important results can be obtained from scenarios that involve flat prior distributions. Suppose that you observe five dots all moving downward, as indicated by the arrows in figure 2.5. Most people would perceive such a stimulus as a single group or object moving downward. The traditional account of the percept is that the brain has a tendency to group the dots together because of their common motion. This is captured by the Gestalt principle of "common fate." Gestalt principles or Gestalt laws, originally formulated by Wertheimer [200], describe how humans group together disparate elements according to their locations or features. Such principles, however, are merely narrative summaries of the phenomenology. A Bayesian model can provide a deeper *explanation* of the percept and, in some cases, even make quantitative predictions.

Let us take a Bayesian approach to understanding the five-dot case. The retinal image of each dot serves as a sensory observation. We will denote these five retinal images by I_1, I_2, I_3, I_4, and I_5.

Step 1: Generative model The first step in Bayesian modeling is to formulate a generative model: a graphical or mathematical description of the possibilities that could have produced the sensory observations. Let's say that the brain considers only two possibilities:

Hypothesis 1: All dots are part of the same object, and they therefore always move together. They move together, either up or down, each with probability $\frac{1}{2}$.

Hypothesis 2: Each dot is an object by itself. Each dot independently moves either up or down, each with probability $\frac{1}{2}$.

(We are assuming that dots are allowed to move only up and down, and that speed and position do not play a role in this problem.) The generative model diagram below shows each hypothesis in a big box. Inside each box, the bubbles contain the variables and the arrows represent dependencies between variables, like before.

Exercise 2.5 Put the following variable names in the correct boxes below, for hypothesis 1 (left) and hypothesis 2 (right): retinal images I_1, I_2, I_3, I_4, and I_5, and motion directions s (a single motion direction), or s_1, s_2, s_3, s_4, and s_5. The same variable might appear more than once.

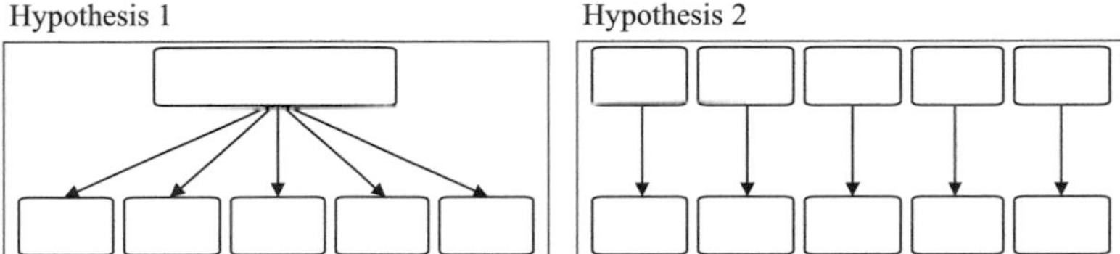

Step 2: Inference Consider now the specific observed configuration {down, down, down, down, down}, that is, all five dots are moving down, which we will denote by I_{obs}. The brain's challenge is to determine based on I_{obs} which of the two hypotheses (1 or 2) , which we will denote by H_1 and H_2, is the right one. As before, the *likelihood* of each hypothesis is the probability of the sensory observations if that hypothesis were true. By the information provided in the problem, the two likelihoods are

$$\text{Likelihood}(H_1) = p(I_{\text{obs}}|H_1) = \frac{1}{2}; \tag{2.41}$$

$$\text{Likelihood}(H_2) = p(I_{\text{obs}}|H_2) = \left(\frac{1}{2}\right)^5 = \frac{1}{32}. \tag{2.42}$$

Note, as we have seen previously, that the two likelihoods do not sum to 1. Note also that the second likelihood is much lower than the first one. This is because, under H_2, quite a coincidence is required for all the dots to happen to be moving downward.

Priors Let us say that hypothesis 1 occurs as often in the world as hypothesis 2. The observer can use these frequencies of occurrence as prior probabilities, reflecting expectations in the absence of specific sensory observations. The prior probabilities of hypotheses 1 and 2 are then

$$\text{Prior}(H_1) = \frac{1}{2}; \tag{2.43}$$

$$\text{Prior}(H_2) = \frac{1}{2}. \tag{2.44}$$

Protoposteriors The protoposteriors of the two hypotheses are

$$\text{Protoposterior}(H_1) = \text{Prior}(H_1) \cdot \text{Likelihood}(H_1) = \frac{1}{2} \cdot \frac{1}{2} = \frac{1}{4}; \tag{2.45}$$

$$\text{Protoposterior}(H_2) = \text{Prior}(H_2) \cdot \text{Likelihood}(H_2) = \frac{1}{2} \cdot \left(\frac{1}{2}\right)^5 = \frac{1}{64}. \tag{2.46}$$

Normalization The normalization is the sum of the protoposteriors:

$$\text{Normalization} = \text{Protoposterior}(H_1) + \text{Protoposterior}(H_2) = \frac{1}{4} + \frac{1}{64} = \frac{17}{64}. \tag{2.47}$$

Posteriors The posterior probabilities are obtained by dividing each protoposterior by the normalization:

$$\text{Posterior}(H_1) = p(H_1 | I_{\text{obs}}) = \frac{\text{Protoposterior}(H_1)}{\text{Normalization}} = \frac{\frac{1}{4}}{\frac{17}{64}} = \frac{16}{17}; \tag{2.48}$$

$$\text{Posterior}(H_2) = p(H_2 | I_{\text{obs}}) = \frac{\text{Protoposterior}(H_2)}{\text{Normalization}} = \frac{\frac{1}{64}}{\frac{17}{64}} = \frac{1}{17}. \tag{2.49}$$

Percept In this case, the MAP percept is H_1, that is, that the dots move down together. This is consistent with the law of common fate, but we have arrived at our conclusion by calculating probabilities. A similar logic applies to the other Gestalt principles. In section 10.5.1, we will explore the Gestalt principle of good continuation, which states that contours with smooth edges are more likely to be perceived as continuous than contours with edges that have abrupt or sharp angles.

2.7 Optimality, Evolution, and Motivations for Bayesian Modeling

The notion of Bayesian inference is closely tied to that of *optimal* behavior. Discussing optimality requires specifying objective function—a function that specifies how good or bad a response is, given a true state of the world. In perception, the objective function is often accuracy (correctly reporting the true state of the world) or error (amount of deviation from the true state of the world). In other decisions, external rewards and costs play a role. An observer could be optimal *with respect to* a specific objective function. For instance, considering further an example from chapter 1, an observer might calculate a posterior probability of "only" 0.1 that a lion is hiding in the grass; nevertheless, it may be optimal to leave the area because the cost of remaining could be extremely high if the lion is present. Given a generative model and an objective function, computing the posterior is always a part of deriving the optimal solution. This is because the posterior represents whatever there is to know about the world state of interest given the observations. We will return to the notion of optimality in chapters 4 and 13.

A caveat on the link between Bayesian inference and optimality is that if the observer or agent uses an incorrect generative model in inference, their behavior will not be optimal. Using an incorrect generative model in inference is a form of *model mismatch*, which we will discuss in chapter 3.

The link between Bayesian inference and optimality serves as a potential motivation for building Bayesian models. The argument would be that through evolution, some common and important brain functions might to a large extent have been optimized. Therefore, it is likely that in some tasks, behavior close to optimal will be found. This argument is more plausible for evolutionarily old functions—perception and movement—than for more recent functions, such as higher cognitive functions. However, one does not need to accept this argument in order to motivate Bayesian modeling. One could also view a Bayesian model as a starting point for building alternative, suboptimal models.

The fact that the posterior distribution is a part of *deriving* the optimal strategy should not, however, be taken to mean that optimal behavior implies that the brain represents posterior distributions. It could simply mean that the mapping from observations to response is the same as the one obtained using the Bayesian recipe. This is sometimes called "as if" Bayesian behavior. Demonstrating that the brain actually *represents* priors, likelihoods, and posteriors is more involved, and we will discuss it in chapter 15.

2.8 Summary

In this chapter, we introduced the precise formulation of Bayes' rule and applied it to a range of discrete estimation problems. We have seen how Bayes' rule makes concrete, meaningful statements about probabilities possible. Regarding the use of Bayes' rule, we learned the following:

- Bayesian modeling starts with a model of the statistical structure of the world and the observations: the generative model.
- Conditional probabilities are not symmetrical. In general, $p(A|B) \neq p(B|A)$.
- Bayes' rule calculates the probabilities of hypotheses conditioned on the observation—the posterior probabilities, $p(H|\text{obs})$—from the probabilities of the observation conditioned on the hypotheses—the likelihoods, $p(\text{obs}|H)$—and the prior probabilities, $p(H)$.
- $p(\text{obs})$ in Bayes' rule normalizes the probability and can be rewritten as $p(\text{obs}) = \sum_H p(\text{obs}|H)p(H)$.
- Priors do not always play the main role in Bayesian models. In some problems, such as the Gestalt example, the prior is relatively unimportant and the likelihood dominates.
- In some scenarios, priors and/or likelihoods may change over time.
- Bayesian inference is a component of deriving the optimal strategy in any inference task. However, not all Bayesian inference is optimal.
- Finding evidence for (near-)optimal behavior does not necessarily mean that the components of the Bayesian model are internally represented in the brain.

2.9 Suggested Readings

- Gar Ming Chan. "Bayes' Theorem, COVID19, and Screening Tests." *American Journal of Emergency Medicine* 38, no. 10 (2020): 2011–2013.

- Norman Fenton. "Improve Statistics in Court." *Nature* 479, no. 7371 (2011): 36–37.
- Wilson S. Geisler and Jeffrey S. Perry. "Contour Statistics in Natural Images: Grouping across Occlusions." *Visual Neuroscience* 26, no. 1 (2009): 109–121.
- Thomas L. Griffiths and Joshua B. Tenenbaum. "Optimal Predictions in Everyday Cognition." *Psychological Science* 17, no. 9 (2006): 767–773.
- Sharon Bertsch McGrayne. *The Theory that Would Not Die*. New Haven, CT: Yale University Press, 2011.
- Jason Rosenhouse. *The Monty Hall Problem: The Remarkable Story of Math's Most Contentious Brainteaser*. Oxford: Oxford University Press, 2009.
- James V. Stone. *Bayes' Rule: A Tutorial Introduction to Bayesian Analysis*. Sebtel, 2013.
- Max Wertheimer. "Gestalt Theory." In *A Source Book of Gestalt Psychology*, edited by Willis D. Ellis, 1–11 Kegan Paul, Trench, Trubner, 1938.

2.10 Problems

Problem 2.1 Think of three daily-life examples of random variables A and B for which intuitively, $p(A|B) \neq p(B|A)$. In each case, state which probability is greater, and explain why.

Problem 2.2 Imagine you have collected data about reported sightings of the dodo throughout history. We will call these data S. Suppose you are interested in the time the dodo went extinct, denoted E. Then the likelihood function of interest to you is

(a) $p(E|S)$ as a function of S

(b) $p(E|S)$ as a function of E

(c) $p(S|E)$ as a function of S

(d) $p(S|E)$ as a function of E.

Incidentally, a paper has calculated this likelihood [153]:

The dodo (Raphus cucullatus); *Frederick William Frohawk*, 1905.

Problem 2.3 In early July 2021, following large gatherings associated with Independence Day festivities, hundreds of people in Provincetown, Massachusetts, became infected with the virus that causes COVID-19. According to an article published in the *Washington Post* [85], "A sobering scientific analysis . . . found that three-quarters of the people infected . . .

were fully vaccinated." The article goes on to emphasize that infected fully vaccinated individuals are very unlikely to suffer severe illness. Nevertheless, the quoted statement understandably alarmed many readers, as it suggested that the vaccines against COVID-19 were ineffective at preventing infection. A crucial piece of information, not provided in the article, was that a large majority of people in Provincetown were fully vaccinated against COVID-19. Interpret these data, and the alarm readers may have felt, in light of the prosecutor's fallacy.

Problem 2.4 At a particular university, 15 percent of all students are in humanities, 55 percent of all students are undergraduates, and 18 percent of undergraduates are in humanities. What is the probability that a random humanities student is an undergraduate?

Problem 2.5 Bayes' rule plays an important role in medical diagnosis. We will illustrate that in this problem. 1 percent of the population suffers from disease D; this rate is also called the *prevalence* of the disease. A diagnostic test for D is being piloted. The probability that someone without D tests positive (false-alarm rate) is 2 percent. The probability that someone with D tests negative (miss rate) is 3 percent.

(a) Make a quick guess of the probability that someone who tests positive actually has D.

(b) Calculate this probability. If it is very different from your answer to (a), what went wrong in your intuition?

(c) This variation was proposed by Huihui Zhang, then a student at Beijing University. Suppose now that there is an extra variable we have ignored, namely whether someone goes to the doctor to have a diagnostic test done. This probability is higher if someone has the disease (because there will likely be symptoms) than if someone does not have the disease. Assume a 5-to-1 probability ratio for this. Now recalculate the probability that someone who tests positive actually has D. Is it closer to your original intuition?

For more practice, see problem B.7.

Problem 2.6 Explain intuitively why likelihoods do not need to sum to 1, whereas priors and posteriors do.

Problem 2.7 Prove, using Bayes' rule, that when the likelihood function is perfectly flat (has the same value for all hypotheses), the posterior distribution is identical to the prior distribution.

Problem 2.8 We build on the luggage carousel problem from section 2.5. Prove, using Bayes' rule, that if you see a bag on the luggage carousel that does not match yours (for instance, a small red bag, when yours is large and black), the posterior probability that it is yours is zero.

Problem 2.9 We again build on the luggage carousel problem from section 2.5. You are one of 100 passengers waiting for your bag at an airport luggage carousel. Your bag looks the same as 5 percent of all bags. Derive a general expression for the probability that the bag you are viewing (which matches your bag visually) is your own, as a function of the number of bags you have viewed so far. How many bags must you view (without finding your own) before the posterior probability that the bag you are viewing (which matches your own visually) is greater than 70 percent?

Problem 2.10 This problem will reveal a central feature of Bayesian inference; namely, that an observation increases or decreases support for a hypothesis not through the likelihood of the hypothesis but rather through the ratios of likelihoods among hypotheses. As a consequence, we can omit irrelevant details (i.e., details that scale the likelihoods of all hypotheses equally) from our definition of the observation.

In the luggage carousel example (section 2.5.), we defined the visual observation as the shape, size, and color of the bag, and we therefore took $p(\text{observation}|H_1)$ to equal 1 when the observation matched the shape, size, and color of your bag. But of course, the exact "look" of a bag on a luggage carousel involves much more than just its shape, size, and color. For instance, as the bag enters the carousel, it may come to rest at any one of many different orientations. Suppose we were to expand our definition of the observation to include the bag's *orientation* as well as the other three features. To keep things simple, let assume that there are 360 possible angles (one for each degree around the circle) and two possible sides (right side up or upside down), for a total of 720 possible orientations with which a bag may come to rest on the carousel.

(a) If we further assume that each orientation is equally probable, then the probability of the observation given hypothesis 1 is no longer 1, but rather $\frac{1}{720}$. Similarly, the probably of the observation given hypothesis 2 would no longer be 0.05, but rather $\frac{0.05}{720}$. Since the likelihoods have changed, must not the posterior distribution change as well? Explain why or why not.

(b) Now suppose that the 720 orientations do not all have probability $\frac{1}{720}$ but that every orientation, taken individually, still has the same probability under both hypotheses. For instance, every bag (whether yours or not) comes to rest upright and aligned parallel to the edge of the carousel with probability 0.2. Again, would your inference be affected by the orientation of the observed bag?

Problem 2.11 Suppose you are waiting to catch a particular bus in a city that has just ten bus routes; the route followed by each bus is indicated by an integer in the corner of its front display. You see the bus below from a distance, and naturally wonder whether this is the bus you are waiting for.

(a) Based on the visual image of the difficult-to-discern bus route number, and your intuitive understanding of how different route numbers might appear, construct a plausible likelihood function that plots p(visual observation | hypothesized bus route), for all numbers from 1 to 10. (Assume that the written destination does not mean anything to you.)

(b) As it turns out, you happen to know that only buses 3, 4, 5, and 6 travel down the street you are on. Furthermore, you know that buses 3 and 4 come twice as frequently as buses 5 and 6. Based on this background knowledge, construct your prior distribution for the bus number.

(c) Use Bayes' rule to calculate your posterior distribution for the number of the bus.

Problem 2.12 This problem was proposed by Jonathan Gornet when he was an undergraduate at New York University. You are a student in a math class. The professor writes a symbol on the board that looks like a "u" or a "v" and you try to determine which one it is. The sensory observations consist of the retinal image of the handwritten letter, which we denote by I_{obs}. We make the following assumptions:

- There are no other possible letters.
- There is no relevant context.
- "u" occurs 1.5 times as often as "v".
- The probability that a "u" produces I_{obs} is 0.0008, whereas the probability that a "v" produces I_{obs} is 0.0010.

(a) Explain why the probabilities in this last line are so low.

(b) Calculate the posterior probabilities $p(\text{"u"}|I_{\text{obs}})$ and $p(\text{"v"}|I_{\text{obs}})$.

(c) What would you conclude?

(d) If there were other possible symbols, what can you say about their likelihoods and/or priors? Explain.

Problem 2.13 Here, we start to examine the central problem of color vision in Bayesian terms. Ironically, this problem can already be illustrated in colorless vision. In section 1.3, we alluded to the fact that a particular light intensity entering your eyes is consistent with multiple combinations of the true shade of the surface and the intensity of the illuminating light source. To make this more precise, we use the following formula for the intensity measured by the retina:

$$\text{Retinal intensity} = \text{surface shade} \cdot \text{light intensity}. \tag{2.50}$$

Surface shade should be interpreted as the proportion of the incident light that is reflected; as such, it is a number between 0 and 1. We also take light intensity to be between 0 and 1. As an example, if the surface shade is 0.5 (mid-level gray) and the light intensity is 0.2 (very dim light), then the retinal intensity is $0.5 \cdot 0.2 = 0.1$.

(a) Suppose your observed retinal intensity is 0.3. Connect in the diagram below all combinations of hypothesized surface shade and hypothesized light intensity that could have produced this retinal intensity. You should get a curved, not a straight line.

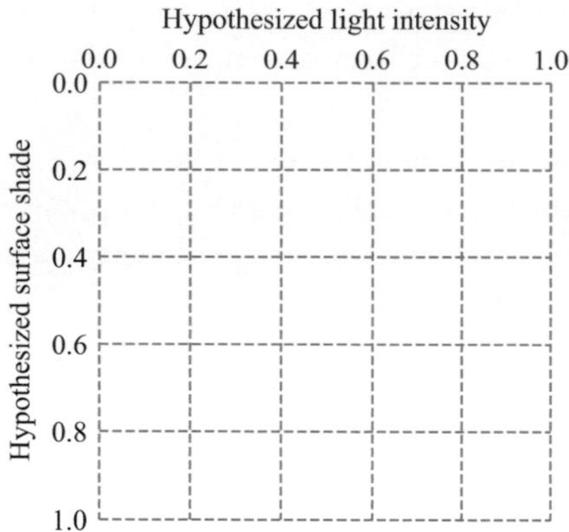

(b) Explain the statement "The curve that we just drew represents the combinations of surface shade and light intensity that have a high likelihood."

(c) Explain how this curve relates to the notion of ambiguity in perception.

Problem 2.14 Imagine you live in a very boring world consisting of a 2×10 grid of squares:

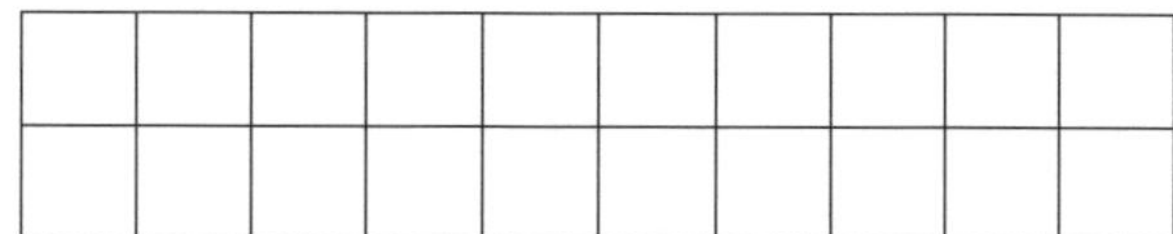

Only two things ever happen in this world: With a probability of 40 percent, a vertical bar will appear in this world, consisting of two black squares in a column, chosen so that each possible column is equally probable. With a probability of 60 percent, one black square will appear in a random position in the top row, and another black square will appear in a random position in the bottom row. When doing inference, we will refer to these possibilities as hypotheses 1 and 2, respectively.

(a) Suppose you have the following retinal image:

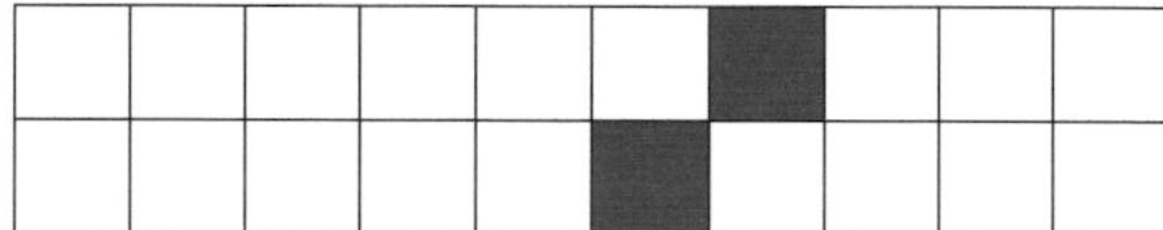

Calculate that based on this retinal image, the posterior probability of hypothesis 1 is 0 percent. Write out all steps in your reasoning, with a brief explanation accompanying each step.

(b) Suppose you have the following retinal image:

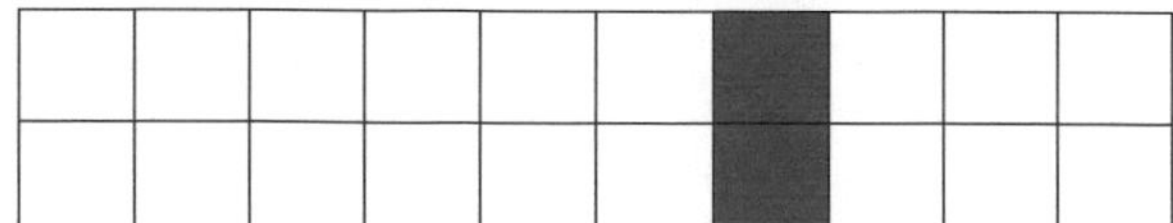

Calculate that based on this retinal image, the posterior probability of hypothesis 1 is roughly 87 percent. Write out all steps in your reasoning, with a brief explanation accompanying each step.

(c) How does your answer to (b) explain why human observers tend to perceive the second image as containing a single object?

Problem 2.15 Imagine you live in only a slightly more interesting world consisting of a 5×5 grid of squares. Again, only two things ever happen in this world:

"Big box": With a probability of 50 percent, a big box consisting of 2×2 small black squares will appear in a random position.

"Small boxes": With a probability of 50 percent, 4 independent black boxes will appear, each occupying one square and again in random places. The boxes will appear one by one, and no box can occupy a location of a box that is already present.

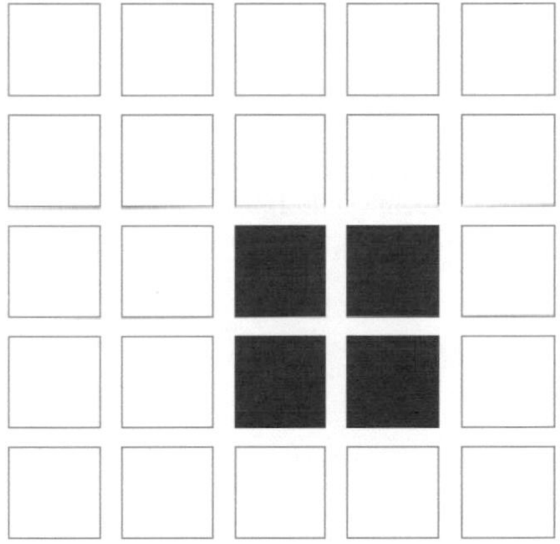

When doing inference, we will refer to these possibilities as hypotheses 1 and 2, respectively.

(a) Draw a generative model diagram. Explain why you drew the diagram the way you did.

(b) We will first consider hypothesis 1. How many possibilities exist within this 5×5 grid to place a 2×2 object? Assuming equal probabilities of possible locations of appearance, what is the probability that it appears exactly in the observed location?

(c) Is the answer to part (b) the likelihood, prior, or posterior of the hypothesis that there is a single object? Explain.

We will now consider hypothesis 2.

(d) Fill in: to place one of these four small boxes, there are ____ out of ____ locations on the grid that are consistent with the observations. This gives a probability of ______. To then place the second small box, there are ____ out of ____ remaining locations on the grid that are consistent with the observations. This gives a probability of ______.

(e) Repeat for the third and fourth small boxes.

(f) Multiply the four probabilities that you found in (d) and (e).

(g) Is the answer to part (f) a likelihood, a prior, or a posterior of the hypothesis? Explain.

(h) How many times bigger is the probability you found in (b) than the one you found in (f)?

(i) How does your answer to (h) explain why human observers tend to perceive the image as containing a single big box.

Problem 2.16 This is a continuation of problem 1.12 about a Bayesian account of motion sickness, following Treisman's hypothesis. In that problem, you drew bar diagrams without numbers. Now that we know how Bayes' rule works, we can do the same calculation more quantitatively.

(a) Make up reasonable numbers for the prior probabilities of hypotheses 1, 2, and 3. They have to sum to 1.

(b) Again assume that you experience a big discrepancy between your visual and vestibular observations. Make up reasonable numbers for the likelihoods of hypotheses 1, 2, and 3. (Only their relative magnitudes matter.)

(c) Based on (a) and (b), calculate the posterior probabilities of the three hypotheses. Verify that they sum to 1. Also verify that they capture Treisman's hypothesis.

Problem 2.17 In the study of the perception of three-dimensional objects, the *generic-view principle* [56, 96] states that the observer is not in a special position relative to the scene. This principle explains why the image in (*A*) tends to be perceived as depicting a square, instead of as the cube in (*B*) viewed from a special viewing angle.

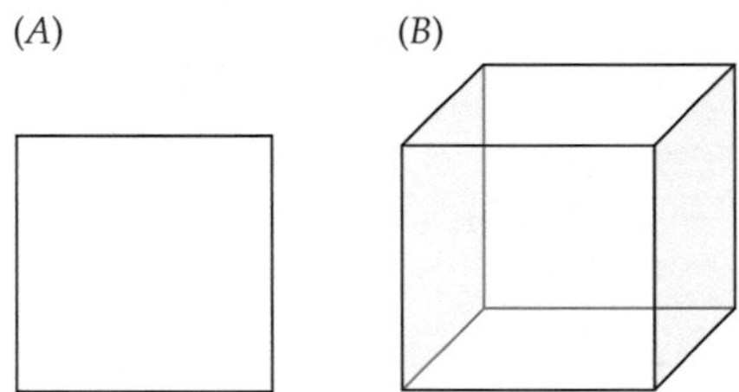

Discuss the generic-view principle in Bayesian terms.

3

Bayesian Inference under Measurement Noise

How does a Bayesian observer infer the state of the world from a noisy measurement?

In chapter 1, we introduced the concept of inference through a variety of daily-life examples. In chapter 2, we examined how to calculate with Bayes' rule. However, the examples we used there involved inferences of categorical variables. The use of categorical variables makes Bayesian calculations easy, but in practice, many variables are continuous rather than categorical. World state variables that the brain may want to infer include the orientation of a line segment, the location of a sound source, the speed of a moving object, the color of a surface, or the time elapsed between two events. In this chapter, we discuss Bayesian models for such continuous variables.

Plan of the Chapter

In this chapter, we mostly consider real-valued world state variables; such variables can take values from negative infinity to infinity. We consider an observer who infers such variables from noisy sensory observations. We will introduce the concept of a measurement, which is an abstraction of the sensory observations that lives in the same space as the world state variable itself. Imperfections in the internal representation of a stimulus can then conveniently be modeled as Gaussian noise in the measurement. We then discuss how the observer combines a prior with a likelihood to compute a posterior.

3.1 The Steps of Bayesian Modeling

Psychophysics is the study of how controlled stimuli are perceived or acted upon by organisms. For example, an experimenter might show you two lines on a computer screen and ask you which one is longer. When the lines are very similar in length, this is a difficult task and you will make mistakes. These mistakes can tell the experimenter about the way you are solving the task. Ever since Gustav Theodor Fechner, around 1860, researchers have used psychophysical methods to probe the nature of perceptual processing. The psychophysics task we will use as the leading example in this chapter is an auditory localization task. Imagine that you are facing a projection screen that displays a horizontal line stretching across the width of the screen. Located behind the screen, at the same elevation as the line, is a densely spaced array of many very small loudspeakers. A tone will originate from one

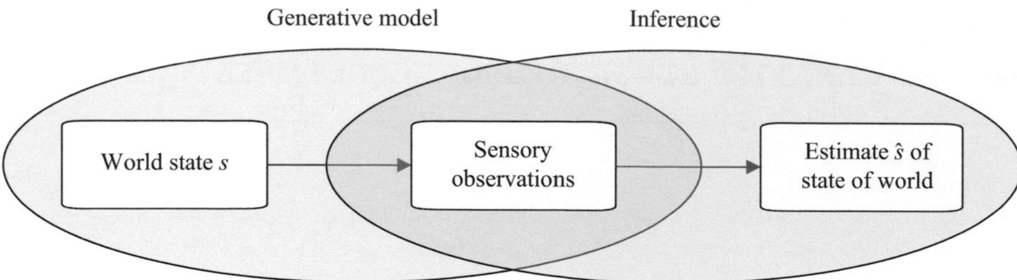

Figure 3.1
Schematic of a Bayesian model. The probabilities of world states s and of sensory observations given world states constitute the generative model. Specifying these distributions is step 1. On each trial, the observer performs inference to obtain an estimate of the world state. Specifying an expression for this estimate is step 2. Across many trials, the estimate itself follows a distribution for a given true s. Specifying this distribution is step 3.

of these speakers. Your job as a subject is to report with a cursor the location from which you perceived the tone to emanate. You do this many times; each stimulus-response pair constitutes a *trial*.

In this experiment, the subject is estimating a continuous quantity, namely the position of a sound source along a line. The subject has sensory observations, but possibly also prior knowledge. This sets up the Bayesian inference problem. The first steps involved in building a Bayesian model for this simple task are conceptually the same as those in section 2.1, but we add a third step here (figure 3.1; see also the inside cover).

Step 1 consists of specifying the *generative model*: the probability distributions associated with world state variables and sensory observations. The generative model is a full statistical description of what is happening in a task. In the shiny floor example of section 2.1, the generative model specified the probability that the floor is wet, the probability that the floor is shiny given that it is wet, and the probability that the floor is shiny given that it is dry. Many of the distributions in the generative model are specified by the experimental design. For instance, the probability of a sound occurring at a particular location is specified in the experimental design. However, we need to make an assumption about the distribution of the sensory observations. Many Bayesian models allow for the possibility that sensory observations are noisy. "Noise" has a diverse set of meanings across distinct scientific fields, but in this book we mean that the same stimulus does not always produce the same internal representation in the brain. Noise thus implies random variability of the sensory observations from trial to trial. Noise can be due to a wide variety of factors, both external (in the world) and internal (in the brain).

Step 2 describes how the observer performs inference, that is, how they estimate the state of the world based on sensory observations and prior expectations. As in chapter 2, step 2 consists of two substeps. Step 2a uses Bayes' rule to compute the observer's posterior distribution, that is, the observer's probability distribution over the world state of interest, given the sensory observations. The observer's inference process "inverts" the generative model. In the shiny floor example, the inference process consisted of computing the probability that the floor is wet from the sensory observations and prior information. The generative model as assumed by the observer, along with the sensory observations, completely defines the posterior distribution; no additional information is needed. Step 2b specifies how the observer obtains an estimate of the world state from the posterior distribution. This could, for example, be the mode or the mean of the posterior distribution.

In the presence of noise in the observations, there is a step 3 to Bayesian modeling. The inference process describes the observer's calculation of a posterior probability distribution over world states, and selection of a world state estimate; in the end, the inference process is summarized as an input-output relationship between the sensory observations and the world state estimate. Step 3 derives the probability distribution of the estimate given the stimulus, which is done by combining the input-output relationship with the distribution of the sensory observations. In this chapter, we discuss steps 1 and 2. In the next chapter, we discuss step 3 for the same problem.

3.2 Step 1: The Generative Model

The generative model is a description of the statistical structure of the task. In the auditory localization task, the world state the observer tries to infer is a single property of the stimulus, namely its horizontal position along a continuum; the sound's loudness, frequency, or other characteristic is not of interest in this task. We will often call the task-relevant property of a stimulus, denoted by s, simply "the stimulus." The sensory observations generated by the sound location consist of a complex pattern of auditory neural activity. For the purpose of our model, and reflecting common practice in the modeling of psychophysical data, we reduce the sensory observations to a single scalar, namely a noisy *measurement*, denoted by x. The measurement "lives" in the same space as the stimulus itself and uses the same units as the stimulus. We will now elaborate on this concept.

3.2.1 The Measurement: An Abstracted Sensory Representation

A physical stimulus elicits activity in the nervous system. This activity will vary randomly from trial to trial even when the physical stimulus itself is identical each time. Such variability, or noise, has many sources. Our sensors are subject to random variability due to intrinsic stochastic processes. For instance, thermal noise affects the responses of hair cells in the inner ear that sense sound waves. The transduction process by which the nervous system captures physical energy and converts it into an electrical response—the absorption of photons by photoreceptors—is also stochastic. At the subcellular level, neurotransmitter release and ion channel opening and closing are stochastic processes. We may expect there to be behavioral consequences of noise. For example, if we place the index finger of our right hand on top of a table and try to place the index finger of our left hand at the matching location underneath the table, we often observe quite a difference (typical variability is about 2 cm in this task). This indicates noise in our internal proprioceptive representations of limb location. Similarly, it is difficult to estimate whether one object is heavier than another based on our sense of force because the internal measurement of force is noisy—necessitating the use of scales to compare weights. These examples suggest that the relationship between stimulus and sensor response is stochastic.

In this book, as in most behavioral models, we do not directly model the neural representation of a stimulus, because doing so would be both underconstrained and unnecessary when accounting for behavioral data only. Instead, we define a measurement as an abstraction or reduction of the neural representation in stimulus space itself. For example, if the true location s of the sound is 3° to the right of straight ahead, then its measurement x could be 2.7° or 3.1°. The terminology "measurement" stems from an analogy with making physical measurements. If a stick is 89.0 cm long, you might measure its length to be 89.5, 88.1, 88.9 cm, or so on. We say that the measurement "lives" in the same space as the stimulus, because it uses the same units as the stimulus. A measurement of temperature is itself a temperature, a measurement of a color is itself a color, and so on.

In our case study, the common space of stimulus and measurement is the real line, but it could be many other things, such as the positive real line (for a stimulus such as length or weight), the circle (example: motion direction), or a high-dimensional space.

3.2.2 Graphical Model

Our case study contains two variables: the world state variable or a stimulus (true sound location, s) and the observer's measurement of the stimulus, x. These two variables appear in the generative model, which is graphically depicted in figure 3.2. As a reminder, the graphical model consists of nodes that contain the random variables and arrows that represent stochastic dependencies between variables. Each node is associated with a probability distribution. The variable at the end of an arrow has a probability distribution that depends on the value of the variable or variables at the origin of the arrow. In other words, an arrow can be understood to represents the influence one variable has on another. No arrow points to s, and therefore the distribution of s is a regular distribution $p_S(s)$. This distribution represents the overall frequency of occurrence of each possible value of the stimulus. The arrow pointing from s to x indicates that the distribution of x depends on the value of s. Mathematically, this is expressed as a conditional probability distribution $p_{X|S}(x|s)$. Conditional probability distributions are formally defined in section B.11.3. If you are not familiar with probability distributions, or if you have not worked with them recently, this would be a good moment to read appendix B. We will now describe the components of the generative model in detail.

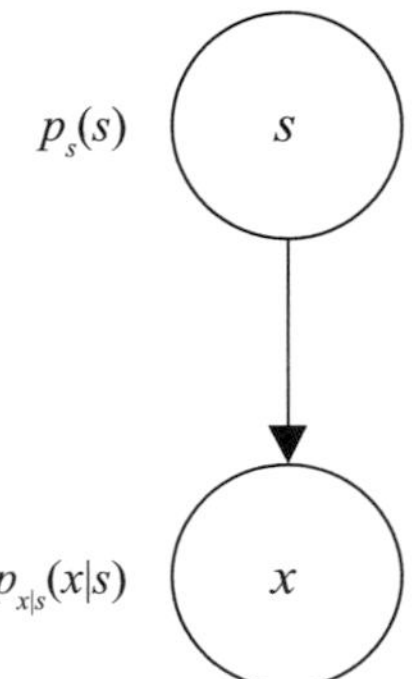

Figure 3.2
Schematic of a Bayesian model. The first step in Bayesian modeling is to define the generative model. This diagram is a graphical representation of the generative model discussed in this chapter. Each node represents a random variable, each arrow an influence. Here, s is the true stimulus and x the noisy measurement of the stimulus.

Box 3.1
Notation for Probability Distributions

Strictly speaking, a probability should be labeled by both the random variable and its value. In other words, $p_s(2)$ would denote the probability distribution of the random variable s evaluated at the value 2. The value is often generic, which leads to somewhat redundant notation, such as $p_s(s)$. Therefore, we typically leave out the subscript indicating the name of the random variable and instead assign this name to the value. This shorthand notation is virtually always unambiguous. Occasionally, it is necessary to include the subscript, for example when a specific value gets substituted and one has to keep track of which distribution is being considered. For more details, see appendix A. In the current chapter, we will keep the subscripts for didactic purposes.

3.2.3 The Stimulus Distribution

The distribution associated with the stimulus s is denoted by $p_s(s)$. This world state distribution, or, in our current example, stimulus distribution, reflects how often each possible value of s occurs. In our case study, the experimenter has programmed a computer to draw the stimulus on each trial from a Gaussian or normal distribution with a mean μ and variance σ_s^2. This distribution is defined by the following probability density function (see box 3.2):

$$p_s(s) = \frac{1}{\sqrt{2\pi\sigma_s^2}} e^{-\frac{(s-\mu)^2}{2\sigma_s^2}}. \tag{3.1}$$

An example of this density is shown in figure 3.3A. The mean being zero implies that the experimenter more often presents the tone straight ahead than at any other location.

Box 3.2
The Gaussian (Normal) Distribution

The most frequently used continuous probability distribution is the Gaussian or normal distribution. Its density function is

$$p(y) = \frac{1}{\sqrt{2\pi\sigma^2}} e^{-\frac{(y-\mu)^2}{2\sigma^2}}. \tag{3.2}$$

This is the famous "bell-shaped" distribution (or bell curve). We will sometimes use the notation

$$p(y) = \mathcal{N}(y; \mu, \sigma^2) \tag{3.3}$$

as shorthand for equation (3.2). The parameters μ and σ turn out to be the mean and standard deviation of the random variable, respectively. The factor $\frac{1}{\sqrt{2\pi\sigma^2}}$ is needed so that the total probability—the integral of $p(y)$—is equal to 1; such a factor is called a *normalization constant* (see section B.5.6). The exponent, $-\frac{(y-\mu)^2}{2\sigma^2}$, has a maximum value of zero at $y=\mu$, which is therefore the mode of the Gaussian distribution. From this maximum outward, the exponent decays. It will be -1 once the difference between y and μ has reached $\sigma\sqrt{2}$. There, the Gaussian will have decreased by a factor of e. Gaussian distributions result when many randomly occurring fluctuations can affect the variable of interest. The more formal version of this statement is called the *central limit theorem*. A typical example is the height of people, which follows a roughly Gaussian distribution, presumably because many factors contribute to height. Suppose that the average height for females is 165 cm, with a standard deviation of 10 cm. In this case we would find many females between 155 and 175 cm (within one standard deviation from the mean), fewer between 145 and 155 and between 175 and 185 cm, and very few above 185 cm or below 145 cm (more than two standard deviations away from the mean).

Gaussian distributions are convenient for analytical calculations; for example, multiplying two Gaussians produces another Gaussian. They are also convenient for simulations; for example, to draw samples from a Gaussian distribution with mean μ and variance σ^2, one can draw samples from one with mean 0 and standard deviation 1 (a *standard normal distribution*), multiply them by σ, and add μ.

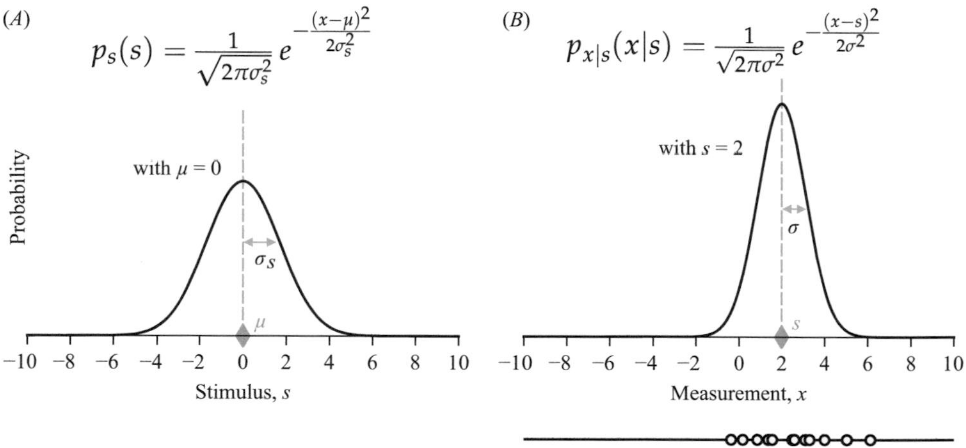

Figure 3.3
The probability distributions that belong to the two random variables in the generative model. (*A*) A Gaussian distribution over the stimulus, $p_s(s)$. Since s is continuous, this distribution is a probability density function. Its mean is $\mu = 0$ and its standard deviation is $\sigma_s = 1.5$. We are not using units in our case study, but if you prefer to be concrete, you can think of the unit being cm, inches, or degrees of visual angle. (*B*) Suppose we now fix a particular value of s, such as 2. Then we assume that the measurement x follows a Gaussian distribution around that value, with standard deviation σ, such as 1. The diagram at the bottom shows a few samples of x.

Exercise 3.1 If one substitutes $y = \mu$ and $\sigma = 0.1$ in equation (3.2), one finds $p(y) = 3.99$. How can a probability be larger than 1? If the answer to this question is not immediately clear, read appendix section B.5.4, on the difference between probability mass and density functions.

3.2.4 The Measurement Distribution

In the next few subsections, we step back from the example of auditory localization; our discussion will be general. The measurement distribution is the distribution of the measurement x for a given stimulus value s. This conditional distribution, $p_{x|s}(x|s)$, describes the probability of occurrence of values of the measurement when the same stimulus value s is repeated many times. If many sources contribute to the variability of the measurement, we will end up with a measurement distribution that is roughly Gaussian. This assertion is—loosely—a consequence of the central limit theorem (see box 3.2). While the Gaussian form of the stimulus distribution, equation (3.1) was chosen for convenience, the

Gaussian form of the measurement distribution is quite fundamental, independent of the experimental design, and common to most Bayesian models we discuss in this book.

The equation for the measurement distribution is

$$p_{x|s}(x|s) = \frac{1}{\sqrt{2\pi\sigma^2}} e^{-\frac{(x-s)^2}{2\sigma^2}}, \tag{3.4}$$

where σ is the standard deviation of the noise in the measurement, also called *measurement noise level* or *sensory noise level*. This Gaussian distribution is shown in figure 3.3B for one value of s. The higher σ, the noisier the measurement and the wider its distribution. The inverse of the variance of the measurement, $\frac{1}{\sigma^2}$, is sometimes called the *reliability* or the *precision* of the measurement.

The level of measurement noise is determined by a variety of internal and external factors that depend on the nature of the stimulus and the sensory modality used to measure it. When visually measuring the spatial location of a small stimulus, σ is affected by factors including retinal eccentricity (distance of a stimulus to the point at which the eye fixates), contrast, and blur (see also figure 1.5). When the location is measured auditorily, σ is affected by factors including signal-to-noise ratio, the angle relative to straight ahead, and pitch. Regardless of modality, σ will also be affected by presentation time and level of attention.

Box 3.3
Noise and Ambiguity

As explained in chapter 1, there are many sources of uncertainty in perception. In this chapter, we consider uncertainty that arises from noise in the observer's sensory measurement. Because our sensory systems are universally subject to measurement noise, this form of uncertainty is always present to some degree in perception. However, uncertainty can additionally arise from ambiguity in the stimulus itself: different world states can produce the same sensory stimulus. An example of an ambiguous image is a shiny floor, which may or may not be wet (section 2.3). Later in the book, we will encounter further examples of ambiguous images (e.g., size-distance ambiguity in section 9.2). Whether uncertainty is caused only by sensory noise, or also by stimulus ambiguity, the end result is that the observation has nonzero probability given more than one hypothesized world state.

3.2.5 Joint Distribution

Together, the two distributions $p_s(s)$ and $p_{x|s}(x|s)$ completely specify the generative model. One could combine them into a single, *joint* distribution which expresses the probability of occurrence of every combination of s and x:

$$p_{s,x}(s,x) = p_s(s)p_{x|s}(x|s). \tag{3.5}$$

This mathematical identity specifies the joint distribution of all variables in the task.

3.3 Step 2: Inference

Organisms do not have direct knowledge of world states. The observer's brain has to infer the value of a world state of interest based on the observations. If the observations take the form of a measurement, that means inferring the value of s from an observed measurement,

Table 3.1
Comparison between step 1 (generative model) and step 2 (inference).

	Step 1: Generative model	Step 2: Inference (decision model)
Point of view of	the world	the observer (decision-maker)
Is realized	across trials	on any single trial
Prior, likelihood, and posterior	play no role	are central
Nature of distributions (see box 3.4)	objective (probability of occurrence)	subjective (degrees of belief)
Outcome:	expressions for the distributions of all variables, including the observations	a rule to map observations to a decision (estimate of world state)

which we will denote by x_{obs}. The observer entertains different hypotheses about what s could be; we denote the hypothesized stimulus variable by s_{hyp}. It is important to distinguish this from the true stimulus s. On a given trial, there is only one true stimulus value, corresponding to the fact that there is only a single objective reality. By contrast, s_{hyp} takes on a range of values even on a given trial; these values represent the different hypotheses that the observer entertains about the stimulus. Then, the observer computes the prior distribution, the likelihood function, and the posterior over s_{hyp}. Finally, the observer "reads out" the posterior distribution to obtain an estimate of the stimulus. Table 3.1 summarizes differences between step 1 and step 2 of the modeling process.

3.3.1 The Prior Distribution

We now return to our example of auditory localization. In section 3.2, we introduced the stimulus distribution $p_s(s)$, which reflects—in the sense of a density function—how often each auditory location occurs in the experiment. Suppose that the observer has learned this distribution through extensive training. Then, the observer will already have an expectation about the stimulus before it even appears, namely that $s=\mu$ will be most probable, and that the probability falls off according to the learned Gaussian curve. The expectation that the observer holds about the stimulus without having received any evidence on the given trial constitutes prior knowledge. The prior probability density of a hypothesized stimulus value s is obtained by substituting that value in the stimulus distribution p_s; thus, the prior probability is equal to $p_s(s)$. In the inference process, $p_s(s)$ is referred to as the *prior distribution* (figure 3.4A). Unlike the stimulus distribution, the prior distribution exists on an individual trial: it reflects the observer's beliefs on that trial. It is therefore an example of a subjective distribution (box 3.4): probability is interpreted as *degree of belief* rather than as probability of occurrence.

3.3.2 The Likelihood Function

Intuitively, the likelihood function represents the observer's belief about a variable given the measurements only—absent any prior knowledge. The likelihood function contains all information about the variable that can objectively be obtained from the measurement: no more information can be obtained, and any different information would be incorrect.

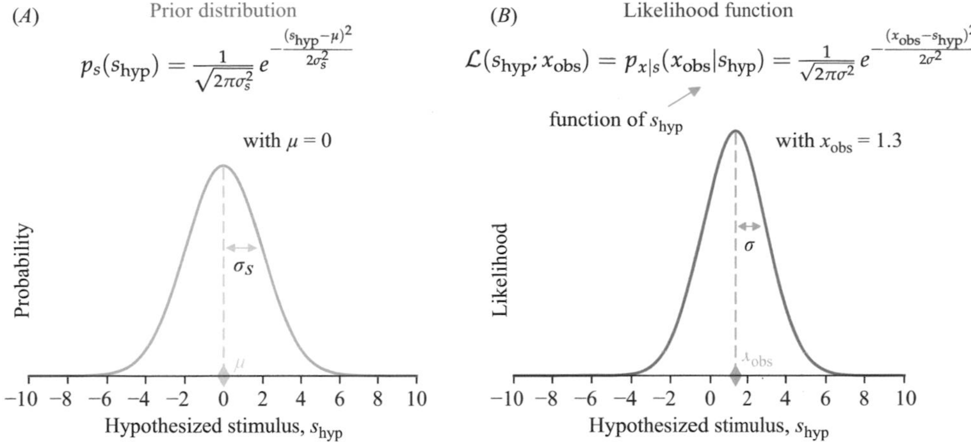

Figure 3.4
Consider a single trial on which the measurement is x_{obs}. The observer is trying to infer which value of the stimulus s produced this measurement. The two functions that play a role in the observer's inference process (on a single trial) are the prior and the likelihood. The argument of both the prior distribution and the likelihood function is s_{hyp}, the hypothesized stimulus. (*A*) Prior distribution. This distribution reflects the observer's beliefs about different possible values the stimulus can take. (*B*) The likelihood function over the stimulus based on the measurement x_{obs}. The likelihood function is centered at x_{obs}.

Box 3.4
Objective and Subjective Probabilities

A distinction is sometimes made between *objective* and *subjective* probability distributions. Objective probabilities reflect theoretical frequencies of occurrence, while subjective probabilities are tied to an observer and reflect degrees of belief. Of the three steps in Bayesian modeling, the second one (inference) deals with subjective probability distributions, because all distributions in that step represent the beliefs the observer holds about world states on a given trial. The first and the third steps deal with objective probability distributions, since sensory observations and estimates can (in principle) be counted. The distinction between objective and subjective probability is discussed further in appendix section B.1. This distinction is not important for calculations, only for interpretation.

The likelihood function derives its mathematical form from the measurement distribution (or more generally, from the distribution of the observations given the world state). In our current example, the measurement distribution is $p_{x|s}(x|s)$. This means that for a given observation, which we denote by x_{obs}, we know the likelihood function over s_{hyp}, which we will denote by $\mathcal{L}(s_{\text{hyp}}; x_{\text{obs}})$:

$$\mathcal{L}(s_{\text{hyp}}; x_{\text{obs}}) \equiv p_{x|s}(x_{\text{obs}}|s_{\text{hyp}}), \tag{3.6}$$

where we use $\equiv$ to indicate a definition. The somewhat complicated notation reflects the fact that we are substituting a specific observation, x_{obs}, and a specific hypothesized state of the world, s_{hyp}, into the measurement distribution that we have from step 1, $p_{x|s}$. Specifically, for the measurement distribution given by equation (3.4), the likelihood function is

$$\mathcal{L}(s_{\text{hyp}}; x_{\text{obs}}) = \frac{1}{\sqrt{2\pi\sigma^2}} e^{-\frac{(x_{\text{obs}} - s_{\text{hyp}})^2}{2\sigma^2}}. \tag{3.7}$$

At first sight, our definition of the likelihood might seem strange: why would we define $p(x|s)$ under a new name? The key point is that the likelihood function is a function of s_{hyp}, not of x_{obs}. The interpretation of the likelihood function is in terms of hypotheses. When an observer is faced with a particular measurement, what is the probability of that measurement when s_{hyp} takes on a certain value? Each possible value of the world state is a hypothesis, and the likelihood of that hypothesis is the observer's belief that the measurement would arise under that hypothesis. Stated otherwise, the likelihood of a hypothesis is the observer's sense of how compatible the measurement is with the hypothesis. Thus, a fundamental difference between the measurement distribution and the likelihood function is that the former is an objective probability distribution, while the latter represents the subjective beliefs of an observer (box 3.4). This is analogous to the distinction between the world state distribution and the prior distribution. In the context of our auditory localization task, the likelihood function in figure 3.4B reflects the observer's belief that the measurement would arise from each hypothesized sound location.

Box 3.5
The Likelihood of What?

> A likelihood function is numerically equal to a conditional probability, but is always a function of the variable after the | sign (for us the world state). It is common but incorrect to say "the likelihood of the measurements" or "the likelihood of the observations." The correct terminology is "the probability of the measurements (given a world state)" and "the likelihood of the world state (given the measurements)."

The likelihood function is general not normalized, that is, does not generally integrate to 1. The reason is that it is a function of the variable after the "given" sign, not of the one before it. This is why the likelihood function is called a function and not a distribution (a distribution is always normalized). The likelihood function in our example, shown in figure 3.4B, happens to be normalized (over s_{hyp}) since, in a Gaussian distribution, argument and mean can be interchanged without changing the distribution. However, we already know from chapter 2 that the likelihood does not need to be normalized.

The maximum-likelihood estimate Using the likelihood function in equation (3.7), one could make a best guess of the value of the stimulus. This best guess is called the maximum-likelihood estimate (MLE) of s, and we denote it by $\hat{s}_{\text{ML}}$; the hat denotes an estimate. Formally, the definition of the MLE is

$$\hat{s}_{\text{ML}} \equiv \underset{s_{\text{hyp}}}{\operatorname{argmax}}\, \mathcal{L}(s_{\text{hyp}}; x_{\text{obs}}). \tag{3.8}$$

Here, "argmax" stands for "the argument of the maximum": the value of the variable written below it for which the function following it takes its largest value. In our case study, the MLE is simply equal to the measurement, x_{obs}. This means that the location of a sound source that would with highest probability produce the measurement x_{obs} is x_{obs} itself.

The width of the likelihood function The width of the likelihood function is interpreted as the observer's level of uncertainty based on the observations alone. A narrow likelihood means that the observer is certain, a wide likelihood that the observer is uncertain. Although it follows from equation (3.7) that the width of the likelihood function is identical to the width of the measurement distribution, these widths have different interpretations. The latter quantifies the spread of the measurements, the former the level of uncertainty based on a single measurement.

Note on terminology: We often talk about the mean and variance of the likelihood function. This is somewhat imprecise, because means and variances are associated with probability distributions. However, we can normalize the likelihood function so that it becomes a probability distribution, and the mean and variance we refer to are the mean and variance of the normalized likelihood function.

3.3.3 The Posterior Distribution

A Bayesian observer computes a posterior distribution over a world state from observations, using knowledge of the generative model. In our case study, the posterior distribution is $p(s_{hyp}|x_{obs})$ the probability density function over hypothesized stimulus s_{hyp} given a measurement x_{obs}. Bayes' rule takes the form

$$p_{s|x}(s_{hyp}|x_{obs}) = \frac{p_{x|s}(x_{obs}|s_{hyp})p_s(s_{hyp})}{p_x(x_{obs})}. \tag{3.9}$$

It is also commonly written as

$$p_{s|x}(s_{hyp}|x_{obs}) \propto p_{x|s}(x_{obs}|s_{hyp})p_s(s_{hyp}), \tag{3.10}$$

or, using equation (3.6), as

$$p_{s|x}(s_{hyp}|x_{obs}) \propto \mathcal{L}(s_{hyp};x_{obs})p_s(s_{hyp}). \tag{3.11}$$

This states that the posterior distribution is proportional to the product of the likelihood and the prior. This product is what we called the *protoposterior* in chapter 2. In the last two expressions, we used the proportionality sign ($\propto$) to stand in for the factor $\frac{1}{p_x(x_{obs})}$, since this factor simply acts as a normalization constant. If we do not know the normalization factor, we still know the full shape of the posterior probability distribution. We refer to box 3.6 for a more detailed explanation. Table 3.2 summarizes the relationships between the likelihood function, the normalized likelihood function, the protoposterior, and the posterior distribution.

Case study We will now compute the posterior in our case study, under the assumptions we made in the previous section about the stimulus distribution and the measurement distribution. Upon substituting the expressions for $\mathcal{L}(s_{hyp};x_{obs})$ and $p_s(s_{hyp})$ into equation (3.10), we see that in order to compute the posterior, we need to compute the product of two Gaussian functions. Multiplying two Gaussian functions over the same variable, followed by normalization, is a common occurrence in Bayesian models of perception. The result is a new Gaussian distribution (box 3.7 and problem 3.4).

Box 3.6
Why That Proportionality Sign?

It is common to see the form of Bayes' rule in equation (3.10), with a proportionality sign. This form is justified because the denominator of Bayes' rule, here $p_X(x_{\text{obs}})$, does not depend on the argument of the posterior, here the world state of interest s. Thus, it simply acts as a multiplicative constant. A multiplicative constant does not change the shape of a function or where that function is maximal. Of course, the multiplicative constant is not an arbitrary number. It has to be such that the total integrated probability equals 1. For this reason, $\frac{1}{p(x_{\text{obs}})}$ is also called a *normalization constant.* We can write $p(x_{\text{obs}})$ as the sum or integral of the numerator over all possible values of the world state:

$$p_X(x_{\text{obs}}) = \int p_{X|S}(x_{\text{obs}}|s)p_S(s)ds. \tag{3.12}$$

This equation is analogous to equation (2.17). There, however, the world state variable was discrete and therefore the normalization consisted of a sum rather than an integral. The common way of dealing with the normalization constant is to first calculate the numerator, $p_{X|S}(x_{\text{obs}}|s_{\text{hyp}})p_S(s_{\text{hyp}})$, and then normalize at the end if desired. There is nothing wrong with explicitly writing $\frac{1}{p_X(x_{\text{obs}})}$. However, this factor would just stand there until the end of the computation. Alternatively, one could keep it and evaluate the inside of the integral along with evaluating the numerator. This would be cumbersome, however, since you would have to write down the same expression twice, once in the numerator, and once inside the integral in the denominator. The effect of working with the proportionality sign is that you first evaluate the entire numerator, and then in the end evaluate the denominator by plugging the final expression of the numerator into the integral. Sometimes that final evaluation of the integral is easy, because the integral has a standard known form.

Table 3.2
Relationships between likelihood, normalized likelihood, protoposterior, and posterior. Ignoring the prior is equivalent to assuming a flat prior.

	Prior ignored	Prior incorporated
Not normalized	Likelihood function	Protoposterior
Normalized	Normalized likelihood function	Posterior distribution

Box 3.7
Multiplying Two Gaussians

Here, we discuss the product of two Gaussian distributions over the same random variable y. One has mean μ_1 and variance σ_1^2, and the other one has mean μ_2 and variance σ_2^2:

$$p_1(y) = \frac{1}{\sqrt{2\pi\sigma_1^2}} e^{-\frac{(y-\mu_1)^2}{2\sigma_1^2}}; \tag{3.13}$$

$$p_2(y) = \frac{1}{\sqrt{2\pi\sigma_2^2}} e^{-\frac{(y-\mu_2)^2}{2\sigma_2^2}}. \tag{3.14}$$

We multiply these distributions just as we would multiply regular functions, and normalize the result (since the product is not automatically normalized). The resulting probability distribution is another normal distribution, now with mean

$$\frac{\frac{\mu_1}{\sigma_1^2}+\frac{\mu_2}{\sigma_2^2}}{\frac{1}{\sigma_1^2}+\frac{1}{\sigma_2^2}} \tag{3.15}$$

and variance

$$\frac{1}{\frac{1}{\sigma_1^2}+\frac{1}{\sigma_2^2}}. \tag{3.16}$$

Since it is a normal distribution, it should get the standard normalization constant of the normal distribution, namely 1 divided by the square root of 2π times the variance. These results are derived in problem 3.4.

Life-saving notation: Expressions arising from multiplying Gaussians involve multiple σ^2 factors, and they appear either in nested fractions or in expressions that are hard to interpret. This can make people miserable. Fortunately, there exists a useful and intuitive simplification of notation. This is to rewrite all expressions in terms of *precision* quantities. Precision is defined as the inverse of variance:

$$J \equiv \frac{1}{\sigma^2} \tag{3.17}$$

with subscripts as needed. In the new notation, the product of the two Gaussians above has a mean equal to $\frac{J_1\mu_1+J_2\mu_2}{J_1+J_2}$ and a precision equal to J_1+J_2. These expressions are substantially simpler. We encourage you to use precision notation whenever Gaussian distributions get multiplied.

Applied to our case study, we find that the posterior is a new Gaussian distribution

$$p_{s|x}(s_{\text{hyp}}|x_{\text{obs}}) = \frac{1}{\sqrt{2\pi\sigma_{\text{post}}^2}} e^{-\frac{(s_{\text{hyp}}-\mu_{\text{post}})^2}{2\sigma_{\text{post}}^2}}, \tag{3.18}$$

where the mean of the posterior is

$$\mu_{\text{post}} \equiv \frac{\frac{x_{\text{obs}}}{\sigma^2}+\frac{\mu}{\sigma_s^2}}{\frac{1}{\sigma^2}+\frac{1}{\sigma_s^2}} \tag{3.19}$$

and its variance is

$$\sigma_{\text{post}}^2 \equiv \frac{1}{\frac{1}{\sigma^2}+\frac{1}{\sigma_s^2}}. \tag{3.20}$$

From here on, we will simplify notation by introducing *precision* (inverse variance) variables (see box 3.7):

$$J_s \equiv \frac{1}{\sigma_s^2}; \tag{3.21}$$

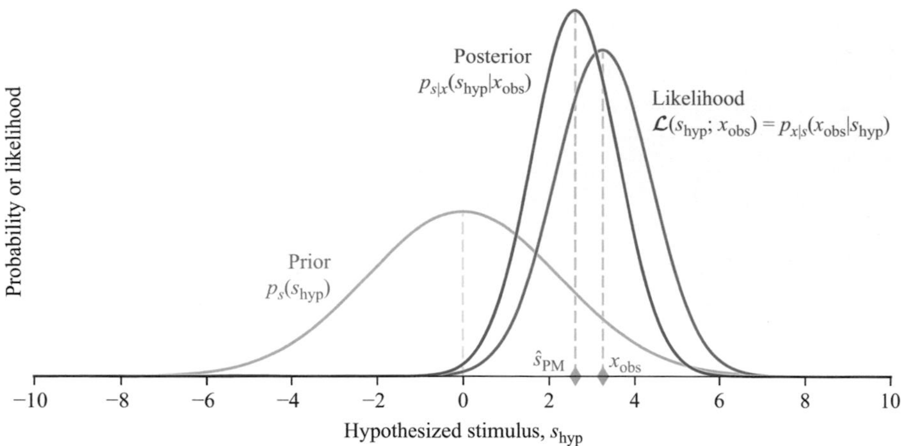

Figure 3.5
The posterior distribution is obtained by multiplying the prior distribution with the likelihood function, and normalizing the resulting function (the protoposterior). The hypothesized stimulus value with the highest posterior probability is the observer's posterior mean estimate of the stimulus, $\hat{s}_{\text{PM}}$.

$$J \equiv \frac{1}{\sigma^2}; \tag{3.22}$$

$$J_{\text{post}} \equiv \frac{1}{\sigma^2_{\text{post}}}. \tag{3.23}$$

In this notation, equation (3.19) can be written as

$$\mu_{\text{post}} = \frac{Jx_{\text{obs}} + J_s\mu}{J + J_s} \tag{3.24}$$

and (3.20) as

$$J_{\text{post}} = J + J_s. \tag{3.25}$$

An example of this posterior is drawn in figure 3.5.

3.3.4 The Posterior Mean

We now look more closely at the mean of the posterior, as given by equation (3.19) or (3.24). The expression of the mean is of the form $wx_{\text{obs}} + (1 - w)\mu$, where w is defined as

$$w \equiv \frac{J}{J + J_s}. \tag{3.26}$$

Since J and J_s are both non-negative, w is a number between 0 and 1. In other words, the posterior mean is a *weighted average* of the observed measurement, x_{obs} and the mean of the prior, μ (see box 3.8). This weighted average will always lie somewhere in between x_{obs} and μ.

Exercise 3.2 Prove this statement mathematically.

Box 3.8
Weighted Averages

> Suppose a student takes a midterm and a final exam and gets grades μ (for *m*idterm) and x (for exam). However, the midterm is less important than the final. Therefore, the teacher weights the final exam grade by a factor $w = 0.7$ and the midterm grade by a factor $1 - w = 0.3$. Then the student's overall grade in the class is the weighted average $wx + (1 - w)\mu$. It will lie in between μ and x.

Where exactly μ_{post} lies is determined by the weights w and $1 - w$ placed on observation and prior mean, respectively. The weights are normalized versions of the precisions of the likelihood function and the prior distribution. If the variance of the likelihood is lower than that of the prior distribution, the precision of the likelihood (i.e., the reliability of the measurement) is higher than that of the prior. As a consequence, the weight to x_{obs} is higher than to μ, causing the mean of the posterior to lie closer to x_{obs} than to μ. Of course, the reverse also holds: if the variance of the likelihood is larger than that of the prior, the mean of the posterior will lie closer to the mean of the prior than to the measurement. To Bayes' rule, the priors and the likelihood are just two pieces of information that need to be combined. Each piece of information has an influence that corresponds to the quality of the information.

Exercise 3.3 In the special case that prior and likelihood have the same variance ($\sigma = \sigma_s$), compute the mean of the posterior.

The intuition behind the weighted average in equation (3.19) is that the prior "pulls the posterior away" from the measurement and toward its own mean, but its ability to pull depends on how narrow it is compared to the likelihood function. If the likelihood function is narrow—which happens when the noise level is low—then the posterior will not budge much: it will be centered close to the mean of the likelihood function, that is, to the measurement. This intuition is often still valid even if the likelihood function and the prior are not Gaussian.

3.3.5 Width of the Posterior

So far, we have discussed only the mean of the posterior. The variance of the posterior is given by equation (3.20). Its square root, the standard deviation, is a measure of the width of the posterior. It is interpreted as the overall level of uncertainty that the observer has about the stimulus after combining the measurement with the prior. It is different from both the variance of the likelihood function and the variance of the prior distribution.

Exercise 3.4

(a) Show that the variance of the posterior can also be written as $\frac{\sigma^2\sigma_s^2}{\sigma^2+\sigma_s^2}$. Note that this is not our favorite way of writing it, as it is harder to interpret than equation (3.20) and in particular than equation (3.25).

(b) Show that the variance of the posterior is smaller than both the variance of the likelihood function and the variance of the prior distribution.

The interpretation of the posterior variance being smaller than the variances of likelihood and prior individually is that combining a measurement with prior knowledge makes an observer more certain about the stimulus than when the observer has only the measurement or only prior knowledge.

Exercise 3.5 What is the variance of the posterior in the special case that $\sigma = \sigma_s$? What are the mean and the variance of the posterior when $\frac{\sigma}{\sigma_s}$ is very large or very small? Interpret your results.

3.3.6 The Posterior Mean Estimate

A Bayesian observer uses the posterior distribution to obtain an estimate of the world state of interest, here the stimulus s. The Bayesian observer does this by minimizing the expected value of some quantity. For continuous variables such as s here, a common assumption is that the observer minimizes the expected squared error. As we will see in chapter 13, this is equivalent to choosing the *mean* of the posterior distribution. This is called *posterior mean* estimation.[1]

In general, the posterior mean estimate (PME), denoted by $\hat{s}_{\text{PM}}$, is defined as

$$\hat{s}_{\text{PM}} \equiv \int s p_{s|x}(s|x_{\text{obs}}) ds, \tag{3.27}$$

or in other words, as the expected value of s under the posterior distribution $p(s|x_{\text{obs}})$. For a review of expected values, see appendix section B.6.

In our example of combining a measurement with a Gaussian prior, we already calculated the posterior mean in equation (3.24). Thus, we have

$$\hat{s}_{\text{PM}} = \frac{J x_{\text{obs}} + J_s \mu}{J + J_s}. \tag{3.28}$$

Thus, the PME is a weighted average of the observed measurement x_{obs} and the prior mean μ, weighted by the inverse variances of likelihood function and prior, respectively.

3.3.7 The MAP Estimate

We also recall another posterior-based estimate from chapter 2: the mode of the posterior, also called the MAP estimate. It is the value of s_{hyp} for which the posterior probability density is maximal:

$$\hat{s}_{\text{MAP}} \equiv \underset{s_{\text{hyp}}}{\operatorname{argmax}}\, p(s_{\text{hyp}}|x_{\text{obs}}). \tag{3.29}$$

For a Gaussian distribution, the mode and the mean are identical, so $\hat{s}_{\text{MAP}} = \hat{s}_{\text{PM}}$. In general, the MAP estimate has the advantage that it generalizes to discrete (categorical) variables, as we saw in chapter 2. In the continuous case, however, it does not always minimize the expected squared error.

1. Also called Bayes-least-squares estimation. An alternative readout is to minimize expected *absolute* error, instead of expected *squared* error; this would lead to a *posterior median* readout (see problem 13.3). For Gaussian posteriors, the mean and the median are the same.

3.4 Uncertainty and Confidence

Uncertainty and confidence are common terms in daily life, but can be made specific in the context of Bayesian modeling.

3.4.1 Uncertainty

Every belief function in the inference stage (step 2) is associated with a notion of uncertainty: prior uncertainty reflects the quality of the observer's knowledge about the state of the world *before* making any observations, likelihood uncertainty or sensory uncertainty reflects the quality of the observer's knowledge about the state of the world solely based on the observations, and posterior uncertainty reflects the quality of the observer's knowledge about the state of the world *after* making the observations.

In all cases, uncertainty is tied to the observer and therefore a *subjective* quantity (box 3.4). It would be incorrect to talk about uncertainty in the context of the generative model (step 1). In particularly, the phrase "the uncertainty in the measurement" reflects confusion; correct usage would be "the uncertainty about the state of the world based on the measurement."

How, then, is uncertainty computed from a probability distribution or likelihood function over a continuous variable? There is no generally agreed-upon definition, but the intuitive use of the term is captured if we define uncertainty as the standard deviation of the probability distribution or (normalized) likelihood function. For example, a narrow posterior distribution $p(s|x)$ means low posterior uncertainty, and a wide posterior distribution means high posterior uncertainty. One could define uncertainty as any monotonically increasing function of the standard deviation of a belief distribution or likelihood function over a state of the world. In the calculations in this book, we simply use the standard deviation itself. Alternatively, we could define uncertainty as the interquartile range of the distribution or as another metric based on quantiles of the distribution. It should be noted that neither standard deviation–like nor quantile-based definitions of uncertainty would generalize to arbitrary discrete (categorical) distributions.

Exercise 3.6 Why not?

In our case study, all distributions are Gaussian and if uncertainty is standard deviation, its values are:

- prior uncertainty: σ_s
- likelihood or sensory uncertainty: σ
- posterior uncertainty: $\sigma_{\text{post}} = \frac{1}{\sqrt{J+J_s}}$, which can also be written as $\frac{\sigma\sigma_s}{\sqrt{\sigma^2+\sigma_s^2}}$.

We see that likelihood uncertainty happens to have the same numerical value as the sensory noise level, but in more complicated examples, that is not necessarily the case. In neural generative models such as those discussed in chapter 14, sensory noise level is even defined in a completely different space than the measurement space.

For Gaussian priors and likelihoods, posterior uncertainty is always smaller than both prior uncertainty and likelihood uncertainty (see exercise 3.4b). The definition of uncertainty extends to non-Gaussian distributions (see figure 3.6). However, for non-Gaussian priors

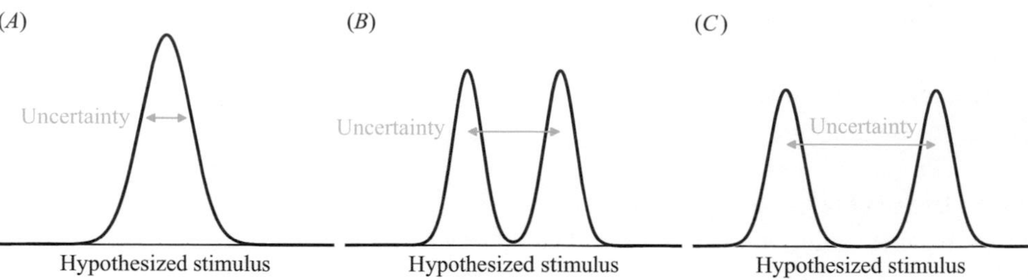

Figure 3.6
Uncertainty when defined as the standard deviation of a random variable. Each of the distributions shown could be the prior, the (normalized) likelihood, or the posterior. The double arrow is two standard deviations long. (*B*) and (*C*) show bimodal (two-peaked) distributions. Uncertainty in (*C*) is higher solely due to the greater separation between the peaks.

and/or likelihoods, it is not necessarily the case that posterior uncertainty is always smaller than both prior uncertainty and likelihood uncertainty (see problem 3.10).

Box 3.9
Terminology: Noise, Uncertainty, Variability

In this book, the term "noise" is reserved for the process by which the observations are generated, that is, it describes the trial-to-trial variability of observations or measurements. Noise is thus part of the generative model. "Uncertainty," on the other hand, reflects the observer's knowledge, or lack of knowledge, about variables in the world. The width of the posterior distribution is a measure of uncertainty, not of noise. Uncertainty is part of the inference process and is *subjective* in the sense of box 3.4. Noise is one possible cause of uncertainty, but not the only one. For example, when an object is partially occluded and there is no direct information about the part of the object behind the occluder, the observer has uncertainty without having noise.

Variability is an encompassing term for anything that varies from trial to trial. Noise is a form of variability and can be called the variability of the measurement. The stimulus estimate is also variable from trial to trial. This can be called "behavioral variability." Uncertainty is *not* a form of variability.

3.4.2 Bayesian Confidence

In daily life, decisions are made with greater or lesser confidence. You might be confident that you can cross the road before a car reaches you, that it is your friend who is approaching you, or that someone's accent is Italian. Confidence naturally fits into a Bayesian framework and is related to the posterior distribution.

That being said, there are different approaches to defining Bayesian confidence. One, which we call *estimate-based Bayesian confidence*, is based on the estimate $\hat{s}$ that the observer makes of the stimulus. Confidence can then be defined as any monotonic function F of the posterior probability density evaluated at the estimate:

$$\text{Bayesian Confidence} \equiv F(p_{s|x}(\hat{s}|x_{\text{obs}})). \tag{3.30}$$

"Monotonic function" here means the same as it did in our definition of uncertainty in section 3.4.1. Estimate-based Bayesian confidence applies equally to continuous and discrete variables.

A second approach, which we call *interval-based Bayesian confidence*,[2] makes sense only for continuous reports of a world state. In this approach, the observer reports some measure of the dispersion of the posterior distribution, either as an interval (starting point and end point) or as the size of such an interval. Measures of posterior uncertainty as defined in the previous subsection, such as posterior standard deviation or the interquartile range of the posterior, would fall into this category. Alternatively, an experimenter can incentivize the subject to set a meaningful confidence interval by rewarding them if the true world state lies within the interval but giving less reward for larger intervals. Then, the Bayesian observer would set an interval that maximizes expected reward. Moreover, the estimate itself could be included in this optimization process instead of being independently set in advance (as, for example, the PME). We will discuss incentivized Bayesian confidence intervals in chapter 13.

We now apply equation (3.30) for estimate-based Bayesian confidence, with F the identity function, to a Gaussian posterior distribution combined with a PME. Then, Bayesian confidence is the maximum of the Gaussian probability density function, which is equal to the normalization factor, $\sqrt{\frac{J_{\text{post}}}{2\pi}}$ or $\frac{1}{\sigma_{\text{post}}\sqrt{2\pi}}$. We further work this out in problem 3.8.

Here, we have examined Bayesian *definitions* of confidence. It is an empirical question whether some form of Bayesian confidence is a good model for *human* confidence ratings. Considerable evidence indicates that human confidence ratings do not simply follow equation (3.30). When a confidence interval rather than a confidence rating is reported, the literature has emphasized that those intervals do tend to follow the predictions of Bayesian models; however, non-Bayesian models for intervals have not been widely tested.

3.5 Model Mismatch in Inference

So far, we have only discussed optimal Bayesian inference. However, that discussion was predicated on the assumption that the observer possesses complete and correct knowledge of the generative model (step 1), and fully utilizes this knowledge during inference (step 2). Instead, it is possible that an observer uses a different, "assumed" generative model to perform inference. This is called *model mismatch* and could have many causes, such as:

- the observer has not yet completed the learning of the generative model,
- the generative model is too complex to learn and the observer approximates it,
- the generative model in the experiment at hand is different from that in the natural environment, and the observer uses the latter for inference,
- the observer holds wrong beliefs about the generative model, for example when a subject is neither instructed about nor trained on the task distributions.

2. To distinguish Bayesian confidence intervals from frequentist ones, Bayesian texts tend to use the term *credible interval*. We do not use that term in this book.

Bayesian observers are optimal when they possess and correctly incorporate full knowledge of all distributions in the generative model, but an observer with a mismatched likelihood or prior will in many cases be suboptimal. In our case study, that would mean having a higher expected squared error than the optimal observer. We will return to this in chapter 4.

3.5.1 Prior Mismatch

A special case of model mismatch is prior mismatch. The prior distribution $p_s(s)$ can be thought of as the observer's belief that the stimulus was s before they receive any sensory information. Within the context of a psychophysics experiment, one cannot blindly assume that subjects learn the stimulus distribution and use it as a prior. In general, the observer's prior might differ from the world state distribution. Subjects might be using a prior that they bring to the experiment, for example one that is based on the world state distribution in the natural world. Well-established examples are a prior that is higher for lower speeds and a prior for light coming from above. Such "natural" priors, acquired over a lifetime of sensory experience, might be hard to override during the relatively short duration of an experiment. Extensive training might be needed to override a natural prior. Of course, subjects might be using a prior intermediate between the "natural world state distribution" and the "experimental world state distribution." The prior might also change over time, as the observer is exposed to more stimuli during the experiment.

Under prior mismatch, the observer would be computing the PME using a different generative model than the correct one:

$$q_{s|x}(s_{\text{hyp}}|x_{\text{obs}}) \propto q_s(s_{\text{hyp}})p_{x|s}(x_{\text{obs}}|s_{\text{hyp}}). \tag{3.31}$$

For example, if the assumed stimulus distribution $q_s(s)$ has mean μ_{assumed} and variance $\sigma^2_{s,\text{assumed}}$, then the PME when the measurement is x_{obs} is

$$\hat{s}_{\text{PM, mismatched}} = \frac{Jx_{\text{obs}} + J_{s,\text{assumed}}\mu_{\text{assumed}}}{J + J_{s,\text{assumed}}}. \tag{3.32}$$

Thus, this observer would be making different trial-by-trial responses than the optimal Bayesian observer, and overall performance would be worse.

We should emphasize that the notion of optimality is usually based on the generative model associated with the experimental statistics. Using a generative model based on natural statistics would be optimal in a different sense, but in practice, the generative model associated with natural statistics is rarely known to researchers (see [57, 63] for notable exceptions).

3.5.2 Improper Priors

A special case of prior mismatch is if the observer uses a constant, or flat, prior on the real line, $q(s) = \text{constant}$. An immediate question is then what value this constant takes. If s were limited to an interval, say $[a, b]$, the answer would be clear: $p(s) = \frac{1}{b-a}$, so that the area under the prior is 1 (figure 3.7). In this chapter, however, the domain of s is the entire real line. A uniform distribution is not properly defined on the entire real line, since the line has infinite length, so the uniform distribution would have value 0. In any practical task, s can of course not grow arbitrarily large in either direction. Therefore, it would be reasonable to cut off its domain at some large value. Choosing this value—and with it, the value of the

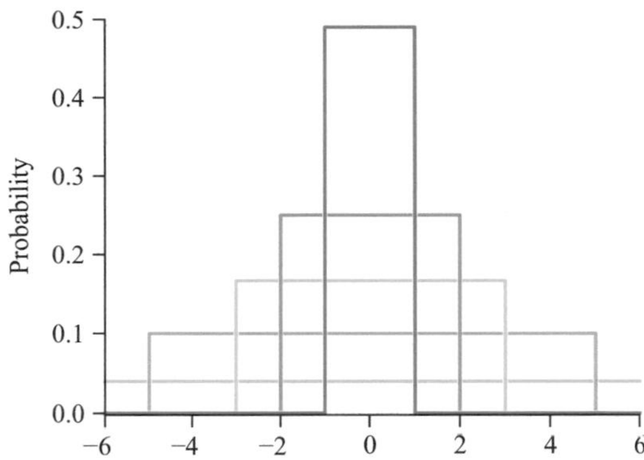

Figure 3.7
Various uniform priors on bounded intervals.

constant prior—would, however, be arbitrary. Fortunately, this conundrum does not need to be solved, since it turns out that in the inference, the value of the constant prior does not play a role. Namely, if this value is c, then the posterior distribution is

$$p_{s|x}(s_{\text{hyp}}|x_{\text{obs}}) \propto c p_{x|s}(x_{\text{obs}}|s_{\text{hyp}}) \propto p_{x|s}(x_{\text{obs}}|s_{\text{hyp}}). \tag{3.33}$$

In other words, since the prior is constant, it gets absorbed by the proportionality constant. After normalizing the posterior distribution, c would drop out. We could assume $p_s(s) = c$ on the entire real line, and the posterior would still be well defined in spite of the prior distribution not being normalized. A nonnormalizable prior distribution (i.e., one whose integral is infinite rather than 1) is called an *improper prior.*

3.6 Heteroskedasticity

In this book, we focus on the case where the measurement noise level σ in equation (3.4) does not depend on the stimulus. In practice, however, it often does. For example, noise in the measurement of visual position increases with retinal eccentricity, and noise in the measurement of auditory position increases with angle away from straight ahead. When the variance of a random variable depends on the mean, we have an example of *heteroskedasticity*. Heteroskedasticity does not prevent us from formulating a Bayesian model. However, heteroskedasticity causes likelihood functions to be non-Gaussian even when the measurement distribution is Gaussian (problem 3.7).

3.7 Magnitude Variables

So far, we have assumed that the domain of the stimulus is the entire real line, from $-\infty$ to ∞. There are many world state variables that have a different domain. In appendix section B.7.6, we discuss variables that are periodic, such as angles. Here, we consider variables that only take positive values and thus have a domain from 0 to ∞, including line length, depth, weight, speed, loudness, temporal duration, and light intensity. Variables that never take negative values can be called *magnitude variables*. One type of probability distribution designed for magnitude variables is the *lognormal distribution*. Since the

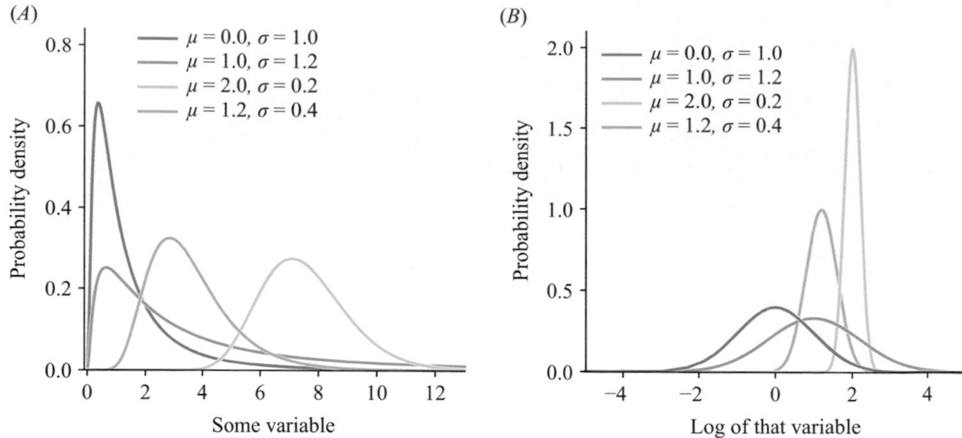

Figure 3.8
Example of lognormal distributions, and their corresponding equivalents on a logarithmic horizontal axis. The lognormal distribution is defined on the positive real line.

domain of s is $[0, \infty)$, the domain of the logarithm of s is the entire real line.[3] Thus, it is possible to define a Gaussian distribution on $\log s$:

$$p_{\log s}(\log s) = \frac{1}{\sqrt{2\pi\sigma_s^2}} e^{-\frac{(\log s - \mu)^2}{2\sigma_s^2}}. \tag{3.34}$$

Transforming this to the original variable s, we obtain

$$p_s(s) = \frac{1}{s\sqrt{2\pi\sigma_s^2}} e^{-\frac{(\log s - \mu)^2}{2\sigma_s^2}}. \tag{3.35}$$

Note the $\frac{1}{s}$ factor, called the Jacobian. Read appendix section B.12.1 if you do not know how to transform probability distributions. Equation (3.35) is a lognormal distribution with parameters μ and σ_s^2, also written as Lognormal$(s; \mu, \sigma_s^2)$. Examples are shown in figure 3.8. Importantly, the parameters μ and σ_s^2 do not correspond directly to mean and variance. The mean of the lognormally distributed variable s is $e^{\mu + \frac{1}{2}\sigma_s^2}$ and its variance is a rather complicated expression that we will not use. The median of s under the lognormal distribution is e^{μ} and its mode is $e^{\mu - \sigma_s^2}$.

A lognormal distribution can also be used for the measurement distribution:

$$p_{\log x | \log s}(\log x | \log s) = \frac{1}{\sqrt{2\pi\sigma^2}} e^{-\frac{(\log x - \log s)^2}{2\sigma^2}}. \tag{3.36}$$

This use has special significance, because an important property of the lognormal distribution is that the standard deviation is proportional to the mean. It turns out that this is empirically a good description of human magnitude judgments—a relation called the *Weber-Fechner law*. For example, the level of noise in the observer's measurement of line length is proportional to the length itself: distinguishing a distance of 1.02 m from a distance of 1 m

3. When we use the notation "log" in this book, we always refer to the natural logarithm (base e).

is about as hard as distinguishing 10.2 cm from 10 cm. We will extensively use lognormal distributions in chapter 9.

3.8 Applications

The Bayesian model described in this chapter has been applied to a wide variety of scenarios within the domains of visual perception, somatosensory perception, auditory perception, and motor control. Here, we mention just a few of the studies from this vast literature.

In visual perception, low-contrast stimuli (e.g., a gray object moving against a background of a different shade of gray) appear to move slower than high-contrast stimuli (for example, a white object moving against the same background) that are in reality moving at the same speed [173]. Weiss and colleagues explain this and many other puzzling visual illusions as resulting from the brain's use of a prior distribution over speed that is higher for lower speeds (a so-called low-speed prior) [199]. A lower-contrast object provides less reliable information, which means that the likelihood function is wider. This results in a posterior estimate that is shifted more toward the mean of the prior. If the prior peaks at 0, then the perceived speed will be shifted toward a lower value when contrast is lower. Subsequently, the Bayesian model for visual motion perception was further refined by estimating the shape of the prior [168].

In the realm of somatosensory perception, a variety of tactile spatiotemporal illusions could be explained using the model in this chapter, again with a prior expectation for low-speed movement [65]. The proposed model provides a unified explanation for several startling tactile illusions in which taps delivered in rapid succession to distinct points on the skin surface are perceived to occur closer together than they actually do. Remarkably, the perceived distance between taps shrinks as the time between taps is reduced, a phenomenon known as perceptual length contraction [65, 66]. Particularly interesting is the confirmed prediction that perceptual length contraction is more pronounced for weaker taps, which evoke broader likelihood functions [178].

In the realm of auditory perception, a Bayesian model explains the observation that owls can process acoustic stimuli remarkably accurately in order to localize a sound source that occurs in the region straight ahead of them, yet they consistently underestimate the position of sound sources that occur in eccentric regions [52]. The model assumes that the owls use a centrality prior, that is, it they expect that sound sources tend to occur more frequently in central locations.

Bayesian models have been applied not only to spatial perception but to temporal perception. For instance, the model of Goldreich [65] provides an explanation for a time perception illusion in which the perceived time between spatially separated stimuli expands as the distance between the stimuli is increased. This time dilation phenomenon (a.k.a kappa effect) occurs in both the tactile and visual domains [36]. Using visual stimuli, Jazayeri and Shadlen [84] show that, when presented with stimuli whose durations are drawn from different distributions, participants' duration estimates are consistent with those of a Bayesian observer model that incorporates the duration distributions as prior probability distributions.

In the domain of movement, Kording and Wolpert [101] study the movements of fingers reaching toward a visual target. In a virtual-reality setup, the target position is laterally displaced by a distance randomly drawn from a Gaussian distribution with a mean displacement of 1 cm. This acts as the prior distribution. Observers are trained until they have learned this distribution. The finger's movement is not visible to the subject; a noisy visual measurement of the finger's position is provided by showing a cloud of dots on a screen halfway through the movement. Importantly, the reliability of the visual measurement is manipulated on a trial-by-trial basis. Observers were found to combine the noisy measurement with their learned knowledge of the prior distribution in the manner described by equation (3.28). In a second experiment, the Gaussian prior was replaced by a bimodal (two-peaked) prior; even this prior was incorporated in good approximation, although this required extensive training. A bimodal prior over orientation was revealed in natural statistics [63]; observers took this prior into account in orientation estimation tasks.

3.9 Percepts

When we model perception in this book, we primarily model decisions that people make in perceptual tasks. Bayesian models provide predictions for those decisions. For example, the posterior mean estimate is what the Bayesian observer should report to minimize squared error in a continuous estimation task. In this example, a different statement, with different philosophical implications, would be that the estimate should be what is being *perceived*, that is, that the observer *experiences* (sees, hears, etc.) the PME of the stimulus. In our view, Bayesian models also provide valid candidate accounts of phenomenological percepts. Perceptual illusions are a domain where compelling percepts arise without prompting, and Bayesian models often provide plausible qualitative explanations of those percepts, as in the case of the hollow-face illusion (problem 1.10).

To us, it is not entirely clear whether or where to draw a line between perceptual decisions and percepts. One distinguishing criterion might be whether a query is made (a question is asked). When you are presented with an anamorphic illusion, you don't need to be asked, "Does this flat drawing look like a three-dimensional scene?"—you simply experience the drawing that way. However, it is possible that even when no explicit query is made, the mind implicitly asks itself questions, such as "What are the identities and locations of the objects in the scene?" This would blur the distinction but would be interesting in itself.

The distinction between perceptual decisions and percepts receives an extra dimension when uncertainty is considered. Bayesian models, in their strong form, would predict that the mind computes a posterior distribution over every world state of interest. In perception, that means that the output of a perceptual computation includes at least a sense of uncertainty or confidence about the state of the world, and at most a full probability distribution. It has been argued that percepts do not come with either [20], even though perceptual decisions might. We believe the jury is still out on this question (see also [43, 128]), and that answering it might empirically be very difficult.

3.10 Summary

We considered inference of a continuous variable based on a noisy sensory observation.

- Sensory noise has many sources.
- We defined the measurement of a stimulus as an abstraction of the noisy internal representation of that stimulus. The measurement lives in the same space as the stimulus itself.

- We defined the first two steps of a Bayesian model in the presence of measurement noise: the generative model and the inference process.
- The generative model consisted of a stimulus distribution (assumed Gaussian for convenience) and a stimulus-conditioned measurement distribution (assumed Gaussian, as justified by the central limit theorem).
- During inference, the observer uses the stimulus and measurement distributions from the generative model to form a prior distribution and a likelihood function, respectively.
- The posterior is the normalized product of prior and likelihood. If prior and likelihood are Gaussian, the posterior is also Gaussian.
- Under the same assumptions, the posterior mean is a weighted average between the observed measurement and the prior mean. In addition, the posterior is narrower than both prior and likelihood, reflecting decreased uncertainty.
- In general, if the objective is to minimize expected squared error, the Bayesian observer estimates the stimulus as the posterior mean.
- Confidence can be defined as the value of the posterior at the posterior mean.
- An observer can be Bayesian but use the "wrong" generative model. This is called model mismatch. Model mismatch might lead to suboptimal behavior.

3.11 Suggested Readings

- Wendy J. Adams, Erich W. Graf, and Marc O. Ernst. "Experience Can Change the 'Light-from-above' Prior." *Nature Neuroscience* 7, no. 10 (2004): 1057–1058.
- Brian J. Fischer and José Luis Peña. "Owl's Behavior and Neural Representation Predicted by Bayesian Inference." *Nature Neuroscience* 14, no. 8 (2011): 1061–1066.
- Daniel Goldreich and Jonathan Tong. "Prediction, Postdiction, and Perceptual Length Contraction: A Bayesian Low-Speed Prior Captures the Cutaneous Rabbit and Related Illusions." *Frontiers in Psychology* 4 (2013): 221.
- Mehrdad Jazayeri and Michael N. Shadlen. "Temporal Context Calibrates Interval Timing." *Nature Neuroscience* 13, no. 8 (2010): 1020–1026.
- Konrad P. Kording and Daniel M. Wolpert. "Bayesian Integration in Sensorimotor Learning." *Nature* 427, no. 6971 (2004): 244–247.
- Alan A. Stocker and Eero P. Simoncelli. "Noise Characteristics and Prior Expectations in Human Visual Speed Perception." *Nature Neuroscience* 9, no. 4 (2006): 578–585.
- Xue-Xin Wei and Alan A. Stocker. "A Bayesian Observer Model Constrained by Efficient Coding Can Explain 'Anti-Bayesian' Percepts." *Nature Neuroscience* 18, no. 10 (2015): 1509–1517.
- Yair Weiss, Eero P. Simoncelli, and Edward H. Adelson. "Motion Illusions as Optimal Percepts." *Nature Neuroscience* 5, no. 6 (2002): 598–604.

3.12 Problems

Problem 3.1 Let s be the stimulus of interest, x the measurement, $p_s(s)$ the stimulus distribution, and $p_{x|s}(x|s)$ the measurement distribution.

(a) Write down the posterior distribution over hypothesized stimulus s, given an observed measurement x_{obs}.

(b) Which of the terms in your expression is called the likelihood function?

(c) What is the difference between the likelihood function and the measurement distribution?

Problem 3.2 In this problem, we numerically calculate a posterior distribution. Suppose the stimulus distribution $p_s(s)$ is Gaussian with mean 20 and standard deviation 4. The measurement distribution $p_{x|s}(x|s)$ is Gaussian with standard deviation $\sigma = 5$. A Bayesian observer infers s from an observed measurement $x_{\text{obs}} = 30$. We are now going to calculate the posterior probability density using numerical methods.

(a) Define a vector of hypothesized stimulus values s: $(0, 0.2, 0.4, \ldots, 40)$.

(b) Compute the likelihood function and the prior on this vector of s values.

(c) Multiply the likelihood and the prior pointwise.

(d) Divide this product by its sum over all s (normalization step).

(e) Convert this posterior probability mass function into a probability density function by dividing by the step size you used in your vector of s values (e.g., 0.2).

(f) Plot the likelihood, prior, and posterior in the same plot.

(g) Is the posterior wider or narrower than likelihood and prior? Do you expect this based on the equations we discussed?

(h) Change the standard deviation of the measurement distribution to a large value. What happens to the posterior? Can you explain this?

(i) Change the standard deviation of the measurement distribution to a small value. What happens to the posterior? Can you explain this?

Problem 3.3 We expand on the previous problem by varying the stimulus distribution. Start with the code from the previous problem. Suppose the stimulus distribution $p_s(s)$ is uniform on the interval $[-15, 25]$ and 0 outside this interval. The measurement distribution $p_{x|s}(x|s)$ is Gaussian with standard deviation $\sigma = 5$. A Bayesian observer infers s from an observed measurement $x_{\text{obs}} = 30$. We are again going to calculate the posterior probability density numerically.

(a) What is the value of $p(s)$ on the interval $[-15, 25]$?

(b) Repeat parts (a)–(f) from the previous problem for this new stimulus distribution.

(c) Is the posterior Gaussian?

(d) Numerically calculate the mean of the posterior.

(e) Numerically calculate the variance of the posterior.

Problem 3.4 We have a closer look at the central inference calculation of this chapter, which involved multiplying two Gaussians. The stimulus distribution $p_s(s)$ is Gaussian with mean

μ and variance σ_s^2. The measurement distribution $p_{x|s}(x|s)$ is Gaussian with mean s and variance σ^2.

(a) Write down the equations for $p_{x|s}(x|s)$ and $p_s(s)$.

(b) A Bayesian observer infers s from a measurement x_{obs}. Use Bayes' rule to write down the equation for the posterior, $p_{s|x}(s_{\text{hyp}}|x_{\text{obs}})$. Substitute the expressions for $p_{x|s}(x_{\text{obs}}|s_{\text{hyp}})$ and $p_s(s_{\text{hyp}})$, but do not simplify yet.

The numerator is a product of two Gaussians. As we discussed in section 3.3.3, the denominator, $p_x(x_{\text{obs}})$, is a normalization factor that ensures that the integral equals 1. For now, we will ignore it and focus on the numerator.

(c) Apply the rule $e^A e^B = e^{A+B}$ to simplify the numerator.

(d) Expand the two quadratic terms in the exponent.

(e) Rewrite the exponent to the form $as^2 + bs + c$, with a, b, and c constants. Importantly, since c is just leading to a constant scaling, no need to calculate it.

(f) Rewrite the expression you obtained in (e) in a simpler form, $e^{c_1(s+c_2)^2+c_3}$, with c_1, c_2, and c_3 constants. Hint: any quadratic function of the form $as^2 + bs + c$ can be written as $a\left(s + \frac{b}{2a}\right)^2 + c - \frac{b^2}{4a}$; this rewriting is known as *completing the square*.

(g) Now rewrite your expression into the form $e^Z e^{-\frac{(s-\mu_{\text{combined}})^2}{2\sigma^2_{\text{combined}}}}$. Express μ_{combined} and σ_{combined} in terms of x, σ, μ, and σ_s.

(h) Recall that $p_{s|x}$ is a probability distribution and that its integral should therefore be equal to 1. However, the expression that you obtained in (g) is not properly normalized because we ignored $p_x(x_{\text{obs}})$. Modify the expression such that it is properly normalized, but without explicitly calculating $p_x(x_{\text{obs}})$ (Hint: does e^Z depend on s?)

Many Bayesian inference problems involve a product of two or more Gaussians, so this derivation will come in handy.

Problem 3.5 The figure below shows a likelihood function and a posterior distribution. Both are Gaussian, with $\sigma_{\text{posterior}} = 1.2$ and $\sigma_{\text{likelihood}} = 1.5$. Assume that the prior is also Gaussian. Which of the following statements is true? Explain. You will need both the plot and the given numbers.

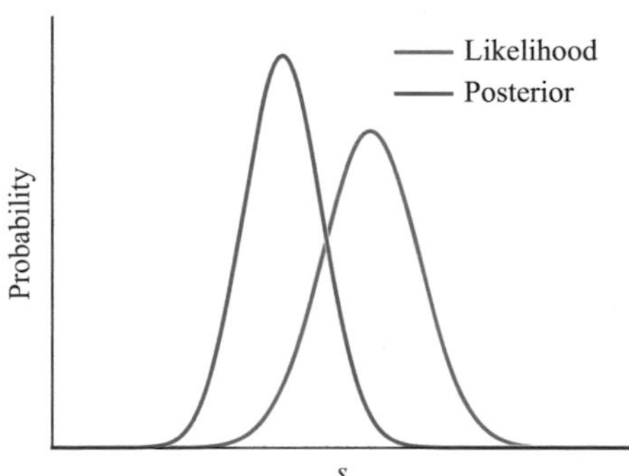

(a) The prior is centered to the left of the likelihood function and is narrower.

(b) The prior is centered to the left of the likelihood function and is wider than it.

(c) The prior is centered to the right of the likelihood function and is narrower than it.

(d) The prior is centered to the right of the likelihood function and is wider than it.

Problem 3.6 Besides Gaussians, there are other distributions for which we can calculate the posterior analytically. Consider a stimulus distribution $p_s(s)$ that is equal to 0 for $s < 0$ and equal to an exponential distribution, $p_s(s) = \lambda e^{-\lambda s}$ with $\lambda > 0$, for $s \geq 0$. For the measurement distribution $p_{x|s}$, we assume the usual: a Gaussian with mean s and variance σ^2. A Bayesian observer infers s from a measurement x_{obs}.

(a) Derive an equation for the posterior mean estimate.

(b) Assume $\lambda = 1$ and $x_{\text{obs}} = 1$. Plot the prior distribution, the normalized likelihood function, and the posterior in a single plot.

Problem 3.7 In this chapter, we assumed that σ is independent of the stimulus s. This assumption is often violated in real problems, leading to *heteroskedasticity*. Assume the measurement distribution

$$p_{x|s}(x|s) = \frac{1}{\sqrt{2\pi\sigma(s)^2}} e^{-\frac{(x-s)^2}{2\sigma(s)^2}}, \tag{3.37}$$

where $\sigma(s)$ is the following function of s: $\sigma(s) = 1 + s^2$.

(a) For $s = 0, 1, 2$, plot the measurement distribution (three curves in one plot, color-coded). All three should look Gaussian.

(b) For $x_{\text{obs}} = 0, 1, 2$, plot the likelihood function over hypothesized s (three curves in one plot, color-coded). None of them should look Gaussian.

(c) Explain how it is possible that the measurement distributions are all Gaussian but the likelihoods are not.

(d) If the prior were Gaussian, would the posterior be Gaussian as well? Explain your answer using math.

Problem 3.8 This problem builds on section 3.4.2.

(a) Show that in our case study, Bayesian estimate-based confidence associated with the PME is $\sqrt{\frac{1}{2\pi}\left(\frac{1}{\sigma^2} + \frac{1}{\sigma_s^2}\right)}$.

(b) Why would an equally legitimate definition of Bayesian confidence in this case be $\frac{1}{\sigma^2} + \frac{1}{\sigma_s^2}$?

(c) Explain, for either expression, why the dependence on σ and σ_s make sense.

Problem 3.9 This problem combines concepts from sections 3.4.2 and 3.5. When the observer uses a wrong prior to compute the posterior (prior mismatch), can decision confidence be higher than when they use the correct prior? If so, give an example. If not, prove mathematically that it is impossible.

Problem 3.10 In section 3.4.1, we remarked that for Gaussian priors and likelihoods, posterior uncertainty—defined as standard deviation—is always smaller than both prior uncertainty and likelihood uncertainty. Construct an example involving non-Gaussian priors or likelihoods in which this property no longer holds.

Problem 3.11 Some stimulus variables are periodic (or circular), taking values between (for instance) 0 and 2π. Examples are orientation, motion direction, and time of day. In appendix section B.7.6, we discuss the Von Mises distribution, which is suitable for such variables. Suppose that motion direction follows a Von Mises distribution with circular mean μ and concentration parameter κ_s

$$p(s) = \frac{1}{2\pi I_0(\kappa_s)} e^{\kappa_s \cos(s-\mu)}. \tag{3.38}$$

Further assume that the measurement distribution is also Von Mises, with circular mean s and concentration parameter κ:

$$p(x|s) = \frac{1}{2\pi I_0(\kappa)} e^{\kappa \cos(x-s)}. \tag{3.39}$$

(a) A Bayesian observer infers s from a measurement x_{obs}. Show that the posterior over s is a Von Mises distribution with (circular) mean μ_{post} given by

$$\cos \mu_{\text{post}} = \kappa_s \cos \mu + \kappa \cos x_{\text{obs}}; \tag{3.40}$$

$$\sin \mu_{\text{post}} = \kappa_s \sin \mu + \kappa \sin x_{\text{obs}}; \tag{3.41}$$

and concentration parameter $\sqrt{\kappa_s^2 + \kappa^2 + 2\kappa\kappa_s \cos(x_{\text{obs}} - \mu)}$.

(b) Compare and contrast with the Gaussian case.

4

The Response Distribution

How does a Bayesian model predict a human observer's responses on a perceptual task?

In chapter 3, we modeled how priors and likelihoods are combined; we now ask how to relate such a model to behavior. In psychophysics experiments, the researcher provides stimuli and measures a participant's responses. Thus, our goal is to predict the probability of each possible response given the stimuli. Our model so far (step 2) predicts the observer's estimate of the world state given the sensory observations, but not given the stimuli. We now add a third modeling step to fill this gap.

Plan of the Chapter

We will discuss why, upon repeated presentations of the same stimulus, a Bayesian observer's posterior mean estimate (the observer's response) is a random variable. We derive the probability distribution of the Bayesian observer's responses. This constitutes step 3 of Bayesian modeling. The response distribution allows investigators to compare the predictions of the Bayesian model to the observer's actual behaviour in the context of a psychophysical experiment. We discuss the bias and variance of the PME, and compare them to those of the MLE. Finally, we return to the topic of optimality.

4.1 Inherited Variability

The experimenter can only control the stimulus. To compare our Bayesian model with an observer's behavior in a psychophysical task, we need to specify what the Bayesian model predicts for the observer's responses when the true stimulus is s. In other words, we need to know the distribution of the stimulus estimate when the true stimulus is s. We denote this distribution $p(\hat{s}|s)$, and call it the *estimate distribution* or the *response distribution*. We first describe the core source of variability in the estimate distribution.

In the most basic form of Bayesian modeling, even when the measurement is noisy, everything that comes after the measurement is a deterministic function of the measurement: the likelihood function is a deterministic function of the measurement, the posterior is deterministically computed from the likelihood, and the estimate (e.g., a posterior mean estimate) is deterministically computed from the posterior. Thus, once we give you the value of the observed noisy measurement, x_{obs}, you can compute everything else up to the observer's

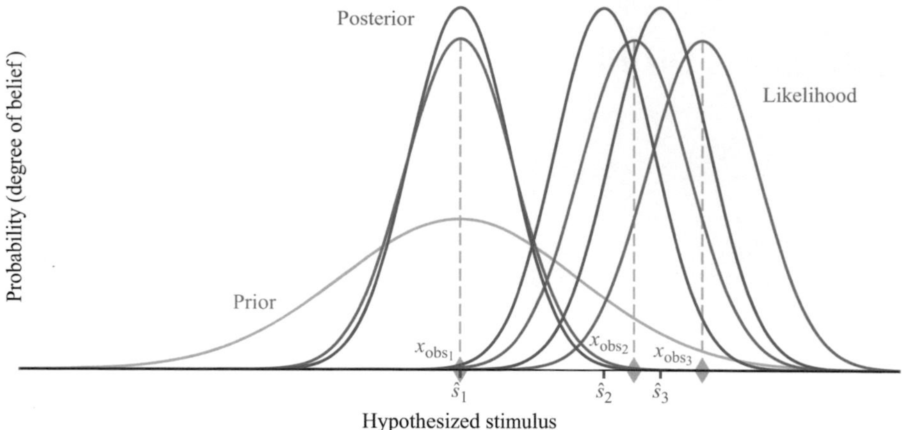

Figure 4.1
Likelihood functions (red) and corresponding posterior distributions (blue) on three example trials. The prior distribution is shown in yellow. The message here is that the likelihood function, the posterior distribution, and the PME are not fixed objects: they move around from trial to trial because the measurement x_{obs} does. Of interest to a Bayesian modeler (in step 3) are the mean and the variance over trials of the PME.

response. However, even when step 2 is fully deterministic, the measurement itself is variable from trial to trial even when the true stimulus is kept fixed. This means that the above quantities (likelihood, posterior, and posterior readout) will all vary along with the measurement. Figure 4.1 illustrates this variation.

In particular, because x_{obs} is a random variable for given s, so is the stimulus estimate. Hence, in response to repeated presentations of the same stimulus, the posterior mean estimate will be a random variable with a probability distribution. In other words, the stochasticity in the stimulus estimate, or the subject's response, is "inherited from" the stochasticity in the measurement.

4.2 The Response Distribution

We now concretely calculate the response distribution in the example of chapter 3. Recall that we derived, for a simple model with a Gaussian stimulus distribution and a Gaussian measurement distribution, that the posterior distribution is also Gaussian, with mean given by equation (3.19),

$$\mu_{\text{post}} = \frac{Jx_{\text{obs}} + J_s\mu}{J + J_s}, \tag{4.1}$$

where $J = \frac{1}{\sigma^2}$ and $J_s = \frac{1}{\sigma_s^2}$. We discussed in section 3.3.6 that a reasonable way for the observer to estimate the stimulus is the mean of the posterior:

$$\hat{s}_{\text{PM}} = \mu_{\text{post}}. \tag{4.2}$$

In our simple model, the estimate is also the *response*, that is, what the observer would respond in an experiment.

How do we go from here to the probability distribution of the response given the stimulus? The estimate $\hat{s}_{\text{PM}}$ is a linear function of x_{obs}. Moreover, we know from step 1 that when the true stimulus is s, x_{obs} follows a Gaussian distribution with mean s and variance σ^2. We can now use the properties of linear combinations of random variables (see box 4.1 and exercise 4.1) to show that when the true stimulus is s, the stimulus estimate $\hat{s}_{\text{PM}}$ follows a Gaussian distribution with mean

$$\mathbb{E}[\hat{s}_{\text{PM}}|s] = \frac{Js + J_s\mu}{J + J_s} \tag{4.3}$$

and variance

$$\text{Var}[\hat{s}_{\text{PM}}|s] = \frac{J}{(J + J_s)^2}, \tag{4.4}$$

or, using the shorthand notation for Gaussian distributions (see equation (3.3)),

$$p(\hat{s}_{\text{PM}}|s) = \mathcal{N}\left(\hat{s}_{\text{PM}}; \frac{Js + J_s\mu}{J + J_s}, \frac{J}{(J + J_s)^2}\right). \tag{4.5}$$

Note on notation: The notation $\mathbb{E}[\hat{s}|s]$ indicates the expected value of $\hat{s}$ under the conditional distribution $p(\hat{s}|s)$. The expected value of a random variable with a conditional probability distribution is also called a *conditional expectation*. We use a subscript under the $\mathbb{E}$, Var, and so on, only to indicate the random variables in case of ambiguity. For details, see appendix A.

Box 4.1
Properties of Linear Combinations of Random Variables

In Bayesian models, we often encounter random variables that are weighted sums of other random variables. The following properties are useful:

1. *General:* If a random variable X has mean μ and variance σ^2, and a and b are constants, then the random variable $aX + b$ has mean $a\mu + b$ and variance $a^2\sigma^2$.
2. *General:* If random variables X and Y are independent and have means μ_X and μ_Y, and variances σ_X^2 and σ_Y^2, respectively, then the random variable $X + Y$ has mean $\mu_X + \mu_Y$ and variance $\sigma_X^2 + \sigma_Y^2$.
3. *Specific to Gaussian distributions:* If a random variable X follows a Gaussian distribution, then $aX + b$ also follows a Gaussian distribution.
4. *Specific to Gaussian distributions:* If random variables X and Y are independent and each follow a Gaussian distribution, then $X + Y$ also follows a Gaussian distributions.

Exercise 4.1 We now use the properties in box 4.1 to derive the response distribution.

(a) Combine the first two properties to show that the random variable $aX + bY$ has mean $a\mu_X + b\mu_Y$ and variance $a^2\sigma_X^2 + b^2\sigma_Y^2$.

(b) Using this result, prove equations (4.3) and (4.4).

The variance of the posterior mean estimate differs from any variance we have encountered so far: from that of the measurement distribution, that of the likelihood function, and that of the posterior distribution. Table 4.1 might help to keep the different distributions apart.

	Mean	Variance
Step 1: Generative model	Stimulus distribution, $p(s)$	
	μ	σ_s^2
	Measurement distribution, $p(x\|s)$	
	s	σ^2
Step 2: Inference	Prior distribution, $p(s)$	
	μ	σ_s^2
	Likelihood function $\mathcal{L}(s_{\text{hyp}}; x_{\text{obs}}) = p(x_{\text{obs}}\|s_{\text{hyp}})$	
	x_{obs}	σ^2
	Posterior distribution $p(s_{\text{hyp}}\|x_{\text{obs}})$	
	$\hat{s}_{\text{PM}} = \frac{Jx_{\text{obs}} + J_s\mu}{J+J_s}$	$\frac{1}{J+J_s}$
Step 3: Response distribution	Response distribution $p(\hat{s}\|s)$	
	$\mathbb{E}[\hat{s}_{\text{PM}}\|s] = \frac{Js + J_s\mu}{J+J_s}$	$\frac{J}{(J+J_s)^2}$

Table 4.1
Means and variances of all distributions discussed in the case study of chapters 3 and 4. We adopt "precision notation," in which $J = \frac{1}{\sigma^2}$ and $J_s = \frac{1}{\sigma_s^2}$.

4.3 Belief versus Response Distributions

We are now ready to address one of the biggest confusions in Bayesian modeling, between belief and response distributions. Priors, normalized likelihoods, and posteriors represent the degree of belief that an observer has in different hypothesized states of the world. These beliefs are defined on each individual trial and are internal to the observer; they are not directly measurable. The response distribution, on the other hand, is a summary of the observer's behavior across many trials. It is directly measurable and exists even if the observer is not Bayesian and does not hold any belief distributions. The horizontal axis of a belief distribution represents a hypothesized world state variable, while the horizontal axis of a response distribution represents an estimate of the world state variable. Table 4.2 puts belief and response distributions side by side.

One way in which the distinction between belief and response distributions manifests is in their variances. Specifically, the variance of the posterior mean estimate, $\text{Var}(\hat{s}_{\text{PM}}) = \frac{J}{(J+J_s)^2}$, is different from the variance of the posterior (from equation (3.20)), $\sigma^2_{\text{post}} = \frac{1}{J+J_s}$. Conceptually, this is possible because they have a completely different meaning: the former is the variability of behavior as measured by the experimenter, the latter is the uncertainty of the observer on a given trial. In general, these two quantities will be different.

Table 4.2
Comparison between belief and response distributions in Bayesian models. It might be useful to consider this table along with table 3.1, which compares steps 1 and 2.

	Belief distributions (prior, likelihood, posterior)	Response distribution
Bayesian modeling step	step 2 (inference)	step 3 (response distribution)
Nature of distributions (see Box 3.4)	subjective (degrees of belief)	objective (probabilities of occurrence)
Point of view of	the observer (decision-maker)	the experimenter
Horizontal axis	hypothesized world state	estimate of world state
Is realized	on any single trial	across many trials
Exclusive to Bayesian models	yes	no

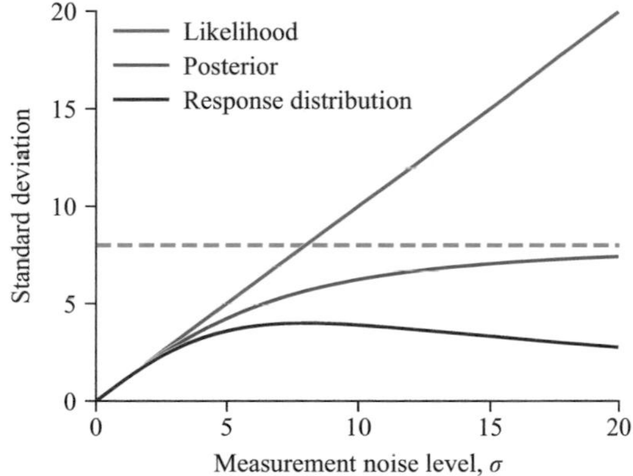

Figure 4.2
The standard deviation of the normalized likelihood function, the posterior distribution, and the response distribution (distribution of the posterior mean estimate) as a function of the measurement noise level, when the stimulus distribution has a standard deviation of $\sigma_s = 8$. As σ grows very large, the standard deviation of the posterior converges to the standard deviation of the prior, σ_s, while the standard deviation of the response distribution eventually decreases to 0.

The difference becomes clear when we plot the corresponding standard deviations as a function of the measurement noise level (figure 4.2). As the measurement noise level increases, the posterior gets wider and wider. Initially, the same holds true for the distribution of the PME. However, when σ grows large enough, the standard deviation of the posterior mean estimate decreases again.

Exercise 4.2 Why does it make intuitive sense that the standard deviation of the posterior mean estimate decreases again?

4.4 Maximum-Likelihood Estimation

Instead of using the posterior mean as the estimate of the stimulus, the observer could instead simply use the measurement itself. Recall from section 3.3.2 that in our case study, the measurement is also the MLE:

$$\hat{s}_{\text{ML}} = x_{\text{obs}}. \tag{4.6}$$

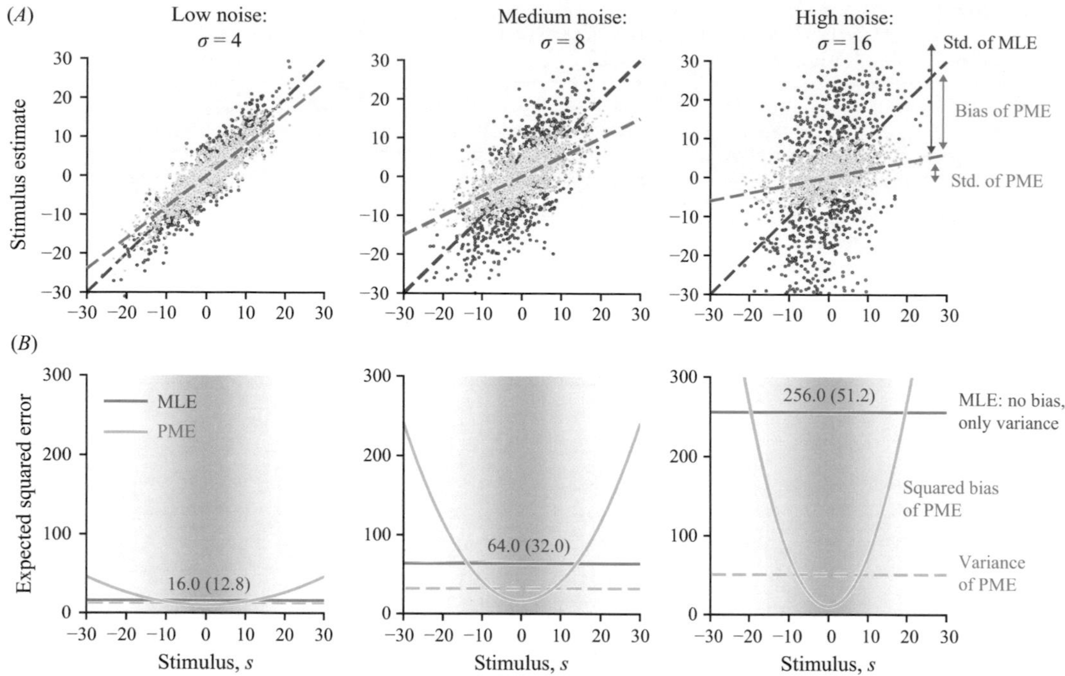

Figure 4.3
Comparison between the PME and the MLE. In this example, the stimulus distribution has $\mu = 0$ and $\sigma_s = 8$. (*A*) Scatterplots of PMEs and MLEs against the true stimulus. Dashed lines indicate the expected values. The larger the noise, the lower the slope of the expected value of the PME. (*B*) Mean squared error as a function of the stimulus for the PMEs and MLEs. Mean squared error (solid lines) is the sum of squared bias and variance. Although the PME is biased, its variance (dashed light blue line) is lower than that of the MLE (green line). The stimuli that occur often according to the stimulus distribution (shading indicates probability) are such that the overall (stimulus-averaged) MSE of the PME (light blue number, in parentheses) is always lower than that of the MLE (green number).

Just as we studied the distribution of the posterior mean estimate $\hat{s}_{\text{PM}}$ for a given s, we can also study the distribution of the maximum-likelihood $\hat{s}_{\text{ML}}$ for a given s. But we already know this distribution, as it is equal to the measurement distribution $p_{x|s}$: Gaussian with mean s and variance σ^2.

Examples of $\hat{s}_{\text{ML}}$ and $\hat{s}_{\text{PM}}$ for stimuli s randomly drawn from a Gaussian stimulus distribution $p_s(s)$ are shown in figure 4.3A. This shows that the MLE is on average equal to the stimulus, whereas the PME tends to lie closer to the mean of the prior. The higher the measurement noise, σ, the closer the PME tends to lie to the mean of the prior.

4.5 Bias and Mean Squared Error

We saw that the PME is a weighted average between the measurement and the mean of the Gaussian stimulus distribution. Therefore, the mean PME is a weighted average between the true stimulus and the mean of the stimulus distribution. As a consequence, the mean of the optimal estimate is not equal to the true stimulus: the posterior mean estimate is *biased* from the stimulus toward the mean of the prior.

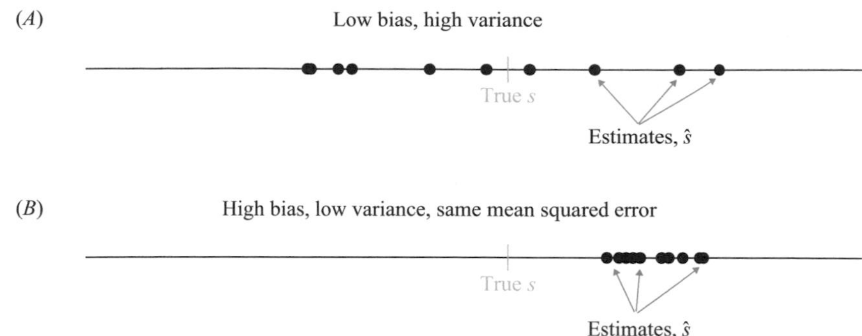

Figure 4.4
The same mean squared error can be produced by an estimate with low bias and high variance (*A*) and by an estimate with high bias and low variance (*B*).

The *bias* of an estimate $\hat{s}$ (which could be the PME, the MLE, the MAP estimate, or any other estimate) is defined as the difference between the average estimate and the true stimulus:

$$\text{Bias}[\hat{s}|s] \equiv \mathbb{E}[\hat{s}|s] - s. \tag{4.7}$$

Exercise 4.3

(a) Calculate the bias of the MLE for a given s.

(b) Calculate the bias of the PME for given s.

Bayesian modelers often say that Bayesian estimation is optimal. But the PME (and pretty much every other Bayesian estimate) is biased—its mean is different from the true stimulus. This seems contradictory: wouldn't it always be better to have an unbiased estimate?

To resolve this paradox, we have to realize that bias is not all that matters. It turns out that the posterior mean estimator is good in the sense that it minimizes the overall *mean squared error* (MSE) between the estimate and the true stimulus. The squared error on a single trial is $(\hat{s}-s)^2$. The MSE of $\hat{s}$ for given s is

$$\text{MSE}[\hat{s}|s] \equiv \mathbb{E}_{\hat{s}|s}[(\hat{s}-s)^2|s]. \tag{4.8}$$

MSE has two components: not only bias, but also variance. Mathematically,

$$\text{MSE}[\hat{s}|s] = \text{Bias}[\hat{s}|s]^2 + \text{Var}[\hat{s}|s]. \tag{4.9}$$

This relation is also called the *bias-variance decomposition of the mean squared error*. Both bias and variance are bad, in different ways. Figure 4.4 shows how the same expected squared error can arise from low bias and high variance, and from high bias and low variance.

The MSE in equation (4.9) depends on s, and we will call it the *stimulus-conditioned mean squared error*. We can average this quantity over all s to obtain an *overall mean squared error*:

$$\text{MSE}[\hat{s}] \equiv \mathbb{E}_{\hat{s},s}[(\hat{s}-s)^2] \tag{4.10}$$

$$= \mathbb{E}_s\left[\mathbb{E}_{\hat{s}|s}[(\hat{s}-s)^2|s]\right] \tag{4.11}$$

$$= \mathbb{E}_s\left[\text{MSE}[\hat{s}|s]\right] \tag{4.12}$$

Intuitively, this is the expected squared error across an entire experiment, obtained by averaging over many trials, each of which features a randomly drawn s and a subject response $\hat{s}$.

It turns out that the PME is optimal in that it minimizes the overall, s-independent, mean squared error MSE[$\hat{s}$]. Let's gain some intuition for that statement in two extreme cases. When the measurement is exceptionally noisy, you will be off by the least, on average, if you ignore the measurement altogether and always pick the mean of the stimulus distribution, μ. Your variance is guaranteed to be zero. The price you pay is a large bias if the stimulus happens to be far from average, but those are exactly the stimulus values that are very rare. On the other extreme, if the measurement x_{ob} is noiseless, you should estimate the stimulus to be x_{obs}. It makes intuitive sense that for intermediate noise levels, a strategy of picking an estimate in between the mean stimulus and x_{obs} will cause you to be off by the smallest possible amount on average. Thus, an optimal strategy would involve a bias toward the mean stimulus. Of course, this argument does not yet uniquely point toward the PME.

We can calculate the stimulus-conditioned MSE and the overall MSE of the PME. To do so, we start with equation (4.9), choose $\hat{s}=\hat{s}_{\text{PM}}$, and substitute equations (4.7), (4.3), and (4.4):

$$\text{MSE}[\hat{s}|s]=\left(\frac{Jx_{\text{obs}}+J_s\mu}{J+J_s}-s\right)^2+\frac{J}{(J+J_s)^2} \tag{4.13}$$

$$=\frac{J_s^2}{(J+J_s)^2}(s-\mu)^2+\frac{J}{(J+J_s)^2}. \tag{4.14}$$

This can be further simplified, but we keep this form so that the first term is still the squared bias and the second term is still the variance. The squared bias grows quadratically with the distance between s and the mean stimulus, while the variance is independent of s. We have plotted both terms as a function of s in figure 4.3B. For comparison, we have also plotted the stimulus-conditioned MSE of the MLE. The MLE is unbiased, since $\mathbb{E}[\hat{s}_{\text{ML}}|s]=s$, and so its stimulus-conditioned MSE is equal to its variance, which has a constant value of σ^2. The plot allows us to understand why the PME is optimal even though it is biased. For any s, its variance is lower than the variance of the MLE. The squared bias gets added to the variance, but their sum still stays below the variance of the MLE as long as the stimulus is close enough to the mean stimulus. That is exactly where most stimuli will be, since the stimulus distribution is Gaussian (shown as a gray shading).

When we evaluate the expected value over s in equation (4.12) to calculate the overall MSE, as one would experience in a sufficiently long experiment, we end up with a lower number for the PME than for the MLE. We work this out in problem 4.9. In short, a strategy in which the prior mean and the likelihood mean are combined will produce greater rewards in the long run. Thus, a bias toward the mean of a prior is actually a sign of an optimal strategy.

As pointed out by E. T. Jaynes, the word "bias," because it carries negative connotations when used in everyday speech, is an unfortunate choice of terminology: "When we call the quantity. . . . 'bias', that makes it sound like something awfully reprehensible, which we must get rid of at all costs. If it had been called instead the 'component of error orthogonal to the variance', . . . it would have been clear to all that these two contributions to the error are on an equal footing. . . . This is just the price one pays for choosing a technical terminology that carries an emotional load, implying value judgments" [13: 514].

4.5.1 An "Inverted Bias" Perspective

Because it is such an important point to understand, we will offer a second answer to the question "how can a biased estimator be optimal?" We can view the scatterplot (figure 4.3A) from a different perspective (figure 4.5). So far, we calculated the overall MSE by taking the expectation first over $\hat{s}$ for a given s, then over s, as in equation (4.12). However, one can also reverse the order of calculations and take the expectation first over s for a given $\hat{s}$, then over $\hat{s}$. We will gain new insights by considering this alternative approach. We will see that the mean difference between the MLE and the stimulus across all trials is zero. The mean difference between the PME and the stimulus across all trials is also zero, but the PME is distributed around the stimulus with smaller variance.

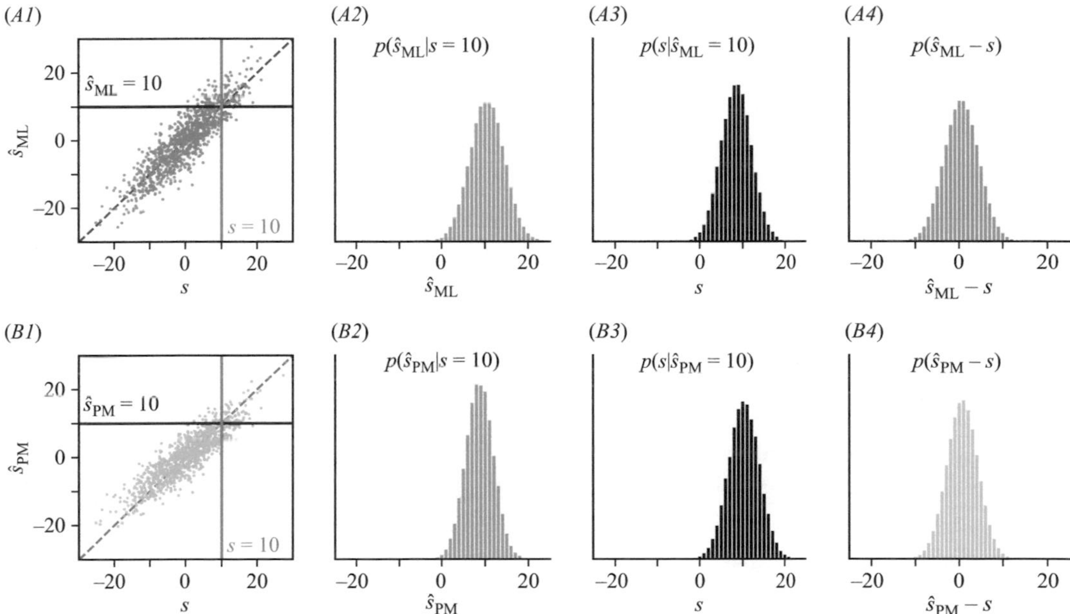

Figure 4.5
The PME ($\hat{s}_{PM}$) is more accurate than the MLE ($\hat{s}_{ML}$). (*A1*) 1,000,000 stimuli were drawn from a normal distribution ($\mu = 0$, $\sigma_s = 8$), each yielding a noisy measurement ($\sigma = 4$). Scatterplot of s vs. $\hat{s}_{ML}$ (subsample of 1000 shown). Diagonal line, $\hat{s}_{ML} = s$; vertical line, $s = 10$; horizontal line, $\hat{s}_{ML} = 10$. (*A2*) At each s, $\hat{s}_{ML}$ is distributed with mean s and variance $\sigma^2 = 16$. The distribution $p(\hat{s}_{ML}|s = 10)$ is shown for illustration. (*A3*) For a given $\hat{s}_{ML}$, the distribution of s is not centered on $\hat{s}_{ML}$. The distribution $p(s|\hat{s}_{ML} = 10)$ is shown for illustration; this is the posterior distribution over s given $x = 10$. (*A4*) For every s, $p(\hat{s}_{ML} - s|s)$ has mean 0 and variance 16. Therefore, the overall error distribution, $p(\hat{s}_{ML} - s)$, also has mean 0 and variance 16. Because this error distribution has mean 0, its variance is MSE[$\hat{s}_{ML}$]. (*B1*) Scatterplot of s vs. $\hat{s}_{PM}$ for the same data points shown in (*A*). Diagonal line, $\hat{s}_{PM} = s$; vertical line, $s = 10$; horizontal line, $\hat{s}_{PM} = 10$. (*B2*) The PME is biased toward the prior mean; $p(\hat{s}_{PM}|s = 10)$ is shown for illustration. (*B3*) $p(s|\hat{s}_{PM})$ is the posterior distribution over s, given the measurement that resulted in $\hat{s}_{PM}$. Thus, $p(s|\hat{s}_{PM})$ has mean $\hat{s}_{PM}$ and variance 12.8; $p(s|\hat{s}_{PM} = 10)$ is shown for illustration. (*B4*) For every $\hat{s}_{PM}$, $p(\hat{s}_{PM} - s|\hat{s}_{PM})$ has mean 0 and variance 12.8. Therefore, the overall error distribution, $p(\hat{s}_{PM} - s)$, also has mean 0 and variance 12.8. Because this error distribution has mean 0, its variance is MSE[$\hat{s}_{PM}$].

Figure 4.5 plots one million stimuli sampled from a stimulus distribution ($\mu = 0$, $\sigma_s = 8$), each yielding a noisy measurement ($\sigma = 4$). Figure 4.5A shows the statistics for the MLE, and figure 4.5B for the PME. For a given stimulus value, s, the MLE is distributed with mean s (panel *A2*); thus, the MLE is unbiased. By contrast, for a given s, the PME is not distributed with mean s (panel *B2*); thus, the PME is biased. Crucially, however, on trials when the observer reports a particular PME, the stimulus is indeed centered on that estimate (panel *B3*), and this is not the case for trials with a particular MLE (panel *A3*). In essence, the *stimulus* is biased with respect to the *MLE*! In contrast, the stimulus is unbiased with respect to the PME. Most importantly, while the mean difference across all trials between either estimate and the stimulus is zero, the MSE is lower for the PME (panel *B4*) than for the MLE (panel *A4*). Thus, the PME is a more accurate estimator than the MLE.

To further understand this, note that, given a particular MLE (i.e., a particular measurement, x), the distribution of stimuli is the posterior distribution, $p(s|x)$. The posterior distribution is not centered on the measurement; rather, it is shifted away from the measurement and towards the prior mean (equation (4.1)). Consequently, the stimulus is biased with respect to the MLE. In contrast, the stimulus is unbiased with respect to the PME. To understand why, note that each MLE (i.e., each measurement) maps onto a particular PME (equation (4.1)). Consequently, each horizontal row of dots in panel *A1* is simply displaced vertically toward the prior mean (0, in this example) in panel *B1*, such that its position equals the PME. For instance, the reader may verify from equation (4.1) that the row at $\hat{s}_{\text{ML}} = 10$ in panel *A1* is shifted to $\hat{s}_{\text{PM}} = 8$ in panel *B1*. Since panel *B1* plots the posterior mean on the y-axis, it necessary follows that the mean of the distribution of stimuli in each horizontal row (given sufficiently large samples) coincides with the identity line. Following this line of reasoning, we can see that the variance of each horizontal row of stimuli in panels *A1* and *B1* is simply the variance of the posterior distribution. Since the variances of the horizontal rows are equal in panels *A1* and *B1*, but the stimulus is biased only with respect to the MLE (compare panels *A3* and *B3*), it follows immediately that the MSE of the MLE must be greater than that of the PME.

In fact, without doing additional math, we can see that the overall MSE of the PME equals the variance of the posterior distribution, whereas the overall MSE of the MLE equals the variance of the measurement distribution. To see this, let's accumulate the errors ($\hat{s}_{\text{PM}} - s$) from all of the horizontal rows in panel B1. As the distribution of stimuli is centered in each row on the PME (e.g., panel *B3*), the resulting overall error distribution has mean zero (panel *B4*). Consequently, the variance of this distribution (i.e., the variance of the posterior distribution) is the overall MSE of the PME. Following a similar line of reasoning, we can accumulate the errors ($\hat{s}_{\text{ML}} - s$) from all vertical columns in panel *A1*. As the distribution of MLE is centered in each column on s (e.g., panel *A2*), the resulting overall error distribution has mean zero (panel *A4*). Consequently, the variance of this distribution (i.e., the variance of the measurement distribution) is the overall MSE of the MLE. In conclusion, $\text{MSE}[\hat{s}_{\text{PM}}] < \text{MSE}[\hat{s}_{\text{ML}}]$, because the variance of the posterior distribution is less than the variance of the measurement distribution.

4.5.2 All Expenses Paid

So far, we have considered only continuous variables for comparing posterior-based estimation with MLE. How would this comparison play out for discrete (categorical) variables? We consider in particular *nominal* variables, where the categories have labels but no particular ordering. (Ordinal discrete variables are somewhere in between nominal variables and continuous variables.) For nominal variables, the mean of the posterior does not make sense, because there is no underlying space. Instead, as we discussed in chapters 2 and 3, the Bayesian observer would pick the category with the highest posterior probability, or, in other words, perform MAP estimation. Like the PME in the continuous case, the MAP is biased toward more frequent values, and therefore it is equally interesting to ask if this bias is "worth it."

To examine this question, let us consider the following example. A particular country consists of three states s. Table 4.3 shows, for each of the states, the proportions of workers in various occupations, x. These proportions are conditional probabilities $p(x = \text{occupation}|s)$. Now suppose that, as part of a public-relations campaign, a person is randomly selected from the country to win an all-expenses-paid vacation. Given this person's occupation, what state do we think they are from? Maximum-likelihood estimation would lead us to report that, if this person happens to be a teacher, they are from state 1; if they happen to be a farmer, they are from state 3; and if they happen to be a retail worker, they are from state 2 (highlighted, see table 4.3).

It is easy to see what is wrong with this reasoning. The three states have vastly different populations (see second row in the table; numbers are in thousands), and it is more probable, a priori, that a randomly selected citizen will be from a more populous state. On consideration, it is clear that, if a randomly selected citizen is a farmer, in order to maximize the probability of being correct about that person's state of residence, it is not sufficient to consider only the *proportion* of each state's population that are farmers. Instead, we must calculate the *absolute number* of farmers in each state, that is, the proportion of the state's population that are farmers, multiplied by the state's total population. These numbers are computed in table 4.4.

The most probable state of origin for a person of a given occupation is the state that has the greatest number of people in that occupation. This gives a partially different set of decisions (highlighted). Whereas the best guess for the teacher's home state is unchanged, the best guess for the farmer's home state is now state 2, and the best guess for the retail worker's home state is now state 1. This is MAP estimation, and we see in the right column of the table that for the farmer and the retail worker, the MAP estimate is more accurate than the ML estimate.

Although MAP estimation is thus "biased" toward states with higher populations, this bias is entirely rational and indeed optimal. Note also that the MAP estimate does not *always* go with the state with the highest population. In the case of the farmer, the proportion of farmers in state 2 is large enough to overcome the overall population advantage of state 1. This indicates that the observation, acting through the likelihood function, can be strong enough to overcome the prior. This ends our discrete detour.

4.6 Other Estimates

If we had not heard about Bayesian inference, we might have thought that a reasonable estimate of the stimulus would be the average of the mean of the stimulus distribution, μ, and the measurement x_{obs}.

$$\hat{s}_{\text{average}} = \frac{x_{\text{obs}} + \mu}{2}. \tag{4.15}$$

Table 4.3
Percentages of workers in a hypothetical three-state country, computed state by state. Boldfaced per row is the maximum-likelihood decision if a randomly selected person has the indicated occupation.

	$s=1$	$s=2$	$s=3$	
Total population (1000s)	19,771 (79.0%)	3,446 (13.8%)	1,805 (7.2%)	% correct MLE
$x=$ teacher	**1.5%**	0.8%	1.1%	86%
$x=$ farmer	0.4%	2.6%	**3.6%**	28%
$x=$ retail	8.4%	**9.3%**	8.3%	15%
$x=$ other	**89.7%**	87.3%	87.0%	79%

Table 4.4
The same data in absolute numbers. Boldfaced per row is the MAP decision if a randomly selected person has the indicated occupation. The probability that the MAP decision is correct is obtained by dividing the bold number in each row by the sum of numbers in the row. This is also the posterior probability of the MAP estimate. For example, $p(s=2|x=\text{farmer})=\frac{90}{79+90+65}=0.38$.

	$s=1$	$s=2$	$s=3$	% correct MAP estimate
$x=$ teacher	**297**	28	20	86
$x=$ farmer	79	**90**	65	38
$x=$ retail	**1,661**	320	150	78
$x=$ other	**17,735**	3,008	1,570	79

Another way to view this average estimate is that it is the PME under the wrong assumption that σ is equal to σ_s. Indeed, a non-Bayesian estimate can often be interpreted as a PME under a wrong assumption about the generative model. However, no estimate can ever have a lower MSE than the PME, as long as the PME uses the correct generative model; we explore this in problem 4.9.

4.7 Decision Noise and Response Noise

So far, we have described the mapping from the measurement to the stimulus estimate to be deterministic, namely, as defined by the posterior mean. This is optimal, but not necessarily realistic. Noise could be added to this mapping. One form of noise is additive Gaussian noise with constant variance. Another form of decision noise comes from the possibility that the stimulus estimate is sampled from the posterior distribution. However, evidence that the brain samples from the posterior distribution is, in our opinion, sparse.

In addition, we assumed that the observer's response is equal to the stimulus estimate. In practice, every response on a continuum will be subject to some response (e.g., motor) noise, reflecting for instance the accuracy with which the observer is able to position a computer cursor. Moreover, the observer's memory of the estimate might decay slightly between the moment the estimate is made and the moment that the response is submitted. Thus, the observer's response distribution is not necessarily the same as the estimate distribution, and a complete model of the task would include the response noise. For example, the observer's response could be drawn from a Gaussian distribution with mean equal to the stimulus estimate but with some variance σ^2_{motor}. We will treat this case in problem 4.12. However, response noise is not central to the Bayesian formalism.

4.8 Misconceptions

To build a Bayesian model, we must formulate our generative model and do proper inference on it. Some aspects of this inference are counterintuitive, and scientists doing research on Bayesian models occasionally become confused about the resulting relations between variables. Here we address several misconceptions that may arise during Bayesian modeling. We explore more misconceptions in the problems.

The likelihood function is determined by the stimulus One misconception is that a Bayesian observer's likelihood function is determined by the stimulus, and therefore it is always the same function as long as the stimulus is held fixed. In reality, we saw in figure 4.1 that the likelihood function varies from repetition to repetition if the observations are noisy, so a single value of the stimulus gives rise to different likelihood functions. The misconception may arise from confusing the likelihood function with the measurement distribution. The measurement distribution *is* indeed generally assumed to be determined by the stimulus.

A related misconception is that the posterior distribution is determined by the stimulus. In reality, the posterior also varies from trial to trial even when the stimulus is fixed (figure 4.1).

The response distribution is equal to the likelihood function Sometimes one reads something like, "We plot the likelihood of the estimate." This confuses the likelihood from step 2 with the response distribution from step 3 or with the measurement distribution from step 1. The argument of the likelihood function is the hypothesized stimulus, not the estimate. The outcome of step 3 is a distribution of an estimate, not a likelihood of the estimate. And if "estimate" is being confused with "measurement" (which is, after all, the MLE), then the distribution of this measurement is not a likelihood function either, even though the likelihood function is derived from the measurement distribution. Possible corrections would be "We plot the distribution of the estimate" (if the x axis represents the estimate), "We plot a likelihood function over the stimulus on a particular trial" (if the x -axis represents the hypothesized stimulus), or "We plot the distribution of the measurement" (if the x axis represents the measurement).

The response distribution is the product of the measurement distribution and the prior Here is another tempting mistake: "Assuming that an observer reports the PME, to obtain the probability density of an observer's response for a given stimulus, I multiply the measurement distribution $p(x|s)$ with the prior probability distribution. This is correct because, after normalization, it gives me a distribution that is centered between the prior mean and the true stimulus." A more sophisticated version leading to the same conclusion would be: "The MLE of the stimulus is equal to the measurement x [true]. Thus, the distribution of the MLE for a given true stimulus s is equal to the noise distribution, with s [true]. Therefore, the distribution of the PME given a stimulus s can be obtained by multiplying the measurement distribution with the prior [false]." In math, this reasoning would amount to

$$\text{(wrong)} \quad p(\hat{s}|s) \propto p_{x|s}(\hat{s}|s)p_s(\hat{s}), \tag{4.16}$$

where $p_{x|s}(\hat{s}|s)$ is the measurement distribution $p_{x|s}$ evaluated at $\hat{s}$, and $p_s(\hat{s})$ is the prior distribution evaluated at $\hat{s}$.

Both statements suggest an awfully convenient shortcut for directly calculating the response distribution, instead of having to go through steps 2 and 3 of the modeling process. Both statements are also too good to be true. To understand why they are wrong, we could substitute the distributions we used in chapter 3 and the current chapter, $p_s(s) = \mathcal{N}(s; \mu, \sigma_s^2)$ and $p_{x|s}(x|s) = \mathcal{N}(x; s, \sigma^2)$. Then, equation (4.16) would produce a Gaussian distribution with a mean of $\frac{Js+J_s\mu}{J+J_s}$ and a variance of $\frac{1}{J+J_s}$. The mean would be correct, per equation (4.3). The variance, however, would differ from the correct variance of $\frac{J}{(J+J_s)^2}$ from equation (4.4). An intuitive reason why the answer must be wrong can be obtained in the zero-reliability limit ($\sigma \to \infty$). Then, the observer will always estimate the stimulus to be the mean of the prior, and thus the variance of the response distribution should be 0. The wrong argument would give a variance of σ_s^2. More formally, the principal mistake made here is that equation (4.16) is not a correct application of Bayes' rule. First of all, it does not have the mathematically correct form $p(y|x) \propto p(x|y)p(y)$. Moreover, in Bayesian modeling of behavior, Bayes' rule does not act at the level of estimates over many trials, but at the level of a single trial. Therefore, both components on the right-hand side of equation (4.16) are incorrect. The likelihood function should not be a function of the MLE or the measurement, but of the hypothesized value of the stimulus s, again on a single trial. The argument of the prior distribution is not the stimulus estimate, but the hypothesized stimulus, again on a single trial.

When the measurement is equal to the true stimulus, the response distribution is equal to the posterior The next misconception we consider is the following: "For the distribution of the observer's PM estimate for a given stimulus, I can simply use the 'typical' posterior, the one obtained when the measurement happens to be equal to the true stimulus s. It will give me a distribution that is centered between the prior mean and the true stimulus."

This mistake is mathematically identical to the previous one but arrived at through a slightly different line of reasoning. Suppose we correctly calculate the posterior distribution, $p(s|x)$. Now we substitute the true stimulus for x: $p(s|x=s)$. This is a legitimate, though not particularly meaningful, function of s: it reflects the observer's beliefs about the stimulus when the measurement x just happens to coincide with the true stimulus. The final step of the faulty argument would be to regard the distribution $p(s|x=s)$ as the response distribution, $p(\hat{s}|s)$. We further examine this in problem 4.14.

In our Gaussian example, this would again lead to the conclusion that the variance of the response distribution is $\frac{1}{J+J_s}$, while in reality this is the variance of the posterior. Equating the variance of the response distribution and the posterior is a frequent mistake. To correctly describe the relationship of prior, noise distribution and distributions of PMEs, there is no alternative to going through the three steps described in chapter 3 and this chapter.

The prior probability is equal to the overall probability of responding This tempting misconception can take different forms, such as:

- "The prior probability of responding rightward is 0.5."
- "I have a simple way to derive the prior distribution used by the observer directly from the data. I can simply tally up the observer's estimates across all trials in the experiment.

After all, the higher the prior probability of a stimulus value, the more often the observer will report that stimulus value."

Over the course of an experiment, we could keep track of the distribution of the observer's response. It is tempting but incorrect to regard the overall distribution of the observer's response as the observer's prior distribution. In other words, a phrase like "$p(s)$ is the prior probability of responding s" is incorrect. The correct phrase is "$p(s)$ is the prior probability (observer's belief) that the state of the world is s."

To see the distinction between prior distribution and overall response distribution, consider our Gaussian example. Intuitively, if the observer uses the correct prior distribution, then all their responses will be biased toward the mean of the prior and the response distribution will thus be narrower than the prior distribution.

We can demonstrate the difference between the two distributions more formally. When the measurement is x_{obs}, the PME is $\frac{Jx_{\text{obs}}+J_s\mu}{J+J_s}$. This is a random variable because x_{obs} is a random variable. To calculate the distribution of the PME conditioned on the true stimulus s, we used the fact that x_{obs} given s is Gaussian with mean s and variance σ^2. Similarly, we now consider the distribution of the PME across all s. The distribution of x_{obs} across all s is Gaussian with mean μ and variance $\sigma^2+\sigma_s^2$. Then, the PME will have mean μ and variance $\frac{\sigma_s^4}{\sigma^2+\sigma_s^2}$ (see problem 4.11). This expression, which is plotted as a function of σ in figure 4.2, shows that the overall response distribution is not identical to the prior distribution (whose variance would, of course, be independent of σ).

4.9 Reflections on Bayesian Models

The Bayesian model discussed in this chapter, although simple, is in many ways representative of Bayesian modeling in general. The essence of Bayesian observers is that they consider all possible values of a state-of-the-world variable and compute the probabilities of those values. In other words, the Bayesian observer does not commit to a limited set of hypotheses unless so directed by the evidence.

One of the great powers of Bayesian modeling is that it allows one to build a complete model of a perceptual task before having seen any experimental data: the Bayesian model specifies how an observer *should* do the task in order to be optimal. Bayesian modeling is therefore an example of *normative* modeling: the Bayesian model sets the norm—the highest performance that can be achieved by an observer. This stands in contrast to common practice in much psychology, in which modeling, if done at all, is often done *after* having observed certain patterns in the data. In Bayesian modeling, one can write down the model and perform model simulation without having even started an experiment.

4.10 Summary

In this chapter, we wrapped up the prototypical Bayesian model started in chapter 3 by calculating the response distribution for a given stimulus. We learned the following:

- Bayesian modeling consists of three steps: defining the generative model, deriving an expression for the observer's posterior mean estimate (PME), and deriving the distribution of the PME over many trials.

- The mean squared error (MSE) as a measure of the performance of an estimate. We distinguished stimulus-conditioned MSE and overall MSE.
- Stimulus-conditioned MSE is a sum of squared bias and variance.
- Compared with the distribution of the PME, that of the MLE is unbiased, though it has higher variance for frequently occurring stimuli, and it is worse overall as a result.
- It is easy to confuse the functions and distributions in the different steps, as they look similar (in this chapter, they are all Gaussian). They must be distinguished carefully, as we did in table 4.1.
- Many conceptual mistakes can be made if the three steps are not followed. In particular, attempts at calculating the response distribution through shortcuts are bound to fail.
- The PME model is a minimal model. Different forms of decision noise and response noise can be considered as variants and extensions.

4.11 Suggested Readings

- Edwin T. Jaynes. *Probability Theory: The Logic of Science*. Cambridge: Cambridge University Press, 2003.
- Shane T. Mueller and Christoph T. Weidemann. "Decision Noise: An Explanation for Observed Violations of Signal Detection Theory." *Psychonomic Bulletin and Review* 15, no. 3 (2008): 465–494.

4.12 Problems

Problem 4.1 Match the following functions that play a role in Bayesian modeling with the descriptions:

Functions:

(a) Distribution of the posterior mean estimate
(b) Prior distribution
(c) Likelihood function
(d) Posterior distribution
(e) Measurement distribution

Descriptions:

(a) Is the result of inference on an individual trial
(b) Describes how potentially noisy observations are generated
(c) Can be directly compared to human responses in a psychophysical experiment
(d) Is often modeled as a Gaussian function centered at the measurement
(e) May reflect statistics in the natural world

Problem 4.2 In the Gaussian model discussed in this chapter:

(a) Prove mathematically that the variance of the PME, equation (4.4) is smaller than or equal to the variance of the MLE.

(b) How do the two compare when the variance of the prior is much larger than the variance of the measurement? Explain why the answer makes intuitive sense.

Problem 4.3 True or false? If false, explain.

(a) The likelihood function is always equal to the measurement distribution.

(b) The value of the stimulus that maximizes posterior probability is the value of the measurement.

(c) We can obtain the response distribution for a given stimulus by multiplying the measurement distribution with the prior.

(d) The response distribution for a given stimulus is always Gaussian.

(e) If, over the course of an experiment, a Bayesian observer reports one value of the stimulus estimate more often than another value, it means that the prior probability of the former is higher.

Problem 4.4 In problem 3.2, we plotted a posterior distribution along with a likelihood function. Repeat parts (a)–(f) of that problem, but instead of using a single value of the measurement x_{obs}, start with a fixed value of the stimulus, namely $s=10$. From this value of s, draw five values of x_{obs} from the measurement distribution. You should observe that, from trial to trial, the likelihood function and posterior probability density function both "jump around", similar to figure 4.1.

Problem 4.5 Refer to figure 4.2. Prove mathematically that, all else being equal, the standard deviation of the response distribution is highest when $\sigma=\sigma_s$. (Hint: at the maximum of a function, what is the value of the derivative?)

Problem 4.6 An observer infers a stimulus s from a measurement x_{obs}. As in the chapter, the measurement distribution $p(x|s)$ is Gaussian with mean s and variance σ^2. Unlike in the chapter, we use the prior

$$p(s)=e^{-\lambda s}, \tag{4.17}$$

where λ is a positive constant. This is an *improper prior* (see section 3.5.2) but that does not stop us.

(a) Derive an equation for the PME.

(b) Derive an equation for the distribution of the PME for given s.

Problem 4.7 We define *relative bias* as the ratio between bias and the difference between the mean of the stimulus distribution and the true stimulus.

(a) For the PME, derive an expression for relative bias as a function of the ratio $R\equiv\frac{\sigma}{\sigma_s}$. (Hint: the expression will only contain R, no other variables.)

(b) Plot relative bias as a function of R.

(c) Does this plot show what you would expect from the Bayesian observer? Explain intuitively.

Problem 4.8 Consider the estimate $\hat{s}_{\text{prior mean}} = \mu$, which completely ignores the measurement and simply returns the mean of the prior. One of the following questions is a trick question.

(a) Derive an expression for the overall MSE of this estimate.

(b) How large must σ be for this estimate to have a lower MSE than the MLE?

(c) How large must σ be for this estimate to have a lower MSE than the PME?

Problem 4.9 Equation (4.9) represents the MSE of the PME for given s. Using that equation, we now calculate the average of this quantity over all s, that is, the overall MSE in equation (4.12).

(a) Show mathematically that the overall MSE of the PME is equal to $\frac{1}{J+J_s}$. (Hint: $\mathbb{E}[(s-\mu)^2]=\sigma_s^2$.)

(b) Verify that the expression in (a) returns the purple numbers in figure 4.4B.

(c) Consider any estimate that is a linear transformation of the measurement: $\hat{s}_{\text{linear}} = ax_{\text{obs}} + b$, where a and b are constants. Calculate its overall MSE. (Hint: $\mathbb{E}[s^2] = \sigma_s^2 + \mu^2$.)

(d) Optional: Prove mathematically that among these estimates, the PME is the one with the lowest overall MSE. (Hint: take partial derivatives.)

Problem 4.10 In this problem, we will numerically compare the properties of the MAP estimate and the MLE. Refer to section 4.5 and the parameter settings in figure 4.3.

(a) Reproduce figure 4.3A using 1,000 trials.

(b) Reproduce the MLE and PME curves in figure 4.3B by plotting the corresponding mathematical expressions.

(c) For each of the three noise levels, simulate s on 10,000 trials. For each s, compute the squared bias and the variance of the PME using the expressions from (b). Add the two quantities to obtain the stimulus-conditioned MSE. Average across all values of s that you drew to obtain the overall MSE of the PME. The resulting values should be close to the PME-related numbers (12.8, 32.0, 51.2) in figure 4.3B.

Problem 4.11 We saw in section 4.5 that overall MSE is a stimulus average of the stimulus-conditioned MSE. Overall variance, however, is not simply a stimulus average of the stimulus-conditioned variances.

(a) Show that across all trials in an experiment, the MLE has mean μ and variance $\sigma^2 + \sigma_s^2$. (Hint: use box 4.1 on linear combinations of random variables.) We will call this the overall variance of the MLE.

(b) Making use of the variance in (a), show that across all trials in an experiment, the PME has mean μ and variance $\frac{\sigma_s^4}{\sigma_s^2+\sigma^2}$. We will call this the overall variance of the PME.

(c) Show that the overall variance of the PME is always smaller than the overall variance of the MLE.

(d) In the absence of measurement noise ($\sigma = 0$), part (c) predicts that the overall variance of the PME is σ_s^2. Explain why this makes sense.

(e) Perform a similar sanity check corresponding to $\sigma \to \infty$. (Hint: if measurement noise is extremely large, what can be said about the observer's estimate?)

Problem 4.12 We mentioned that an observer's response might be corrupted by response or motor noise. Assume motor noise that follows a Gaussian distribution with mean equal to the posterior mean estimate and with standard deviation σ_m.

(a) What is the distribution of the observer's response when the true stimulus is s?

(b) Think of ways to experimentally distinguish motor noise from measurement noise.

Problem 4.13 A student claims, "To obtain the probability density of an observer's PME for a given stimulus, I multiply the measurement distribution $p(x|s=\mu)$ where μ is the prior mean, with the prior probability density $p(s)$." In other words,

$$\text{(wrong)} \quad p(\hat{s}_{\text{PM}}|s) = p_{x|s}(\hat{s}_{\text{PM}}|\mu)p_s(\hat{s}_{\text{PM}}). \tag{4.18}$$

(a) Although this would produce a distribution that is centered between the prior mean and the true stimulus, this claim is conceptually wrong. Why?

(b) Show mathematically that the variance of the resulting distribution is incorrect.

(c) As a specific example, determine what this student would predict for the variance if measurement noise were extremely large (the limit $\sigma \to \infty$). Explain both answers intuitively.

Problem 4.14 A student claims, "To obtain the response distribution $p(\hat{s}_{\text{PM}}|s)$ in the model, I can simply use the posterior $p(s|x_{\text{obs}})$ obtained when the measurement x_{obs} happens to be equal to s." In other words,

$$\text{(wrong)} \quad p(\hat{s}_{\text{PM}}|s) = p_{s|x}(\hat{s}_{\text{PM}}|s). \tag{4.19}$$

(a) Although this would produce a distribution that is centered between the prior mean and the true stimulus, this claim is conceptually wrong. Why?

(b) Show mathematically that the variance of the resulting distribution is incorrect.

(c) As a specific example, determine what this student would predict for the variance if measurement noise were extremely large (the limit $\sigma \to \infty$), and what the correct calculation would predict. Explain both answers intuitively.

Problem 4.15 A student claims, "To obtain the response distribution $p(\hat{s}_{\text{PM}}|s)$ in the model, I can average the posteriors $p(s|x_{\text{obs}})$ over all x_{obs} for a given s." In other words,

$$\text{(wrong)} \quad p(\hat{s}_{\text{PM}}|s) = \int p_{s|x}(\hat{s}_{\text{PM}}|x_{\text{obs}})p(x_{\text{obs}}|s). \tag{4.20}$$

(a) Although this would produce a distribution that is centered between the prior mean and the true stimulus, this claim is conceptually wrong. Why?

(b) Show mathematically that the variance of the resulting distribution is incorrect.

(c) As a specific example, determine what this student would predict in the infinite-noise limit ($\sigma \to \infty$), and what the correct calculation would predict. Explain both answers intuitively.

Problem 4.16 In this problem, we fit and compare models on synthetic data from a continuous estimation task. Read appendix C if you are not familiar with model fitting and model comparison. Download estimation.csv from https://osf.io/84kpb. The rows correspond to trials (500 trials in total). The first column contains the values of the presented stimulus, the second column the subject's estimates of the stimulus.

(a) Plot the data in a scatter plot of estimate against stimulus, using black dots. Draw a dashed black line to indicate the diagonal. Choose your axis ranges suitably. Label the axes.

(b) By eye, would you say that the observer is doing ML estimation or is taking into account a prior? Why?

We now fit two models to the data. In both models, we assume that the observer's measurement follows a Gaussian distribution with mean equal to the true stimulus and unknown standard deviation σ. We also assume that all trials are independent.

(c) Model 1 states that the observer performs ML estimation. Under this model, write down an equation for the log likelihood function over σ in terms of the stimuli $s_1, \ldots, s_n$ and the estimates $\hat{s}_1, \ldots, \hat{s}_n$. Simplify as much as possible.

(d) Plot the log likelihood function of σ on a grid from 0.1 to 10 in steps of 0.02.

(e) On this grid, what is the MLE of σ?

(f) As an alternative to the grid search, use a built-in optimization algorithm to find the MLE of σ. Justify the algorithm that you chose.

(g) Model 2 states that the observer performs posterior mean estimation using a Gaussian prior with mean 0 and unknown standard deviation σ_{prior}. Under this model, write down an equation for the log likelihood function of the combination $(\sigma, \sigma_{\text{prior}})$. Simplify as much as possible.

(h) Using grids from 0.1 to 10 in steps of 0.02 for both σ and σ_{prior}, plot the log likelihood function as a heat map. Add a color legend. Label the axes; make sure to check which axis corresponds to which variable.

(i) On this grid, what are the MLEs of σ and σ_{prior}?

(j) Use a built-in optimization algorithm to find the MLEs of σ and σ_{prior}.

(k) Which model wins according to the Akaike Information Criterion?

(l) Which model wins according to the Bayesian Information Criterion?

Problem 4.17 In this problem, we fit and compare models on data from a real estimation experiment. The experiment was a sound localization experiment in humans conducted in the lab of Wei Ji Ma. On each trial, a sound was presented at one of five locations on a horizontal line: $-6, -3, 0, 3, 6$ (arbitrary units). The sound sources were hidden and the subject did not know that there were only five locations. The subject reported where they heard the sound coming from; they chose from twenty-one discrete locations: $-10, -9, -8, \ldots, 8, 9, 10$. Download localization.csv from https://osf.io/84kpb. The rows correspond to trials (400 trials in total). The first column contains the values of the true location, the second column the values of the reported location.

(a) **The data.** Separately for each of the five true locations, calculate the proportion of subject responses at each response location. Plot as five solid lines in a single plot, color-coded to distinguish the five true locations.

(b) We first consider a "null model", model 0 (0 parameters). According to this model, the observer chooses a random location (out of the twenty-one possibilities), with equal probabilities. Calculate the log likelihood, AIC, and BIC for model 0. You do not even need the subject responses for this. Explain why not.

We now introduce two more meaningful models:

- Model 1 (one parameter): The observer performs maximum-likelihood estimation of location, that is has a flat prior.
- Model 2 (two parameters): The observer performs MAP estimation of location, with a prior that is Gaussian with mean 0 and unknown standard deviation σ_s.

In both models, we make the usual assumption that the observer makes a noisy measurement of sound location and that this measurement follows a Gaussian distribution with mean equal to the true location and unknown standard deviation σ. Parts (c)–(g) are about models 1 and 2.

(c) **Model predictions I.** Ignore for the moment that the observer makes only discrete responses, and instead imagine that they report a continuous estimate of location, $\hat{s}$. For model 1, write down the equation for the probability density function over $\hat{s}$ when the true stimulus is in location s. Repeat for model 2.

(d) **Model predictions II.** We have to deal with a complication: the response is not continuous, since there are only twenty-one possible response locations. Therefore, we assume that the observer first computes the continuous estimate $\hat{s}$ and then reports the closest possible response location. Thus, the probability of reporting a particular location is the probability that $\hat{s}$ falls in a bin of size 1 around that location. Exception: for the two extreme response locations (-10 and 10), the bin extends infinitely on one side. Suppose that the true location s is -6, and that $\sigma = 3$ and $\sigma_s = 10$. For both model 1 and model 2, compute a vector of twenty-one numbers that gives the probability of reporting each of the twenty-one locations as predicted by the model. Plot both model predictions in the same plot with different colors.

(e) **Model predictions III.** Repeat part (d), except for the plotting, for each of the five true locations and for each combination of parameter values. Choose the possible values of σ and σ_s to range from 0.1 to 15 in steps of 0.1. For model 1, save the results in a matrix of size 21 (responses) $\times$ 5 (true stimulus locations) $\times$ 150 (values of σ), and for model 2, in a matrix of size $21 \times 5 \times 150 \times 150$. To avoid numerical problems, first set probability values that are numerically zero equal to the smallest nonzero value in the matrix.[1] After that, make sure that, separately for every true location and parameter combination, the response probabilities again sum to 1 across the twenty-one response locations.

(f) **Model fitting I.** Assume that trials are independent of each other. Use the "lookup tables" of model predictions from part (e) to calculate the log likelihood of every parameter combination, separately for models 1 and 2. This should give you a vector of length 150 for model 1 and a matrix of size 150×150 for model 2. Plot the former using a line plot and the latter using a heat map.

(g) **Model fitting II.** For models 1 and 2, find the MLE of the parameter(s) on their grids.

(h) **Model comparison.** Report for models 1 and 2 the maximum log likelihood, AIC, and BIC. Combining with part (b), which model wins according to AIC? Which model wins according to BIC? State your overall conclusion about the subject's behavior in a sentence in a way you would do in a paper.

(i) **Model checking.** Separately for each model, add the model fits to the five curves in part (a). Use dashed lines with colors corresponding to those of the data. Plot each model in a separate plot, so that you get three plots, each containing five solid lines (data) and five corresponding dashed lines (model fits). For models 1 and 2, use the MLEs from part (g) and the lookup tables from part (e).

1. This is an easy solution but not the best one. A better solution is to use Inverse Binomial Sampling [187] throughout.

5

Cue Combination and Evidence Accumulation

How can we integrate multiple sensory cues into a single percept?

In chapters 3 and 4, we studied a simple generative model, containing a single stimulus and a single measurement. In this chapter, we study an extension in which there are two measurements, which are based on sensory inputs that are also called cues. The measurements could be an auditory and a visual measurement of an underlying stimulus, such as the location at which a ball drops on the ground. The observer estimates the stimulus value based on both cues. Mathematically, the model is a straightforward extension of the one in chapter 3 and 4. Yet, there are four reasons to study this generative model. First, cue combination occurs very often in daily life. Second, it is a historically early and still prominent domain of application of Bayesian modeling. Third, this generative model is our first example in which the Bayesian observer computes the likelihood function over the world state of interest from two simpler likelihoods. Fourth, a takeaway message of this chapter is that Bayesian inference does not need to involve priors to be interesting. The key aspect of Bayesian decision making is not the prior, it is the reasoning with probability distributions rather than point estimates.

Plan of the Chapter

We begin by discussing the intuitions behind cue combination. We then develop steps 1 through 3 of Bayesian inference for cue combination. We show how integration of sensory evidence over time is mathematically equivalent to cue combination. Finally, we discuss the empirical literature that addresses how well people actually combine cues.

5.1 What Is Cue Combination?

When trying to understand someone's speech, it helps not just to listen carefully but also to simultaneously view the speaker's facial movements and nonverbal gestures. This is an example of cue combination. Combining cues is especially important when an individual cue is noisy, for example when you are trying to understand speech in the presence of background noise. The ability to combine cues enhances performance in perceptual tasks.

In numerous daily perceptual situations, we receive and combine cues from different sensory modalities, yet we do this so effortlessly that we may be unaware it is happening. When tasting food, we may think we are engaging in a purely gustatory activity, but in fact we perceive the flavor of food by combining gustatory, olfactory, thermal, and mechanical (texture) cues.[1] When estimating our acceleration while traveling in a moving vehicle, we may think we are relying on only vision, but we are relying as well on proprioceptive cues conveyed by sensors (muscle spindles and Golgi tendon organs) that signal muscle length and tension, and on vestibular cues conveyed by sensors in our inner ears (semicircular canals and otolith organs) that signal rotational and linear acceleration of the head.

In fact, we combine cues not only across sensory modalities but also within a single modality. Each modality provides an array of distinct cues. In vision, for example, the relative activation of photoreceptors tuned to different wavelengths tells us an object's color, the pattern of reflected light indicates the object's surface texture, and comparison of the images in the two retinae informs us about the object's depth. We effortlessly combine these and other visual cues to infer object identity. Similarly, distinct receptors in the skin provide us with mechanical, thermal, and nociceptive information, and even within each of these somatosensory divisions we obtain multiple cues. For example, different mechanoreceptor subtypes provide information about static pressure (Merkel receptors), skin stretch (Ruffini receptors), low-frequency vibration (Meissner receptors), and high-frequency vibration (Pacinian receptors). When we run a fingertip across an unknown surface, we obtain information about surface texture, friction, hardness, and other qualities by combining cues from these receptors. This allows us to achieve fine perceptual inferences, distinguishing for instance the feel of silk from that of velvet or wool.

Why should the brain combine cues? To answer this question, let us explore the consequences of an obvious alternate strategy: the brain could simply use the single most informative cue that it has at hand, and ignore the rest. This winner-take-all strategy is suboptimal for two reasons. First, our sensorineural responses are noisy, with the consequence that any parameter estimate based even on the most reliable cue is subject to some uncertainty; a strategy that does not include the other, albeit less reliable cues, discards information that can be used to sharpen the precision of the estimate. Second, even when sensorineural noise does not impose serious limitations, an individual cue is often ambiguous; a strategy that does not include all available cues will often fail to overcome ambiguities. To illustrate these points, we consider two examples.

First, let us suppose we wish to infer the location at which a dropped ball hits the ground. This event provides both visual and auditory cues. Now suppose that we base our inference about location entirely on the visual cue because the ball is dropped in a well-lit environment under direct view, a condition in which vision is more informative than audition. Because our photoreceptors and neural responses are noisy, even the estimate based on this most reliable cue will have some uncertainty, as reflected in the width of the posterior distribution over location. We will show below that inclusion of a less reliable cue (e.g., the auditory cue in this example) nevertheless contributes useful information. Thus, by combining cues, we obtain a more precise estimate than the one obtainable from the best cue alone.

Second, let us suppose we wish to infer the identity of a spherical object, placed on a tabletop, in the dark. As we rest our hand on the object, our proprioceptors inform us about its size, but we cannot unambiguously identify an object from its size alone. In this case, our

1. Perhaps even auditory ones [166]!

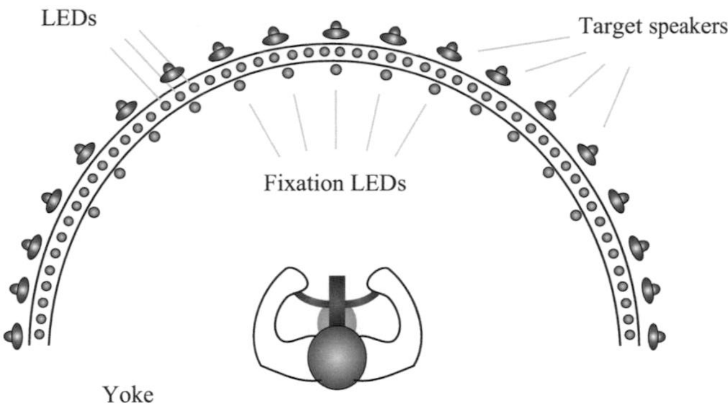

Figure 5.1
Experimental setup for testing auditory or multisensory localization. A speaker produces a brief tone. At the same time, an LED might produce a brief flash. The subject directs a laser pointer at the perceived location of the tone. Figure reproduced from [194].

uncertainty about the object's identity (as opposed to its exact size) does not result primarily from sensorineural noise. Rather, our uncertainty derives from the inherent ambiguity of size as a cue to object identity: even if we knew its size exactly, we would not know the object's identity, because different individual objects (e.g., an apple and an orange) can have the same size. Although size might provide the single most informative cue in this scenario (e.g., greatly narrowing the set of possible objects to those as big as apples and oranges), with further exploration—lifting and manipulating the object—we could glean from our muscle and mechanoreceptors an understanding of the object's weight and surface texture, further narrowing the possible set of objects. In short, we overcome ambiguity by combining cues.

How should we expect the nervous system to combine two pieces of information? Obviously, although the winner-take-all strategy is too extreme, we do expect people to rely more on those cues that are most informative to the task at hand. If, in a particular scenario, vision is more informative than audition, for example when we want to locate a person who is talking in an environment with loud ambient noise, then we should mostly rely on vision. When audition is more informative than vision, we expect the reverse—and indeed at night we often rely primarily on our auditory system.

Over the last decades, many scientists have studied cue combination in the laboratory. In a typical experiment, the subject is surrounded by an array of loudspeakers and light-emitting diodes (LEDs, see figure 5.1). An audiovisual stimulus is produced by the simultaneous occurrence of a brief auditory beep and a light flash. The subject is instructed to indicate the perceived location of the beep. With this apparatus, scientists can probe how visual and auditory cues are combined by the nervous system.

The results of these experiments reveal that when the beep and flash occur at the same—or nearly the same—location, subjects use the visual stimulus to help estimate the location of the auditory stimulus, even when they are instructed to ignore the visual stimulus. The subjects thus naturally and intuitively combine the cues, apparently operating under the assumption that the beep and flash originate from a single source. Indeed, under conditions in which the auditory and visual cues originate from slightly offset locations, subjects are easily led astray as their "auditory localization" estimates are biased by the presence of the visual cue. Importantly, the more precise the visual cue relative to the auditory one, the more strongly subjects rely on the visual cue in formulating their localization estimate.

5.2 Formulation of the Bayesian Model

Here we formulate a Bayesian model for optimal cue combination. In developing our formulation, we use the example of auditory-visual location estimation, but the same approach can be applied to other cue combination scenarios. Our approach follows the same three steps of Bayesian modeling outlined in chapters 3 and 4.

5.2.1 Step 1: Generative Model

Our model consists of three nodes: the stimulus s and two measurements, x_1 and x_2 (see figure 5.2). Associated with the stimulus is a stimulus distribution $p(s)$. Unlike in chapter 3, we assume here that the prior distribution is flat. We do this not only because it is the most common assumption in cue combination studies, but also because it illustrates nicely that a nonuniform prior is not necessary for a Bayesian model to be interesting and important. (For another illustration of the same point, refer to section 2.6.)

Any measurement is noisy and we need to make assumptions about its distribution, just as in section 3.2.4. We denote these distributions by $p(x_1|s)$ and $p(x_2|s)$. The separate arrows pointing to x_1 and x_2 in figure 5.2 reflect a key assumption, namely that these measurements are conditionally independent. Conditional independence of two random variables (see box 5.1) means that they are independent of each other when conditioned on another variable, in this case the stimulus s. Specifically, this means that while vision and audition are noisy, the noise corrupting the two streams is uncorrelated: there is zero noise covariation between the two modalities. It is important to distinguish conditional independence from independence. Our visual and auditory measurements are not going to be independent from one another *without* conditioning on the stimuli. When the stimulus is to the left, both measurements will tend to be to the left, and when the stimulus is

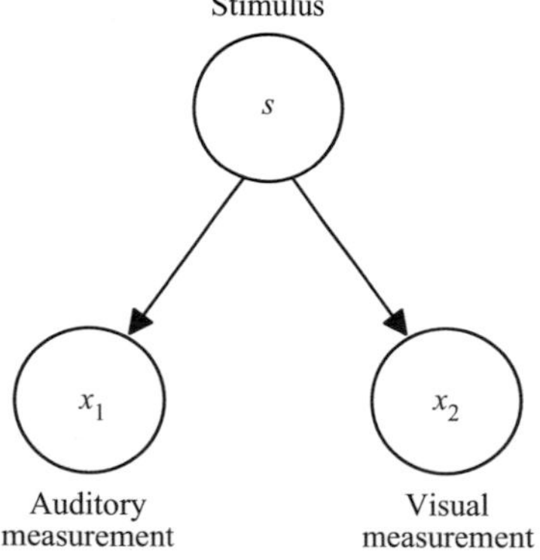

Figure 5.2
Generative model of cue combination. The stimulus affects both auditory and visual measurements, but those are conditionally independent of each other.

to the right, both measurements will tend to be to the right. However, we assume that on repeated presentations of *the same stimulus*, the trial-by-trial variability in the auditory and visual measurements will be uncorrelated. Thus, the visual and auditory measurements are assumed to be independent of one another only when conditioned on s.

The assumption of condiitonal independence is easier to justify when the two measurements come from two different sensory modalities (e.g., audition and vision) than from the same one (e.g., both from vision). That being said, all further calculations in this chapter—and in most of the cue combination literature—are based on an assumption of conditional independence.

Mathematically, the conditional independence of x_1 and x_2 given s is expressed as

$$p(x_1,x_2|s)=p(x_1|s)p(x_2|s). \tag{5.1}$$

Independence would have meant that $p(x_1,x_2)=p(x_1)p(x_2)$, but that is not true here.

Exercise 5.1 What would the generative model look like if the two measurements were not assumed conditionally independent, but instead correlated given s?

Box 5.1
Conditional Independence

Conditional independence occurs when two random variables are independent only given the value of a third one. For example, having Alzheimer's and needing reading glasses are two events that are not independent, because both tend to occur in older people. However, among only eighty-year-olds (i.e., given the age group), the two are probably more or less independent. Another famous example is that homicide rates in a city and ice cream sales are not independent random variables: they both tend to be more probable on hot days. However, given the temperature in the city, the two are conditionally independent.

The intuition is that you condition on the value of the cause of dependence of the two variables. If x, y, and z denote three random variables, then x and y are independent given z if

$$p(x,y|z)=p(x|z)p(y|z).$$

Be careful not to confuse conditional independence with independence, which would be $p(x,y)=p(x)p(y)$.

Another way to interpret equation (5.1) is by starting from the product rule for probabilities:

$$p(x,y|z)=p(x|y,z)p(y|z).$$

Then to get to equation (5.1), we have to make the assumption that $p(x|y,z)=p(x|z)$. In other words, knowledge of z fully specifies our understanding of the probability of x. When we know z, also knowing y does not contribute anything to our assessment of the probability of x.

Exercise 5.2 Think of another real-world example of conditional independence.

For the distribution of each individual measurement, we choose a Gaussian distribution:

$$p(x_1|s)=\frac{1}{\sqrt{2\pi\sigma_1^2}}e^{\frac{-(x_1-s)^2}{2\sigma_1^2}}; \tag{5.2}$$

$$p(x_2|s) = \frac{1}{\sqrt{2\pi\sigma_2^2}} e^{\frac{-(x_2-s)^2}{2\sigma_2^2}}. \tag{5.3}$$

As we did in the previous two chapters, we will use precision notation for convenience. We define two precision variables:

$$J_1 \equiv \frac{1}{\sigma_1^2}; \tag{5.4}$$

$$J_2 \equiv \frac{1}{\sigma_2^2}. \tag{5.5}$$

This concludes the specification of the generative model.

5.2.2 Step 2: Inference

The observer infers the stimulus s from the measurements $x_{\text{obs},1}$ and $x_{\text{obs},2}$. To reduce clutter, we will from now on drop the subscript "hyp" in s_{hyp} that reminded us that in step 2 the observer is considering hypotheses about the stimulus. Instead, you will have to keep track of which step we are on.

The likelihoods over the stimulus are the same expressions as the noise distributions but regarded as a function of s:

$$\mathcal{L}(s; x_{\text{obs},1}) \equiv p(x_{\text{obs},1}|s) = \frac{1}{\sqrt{2\pi\sigma_1^2}} e^{\frac{-(x_{\text{obs},1}-s)^2}{2\sigma_1^2}}; \tag{5.6}$$

$$\mathcal{L}(s; x_{\text{obs},2}) \equiv p(x_{\text{obs},2}|s) = \frac{1}{\sqrt{2\pi\sigma_2^2}} e^{\frac{-(x_{\text{obs},2}-s)^2}{2\sigma_2^2}}. \tag{5.7}$$

We call each of these an *elementary likelihood function*, defined as a likelihood function over a stimulus feature associated with an individual measurement. The posterior distribution over the stimulus is computed from Bayes' rule:

$$p(s|x_{\text{obs},1}, x_{\text{obs},2}) \propto p(s)p(x_{\text{obs},1}, x_{\text{obs},2}|s). \tag{5.8}$$

We have left out the factor $\frac{1}{p(x_{\text{obs},1}, x_{\text{obs},2})}$ for the reason explained in box 3.6: it only acts as a normalization, so if we normalize the distribution in the end, we automatically take this factor into account. Since the stimulus distribution is flat, the prior is flat as well, and the posterior is determined by the likelihood $p(x_1, x_2|s)$ only. To make further progress, we use equation (5.1), the assumption of conditional independence of the measurements. Then, the posterior becomes:

$$p(s|x_{\text{obs},1}, x_{\text{obs},2}) \propto p(s)p(x_{\text{obs},1}|s)p(x_{\text{obs},2}|s) \tag{5.9}$$

$$\propto p(x_{\text{obs},1}|s)p(x_{\text{obs},2}|s). \tag{5.10}$$

This step—making use of the structure of the generative model to express the likelihood in terms of elementary likelihoods—is the only conceptually new element in this chapter compared to chapter 3. Expressing the likelihood over the state-of-the-world variable in terms of elementary likelihoods is at the core of Bayesian inference models for many tasks.

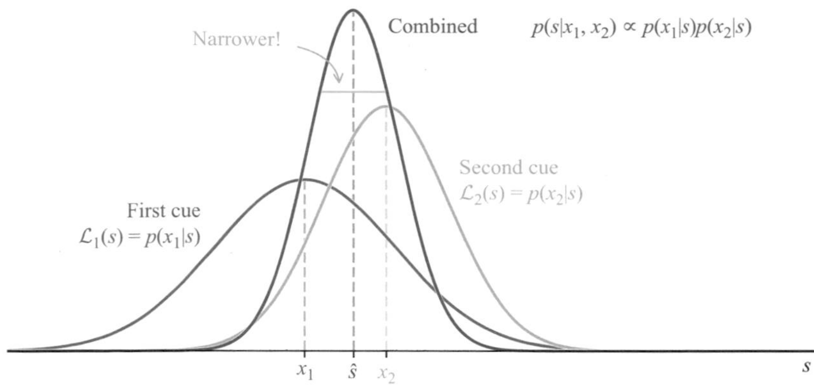

Figure 5.3
Computation of the posterior distribution in cue combination. Note that both red curves are likelihood functions. The prior is flat and not shown.

We can substitute the two Gaussian distributions into this equation and evaluate the expression according to box 3.7. The result is that the posterior is another Gaussian distribution (figure 5.3):

$$p(s|x_{\text{obs},1}, x_{\text{obs},2}) = \frac{1}{\sqrt{2\pi\sigma_{\text{post}}^2}} e^{\frac{-(s-\mu_{\text{post}})^2}{2\sigma_{\text{post}}^2}}, \tag{5.11}$$

with mean

$$\mu_{\text{post}} = \frac{J_1 x_{\text{obs},1} + J_2 x_{\text{obs},2}}{J_1 + J_2} \tag{5.12}$$

and variance

$$\sigma_{\text{post}}^2 = \frac{1}{J_1 + J_2}. \tag{5.13}$$

We can also write the posterior mean as

$$\mu_{\text{post}} = w_1 x_{\text{obs},1} + w_2 x_{\text{obs},2}, \tag{5.14}$$

where the weights are proportional to the precisions:

$$w_1 = \frac{J_1}{J_1 + J_2}; \tag{5.15}$$

$$w_2 = \frac{J_2}{J_1 + J_2}. \tag{5.16}$$

The posterior mean estimate (PME) is

$$\hat{s}_{\text{PM}} = \mu_{\text{post}}. \tag{5.17}$$

As a reminder, the PME is the estimate that minimizes the expected squared error. In this case, the PME is equal to the maximum-a-posteriori estimate (MAP) as well as to the maximum-likelihood estimate (MLE).

You might have noticed that the observer's computation of the posterior is exactly analogous to the computation in chapter 3, where we combined a single cue with a prior. The second cue has now taken the role of the prior, whereas the new prior is flat. This equivalence was already foreshadowed in section 1.5.

The weights in equations (5.16) and (5.17) sum to 1: $w_1 + w_2 = 1$, indicating that the PME is a weighted average of the two measurements. Averaging with weights proportional to the inverse of the variance is by far the most frequently used model in cue combination. An analogy is that of a police investigator trying to reconstruct a crime based on the testimonies of two witnesses. One witness was intoxicated at the time of the crime, the other was not. The testimony of the sober witness would be the result of a low-noise process; of the inebriated witness, the result of a high-noise process. A good investigator would take both testimonies into account, but would more heavily weight the testimony of the less noisy witness.

The variance of the posterior, in equation (5.13), is a measure of the observer's uncertainty. Under the assumptions we made, it is smaller than the variances of each of the elementary likelihood functions. Intuitively, this says that combining cues reduces uncertainty; the observer is more confident in the combined estimate than in the estimate that would be obtained from either cue alone.

Exercise 5.3 Prove mathematically that the variance of the posterior is never larger than that of either likelihood.

5.2.3 Step 3: Estimate Distribution

As the third step in our Bayesian model, we are interested in the distribution of the PME across many trials. The PME is given as a function of the measurements x_1 and x_2 in equation (5.12), but the measurements are themselves random variables—their values vary from trial to trial. As a consequence, the PME varies from trial to trial as well. Since in a behavioral experiment, we never know the measurements on a single trial (they are in the observer's head), we have to compare behavior with the distribution of the PME over many trials. To find the mean and variance of the PME, we apply the rules for linear combinations of normally distributed variables. The means of x_1 and x_2 are both s. Therefore, the model predicts that the mean PME will be $w_1 s + w_2 s = w_1 s + (1 - w_1)s = s$. In other words, for cue combination with a flat prior, the posterior mean estimator is unbiased. (Recall that bias is defined as the difference between the mean estimate and the true stimulus.) The variance of the PME is:

$$\sigma^2_{\text{post}} = \frac{1}{J_1 + J_2}. \tag{5.18}$$

Thus, in this cue combination problem, the variance of the PME distribution happens to be identical to the variance of the posterior. This stands in contrast to chapter 4, where the variance of the estimate distribution was different from the variance of the posterior. This difference arises because the prior was chosen flat in the current chapter. When the prior is Gaussian, then the variance of the PME will differ from the variance of the posterior also for cue combination (see section 5.4).

5.3 Artificial Cue Conflict

We saw that in the Bayesian model for cue combination, the PME is on average equal to the true stimulus. This is not very interesting, since it does not distinguish between the Bayesian cue combination model and a model in which the observer uses only one of the cues. (The variance of the PME does distinguish, but it is better to have two measures than one.) In cue combination experiments, therefore, a common trick is to introduce a small conflict between the true stimuli in the two modalities. In other words, unbeknown to the observer, there is not a single s, but rather two slightly displaced stimuli, s_1 and s_2. Everything else remains the same.

Of course, this requires that the observer still believes that there is a single underlying stimulus, in spite of the discrepancy introduced by the experimenter. The experimenter sometimes explicitly instructs the observer to imagine that the two cues are generated by a single stimulus, for example an auditory and a visual measurement generated by a ball hitting the screen. Although the investigator has deliberately used two discrepant stimuli, the observer nonetheless incorrectly infers a single common stimulus. At first sight, this seems suboptimal. Another way to think about this behavior, however, is to consider that the observer is applying a prior based on natural statistics; in the real world, when the observer simultaneously sees a ball hit the ground and hears a thud, the visual and auditory stimuli nearly always result from the same event and therefore originate from the same location. In the laboratory experiment, the investigator has contrived a situation that would rarely occur in the world and that therefore is easily misinterpreted by the observer. Keep in mind that even when there is truly a single stimulus, the auditory and visual measurements will differ from each other on each trial because of noise (unless somehow the noise is completely correlated). Thus, the mere fact that the two cues differ does not imply that they resulted from two stimuli at different locations. The observer's inference may still be optimal, then, under a real-world prior.

Of course, the observer will only believe in a single stimulus if the discrepancies introduced are small. Otherwise, the observer will notice a conflict. For example, if the sound of bouncing ball originates at a sufficiently large distance from the visual image of the ball, the observer will realize that two separate stimuli were presented. Similarly, if a movie is poorly dubbed, the discrepancy in time between the speaker's mouth movement and voice will be too large to go unnoticed. When the observer does not necessarily believe that there is a single common cause, the observer's inference process changes. This interesting situation will be discussed in section 10.3.

If the observer indeed believes that there is a single common cause, then step 2 above is unchanged. In step 3, however, the means of x_1 and x_2 are no longer both s, but s_1 and s_2, respectively. As a consequence, the mean PME will be:

$$\mathbb{E}\left[\hat{s}_{\text{PM}}|s_1,s_2\right]=w_1s_1+w_2s_2. \tag{5.19}$$

This estimator is biased. For example, the bias with respect to stimulus s_1, which we denote here by Bias_1, is:

$$\text{Bias}_1\left[\hat{s}_{\text{PM}}|s_1,s_2\right]=\mathbb{E}\left[\hat{s}_{\text{PM}}|s_1,s_2\right]-s_1 \tag{5.20}$$

$$=w_1s_1+(1-w_1)s_2-s_1 \tag{5.21}$$

$$=(1-w_1)(s_2-s_1). \tag{5.22}$$

Table 5.1
Overview of the distributions in this chapter. i can be 1 or 2.

Distribution	Argument	Mean	Variance	Precision
Measurement distributions	Measurement x_i	Stimulus s_i	σ_i^2	J_i
Likelihood functions	Hypothesized stimulus s	Measurements x_i	σ_i^2	J_i
Posterior distribution	Hypothesized stimulus s	$\frac{J_1 x_{\text{obs},1}+J_2 x_{\text{obs},2}}{J_1+J_2}$	$\frac{1}{J_1+J_2}$	J_1+J_2
Response distribution	Stimulus estimate $\hat{s}_{\text{ML}}$	$\frac{J_1 x_{\text{obs},1}+J_2 x_{\text{obs},2}}{J_1+J_2}$	$\frac{1}{J_1+J_2}$	J_1+J_2

This predicted bias can be compared against experimental data and is indeed often found to be a good match (see section 5.7).

5.3.1 Distinguishing the Distributions

As in chapter 4, it is important to distinguish between the posterior (single trial, step 2), and the estimate distribution (multiple trials, step 3); see table 5.1. It just happens to be the case that when the prior is flat, as we have assumed so far, the estimate distribution has the same variance as the posterior, but this is not the case in general.

5.4 Generalizations: Prior, Multiple Cues

In chapters 3 and 4, we studied the combination of a Gaussian prior (mean μ, standard deviation σ_s, precision J_s) with a single measurement. In this chapter, we studied the combination of two conditionally independent measurements. The two combinations can easily be combined. The posterior becomes

$$p(s|x_{\text{obs},1},x_{\text{obs},2}) \propto p(s)p(x_{\text{obs},1}|s)p(x_{\text{obs},2}|s) \tag{5.23}$$

and the PME becomes

$$\hat{s}_{\text{PM}} = \mu_{\text{post}} = \frac{J_s\mu + J_1 x_{\text{obs},1} + J_2 x_{\text{obs},2}}{J_s + J_1 + J_2}. \tag{5.24}$$

We can further generalize to multiple cues. Combining N cues with the same underlying stimulus is as easy as combining two. Assume the measurements $x_1, x_2, \ldots, x_N$ are conditionally independent given s. The corresponding generative model is shown in figure 5.4. Then, the posterior is:

$$p(s|x_{\text{obs},1}\cdots x_{\text{obs},N}) \propto p(s)p(x_{\text{obs},1}|s)\cdots p(x_{\text{obs},N}|s) \tag{5.25}$$

$$= p(s)\prod_{i=1}^{N} p(x_{\text{obs},i}|s), \tag{5.26}$$

where in the last line we used product notation. Thus, the prior gets multiplied with the likelihoods derived from the individual measurements. If the measurements are normally distributed with mean s and standard deviations $\sigma_1, \sigma_2, \ldots, \sigma_N$, respectively, then the posterior will have mean

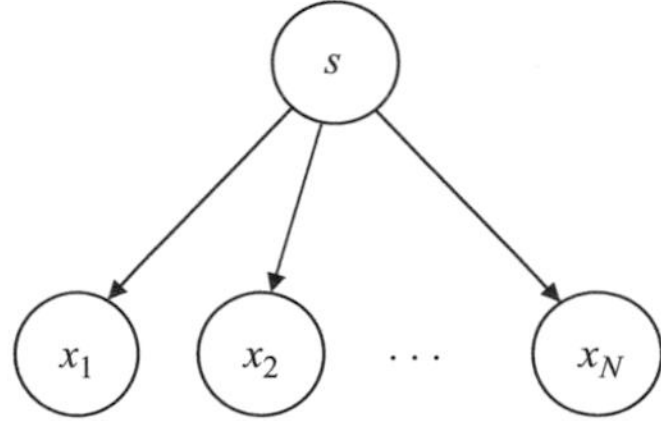

Figure 5.4
Generative model of cue combination with a stimulus s and N conditionally independent cues, corresponding to measurements $x_1, \ldots, x_N$.

$$\hat{s}_{\text{PM}} = \mu_{\text{post}} = \frac{J_s \mu + \sum_{i=1}^{N} J_i x_{\text{obs},i}}{J_s + \sum_{i=1}^{N} J_i} \tag{5.27}$$

and variance

$$\sigma^2_{\text{post}} = \frac{1}{J_s + \sum_{i=1}^{N} J_i}. \tag{5.28}$$

5.5 Evidence Accumulation

A major way in which organisms improve their knowledge of the world is by observing for a longer time. More time allows more evidence to be accumulated. Mathematically, evidence accumulation is cue combination over time and can be described using the same formalism. In fact, evidence accumulation is often regarded as the prototypical example of Bayesian inference, where a posterior gets updated at each time step based on new information.

We now formalize this. Consider an observer who makes a series of conditionally independent, normally distributed measurements $x_1, x_2, \ldots, x_T$, one at each time point. Thus, the generative model is the same as that in figure 5.4, but with cues replaced by time points. We further assume that the measurements all have the same mean, s, and standard deviations $\sigma_1, \sigma_2, \cdots, \sigma_T$, respectively. Under these assumptions, equation (5.26) applies with N replaced by T:

$$p(s|x_{\text{obs},1} \cdots x_{\text{obs},T}) \propto p(s)p(x_{\text{obs},1}|s) \cdots p(x_{\text{obs},T}|s) \tag{5.29}$$

$$= p(s) \prod_{i=1}^{T} p(x_{\text{obs},i}|s). \tag{5.30}$$

Therefore, with the same substitutions, equations (5.27) and (5.28) also apply. In particular, the variance of the posterior will shrink continuously to 0 as more evidence is accumulated.

In the context of evidence accumulation, it makes sense to think of the computation of the posterior as a recursive process, whereby the posterior after obtaining the measurement at a given time point serves the prior used at the next time point. Mathematically, we can rewrite equation (5.30) as an *update equation*

$$p(s|x_{\text{obs},1},\cdots,x_{\text{obs},t+1}) \propto p(s|x_{\text{obs},1},\cdots,x_{\text{obs},t})p(x_{\text{obs},t+1}|s), \tag{5.31}$$

where it is understood that $p(s|x_{\text{obs},1},\cdots,x_{\text{obs},t})$ for $t=0$ is the prior distribution. In other words, the posterior at time t gets multiplied with the likelihood at time $t+1$ to produce, after normalization, the posterior at time $t+1$. This process is called the *Bayesian updating of the posterior*, and it is the most fundamental concept in the application of Bayesian modeling to temporal data.

Modeling evidence accumulation in this way comes with several important caveats:

- Termination. We have not specified how and when the evidence accumulation terminates. This is a difficult modeling problem that has a long history.
- Conditional independence. We assumed that measurements across different time points are conditionally independent. This assumption is easily violated: in many cases, processes with a long time scale (such as slow fluctuations of attention) will cause measurements to be correlated over time. Then, equations (5.27) and (5.28) will no longer apply. In particular, the variance of the posterior might asymptote to a value larger than 0.
- Stationarity. Another important caveat is that we assumed that the true world state, s, does not change over time. Often, the stimulus itself changes as you accumulate evidence. We will discuss an example where it does change in chapter 12.
- Treating different times differently. In our evidence accumulation model, early information is treated on the same footing as later information: time points could be swapped without affecting the inference. This ignores time-dependent effects such as forgetting. A normative way to model forgetting is as the deliberate ignoring of information that is no longer task-relevant. Thus, forgetting naturally emerges in a generative model in which the world state changes over time. We will examine this in chapter 12.

5.6 Cue Combination under Ambiguity

So far, we have considered cue combination under sensory noise. However, as we have seen, uncertainty sometimes arises not from sensory noise but because of inherent ambiguities in the inputs. For instance, in attempting to identify an object in the hand based on its size and weight, we may experience uncertainty not because of sensory noise (given sufficient time to obtain reliable measurements of these variables) but rather because multiple objects can have the same size and weight. Nevertheless, the logic of the inference process is the same in this scenario: the different possible objects are hypotheses (typically discrete ones: apple, orange, etc.), and the measurement of each feature (size, weight, etc.) has a certain probability under each hypothesis. As a function of the hypothesis, this is an individual-feature likelihood function. When the features are independent conditioned on object identity, the likelihood of a particular hypothesis is the product of the individual-feature likelihoods. What complicates matters is that the features do not tend to be conditionally independent. For example, even when restricted to oranges, weight and size tend to correlate strongly. Formally, this is a classification task, not an estimation task. We will discuss classification tasks under ambiguity in chapter 8.

5.7 Applications

There is a long tradition of probing cue combination by analyzing how humans integrate position information from vision and audition. In many cases, vision is very precise (precision ~min arc), while audition is relatively imprecise (precision ~10 degrees). This has the effect that when reliable visual information is available, people generally rely primarily on vision, a behavior predicted from equation (5.19).

To test the more subtle predictions of the Bayesian model, it was necessary to create situations where vision and audition were similarly precise. In a seminal study, Alais and Burr [9] accomplished this by blurring visual inputs. The authors used several levels of blur so that visual precision would change unpredictably from trial to trial. They estimated visual precision by presenting visual stimuli alone, making use of the fact that the prior is flat so the variance of the PME for vision alone will equal the variance of the likelihood function (and the noise distribution). Similarly, they presented auditory stimuli alone to estimate the variance of the auditory likelihood function. Using the resulting estimates for σ_A and σ_V, the authors predicted the weights human subjects would place on vision and on audition when combining these cues. They found that human behavior was well predicted by equation (5.19) with those weights. Moreover, the same model successfully predicted the variance of the PME.

An important technical detail in many cue combination studies is that the experiments usually do not ask for estimates on a continuum (as in figure 5.1) but instead use a so-called two-alternative forced-choice paradigm, in which the subject is presented with stimuli in two intervals and required to make a choice between them. For example, subjects are presented with two sets of auditory-visual stimuli and asked in which of the two the auditory stimulus was more to the left. This allows the investigator to estimate the variances (precisions) of the cues in a way that is unaffected by the subject's prior. We examine the details of this procedure in a later chapter.

The Bayesian model of cue combination has been tested in other sensory modalities as well. A classic study is by Ernst and Banks [47]. The authors studied subjects' estimation of the size of an object that could be both seen and felt. Under normal viewing conditions, vision is often more precise than touch; the authors blurred the visual stimulus in order to reduce its precision. They found that across different visual precisions, the weight subjects placed on vision was close to the value predicted by equation (5.19).

Many experiments have probed the integration of two cues originating from a single sensory modality. One example is the estimation of slant (orientation of a plane) based on visual texture and binocular disparity [95]. In a typical study, subjects are shown a surface with a texture that indicates a given slant. The texture information can be made more or less informative. For example, circles provide a highly informative cue, whereas random white noise provides a very uninformative cue. The disparity cue can also be manipulated, and importantly changed independently of texture. By independently varying the texture and disparity cues, the authors of such studies have generally found that subjects integrate these cues in accordance with the predictions of the Bayesian model. For example, as the texture cue is varied to indicate different slants, it exerts a roughly linear influence on the estimated slant. The slope of that influence, the weight on texture, fits well with the prediction of the Bayesian model.

When we want to estimate the position of our hand in a two-dimensional plane, such as a tabletop, we have to solve a two-dimensional estimation problem. To make this estimate, we can use proprioceptors that signal body posture. We can also use vision. The proprioceptive and visual cues to hand position have different properties. Proprioception is generally noisy, but good at estimating changes in the direction of our smaller joints. Vision is quite good in terms of direction but rather poor at estimating depth. In a seminal study, van Beers and collaborators [184] probed how the nervous system combines visual and proprioceptive cues in this task. It was found that the cue combination proceeds almost exactly as predicted by the Bayesian model.

We discussed artificial cue conflict in auditory-visual localization. A famous example of artificial cue conflict in auditory-visual speech perception is the McGurk effect [125]. When an observer hears the sound of someone saying "baba" while they watch a video of the same person saying "gaga," they may perceive the person saying "dada." Demos of the McGurk effect can be found online. The McGurk effect can be understood as an instance of cue combination in which the observer infers a single, common world state from the auditory and visual observations. The observer is typically not aware of the conflict until they listen with their eyes closed. Auditory-visual speech perception has been modeled using Bayesian models [18, 117].

Finally, cue combination can also take place between individuals. A fascinating study [15] investigated how two individuals combine information about a visual stimulus through verbal communication.

5.8 Summary

In this chapter, we introduced models in which multiple cues have to be combined. We have learned the following:

- Cue combination is a frequent and important perceptual activity that often happens automatically and outside of our conscious control.
- As with combining a prior with a likelihood, all the Bayesian observer needs to do is multiply two probability distributions and normalize.
- Unlike the winner-take-all strategy, the optimal Bayesian solution (posterior mean estimate) is to weight each cue according to its reliability.
- The Bayesian model accounts for human data in a wide variety of settings.
- Cue combination illustrates that the prior can be flat in an interesting Bayesian model.
- Cue combination can take place over time, in which case it is sometimes called evidence accumulation, evidence integration, or decision making. Across subsequent measurements, uncertainty is reduced. The PME is a linear combination of the individual measurements, weighted by their precisions.

5.9 Suggested Readings

- Bahador Bahrami et al. "Optimally Interacting Minds." *Science* 329, no. 5995 (2010): 1081–1085.

- Marc O. Ernst and Martin S. Banks. "Humans Integrate Visual and Haptic Information in a Statistically Optimal Fashion." *Nature* 415, no. 6870 (2002): 429–433.
- Vikranth Rao Bejjanki, Meghan Clayards, David C. Knill, and Richard N. Aslin. "Cue Integration in Categorical Tasks: Insights from Audio-Visual Speech Perception." *PloS One* 6, no. 5 (2011): e19812.
- Anne-Marie Brouwer and David C. Knill. "The Role of Memory in Visually Guided Reaching." *Journal of Vision* 7, no. 5 (2007): 1–12.
- A. L. Yuille and Heinrich H. Bulthoff. "Bayesian Decision Theory and Psychophysics." *Perception as Bayesian Inference*, edited by David C. Knill and Whitman Richards. 123–162. Cambridge: Cambridge University Press, 1996.
- David Alais and David Burr. "The Ventriloquist Effect Results from Near-Optimal Bimodal Integration." *Current Biology* 14, no. 3 (2004): 257–262.
- Robert A. Jacobs. "Optimal Integration of Texture and Motion Cues to Depth." *Vision Research* 39, no. 21 (1999): 3621–3629.
- David C. Knill and Jeffrey A. Saunders. "Do Humans Optimally Integrate Stereo and Texture Information for Judgments of Surface Slant?" *Vision Research* 43, no. 24 (2003): 2539–2558.
- Harry McGurk and John MacDonald. "Hearing Lips and Seeing Voices." *Nature* 264, no. 5588 (1976): 746–748.
- Julia Trommershäuser, Konrad P. Körding, and Michael S. Landy. *Sensory Cue Integration*. Oxford: Oxford University Press, 2011.
- Robert J. van Beers, Anne C. Sittig, and Jan J. van der Gon Denier. "How Humans Combine Simultaneous Proprioceptive and Visual Position Information." *Experimental Brain Research* 111, no. 2 (1996): 253–261.

5.10 Problems

Problem 5.1 An observer combines conditionally independent cues A and B with Gaussian measurement noise. When B becomes more reliable

(a) the observer's estimate will shift toward A;

(b) the observer's estimate will shift toward B;

(c) the observer's estimate will stay the same;

(d) there is insufficient information to answer.

Problem 5.2 True or false? Explain.

(a) In the cue combination model of this chapter, the measurements are assumed to be independent of each other.

(b) Conflicts between two measurements generated by a single source rarely occur in real-world perception.

Problem 5.3 This problem was suggested in 2016 by Nick Johnson, then a PhD student at New York University. Within the model for combining two independent cues with Gaussian measurement noise under a flat prior:

(a) The variance of the posterior is never greater than 50 percent of the largest variance of the single-cue likelihoods. Show this, using equations.

(b) The variance of the posterior is always between 50 percent and 100 percent of the smallest variance of the single-cue likelihoods. Show this, using equations.

(c) Is the statement in (a) still true if the prior is Gaussian rather than flat? If so, prove it. If not, give a counterexample.

(d) Is the statement in (b) still true if the prior is Gaussian rather than flat? If so, prove it. If not, give a counterexample.

Problem 5.4 Suppose $p_1(x), p_2(x), \ldots, p_N(x)$ are Gaussian distributions, where $p_i(x)$ (for every $i = 1, \ldots, N$) has mean μ_i and precision J_i. We multiply these distributions, then normalize:

$$q(x) = kp_1(x)p_2(x)\cdots p_N(x), \tag{5.32}$$

where k is such that $q(x)$ is normalized. We are interested in showing that $q(x)$ is a Gaussian distribution with mean

$$\mu_q = \frac{1}{J_q}\sum_{i=1}^{N} J_i\mu_i \tag{5.33}$$

and precision (inverse variance)

$$J_q = \sum_{i=1}^{N} J_i. \tag{5.34}$$

(a) Show this using direct calculation as in chapter 3.

(b) As an alternative, perform a proof by induction; for $N = 1$, the equations are trivially true. Assume it is true for some N, and show that then it would also be true for $N + 1$.

(c) Making use of the result, derive equation (5.24).

Problem 5.5 This problem builds on section 5.5 about evidence accumulation. An observer infers a stimulus s from a sequence of measurements, $x_{\text{obs},1}, x_{\text{obs},2}, \ldots, x_{\text{obs},T}$, made on a single trial. The stimulus distribution is Gaussian with mean μ and variance σ_s^2. The distribution of the tth measurement, $p(x_t|s)$, is Gaussian with mean s and variance σ^2 (identical for all measurements); we assume conditional independence.

(a) What are the mean and variance of the posterior? You may start with the equations in section 5.4.

(b) For a given stimulus s, we define *relative bias* as the difference between the mean PME and s itself, divided by the difference between the mean of the stimulus distribution and s. Derive an expression for relative bias in terms of μ, σ, σ_s, and T. Simplify the expression as much as you can.

(c) Interpret the expression in (b): explain intuitively how the dependencies on the variables make sense.

(d) Compute the variance of the PME for given s.

(e) Plot this variance as a function of T for all nine combinations of $\sigma_s \in \{1,2,5\}$ and $\sigma \in \{1,2,5\}$. Create one plot for each value of σ_s, for a total of three plots, each containing three curves (color-coded).

(f) Interpret the plots in (e): explain intuitively how the shapes of the functions make sense.

Problem 5.6 In this problem, we examine *suboptimal* estimation in the context of cue combination. Suppose an observer estimates a stimulus s from two conditionally independent, Gaussian-distributed measurements, $x_{\text{obs},1}$ and $x_{\text{obs},2}$. The prior is flat.

(a) We start with a reminder of optimal estimation. Express the PME in terms of the measurements.

(b) What is the variance of the PME across trials?

(c) Now suppose the observer uses an estimator of the form $\hat{s} = wx_{\text{obs},1} + (1-w)x_{\text{obs},2}$, where w can be *any constant*. Show that this estimate is unbiased (just like the PME); this means that the mean of the estimate for given s is equal to s.

(d) What is the variance of this estimate as a function of w? Plot this function. At which value of w is it minimal, and does this value make sense? State your final conclusion in words.

Problem 5.7 In chapter 3 and the present chapter, we were able to derive analytical expressions for the posterior distribution and the response distribution. For more complex psychophysical tasks (e.g., later in this book), however, analytical solutions often do not exist but we can still use numerical methods. To gain familiarity with such methods, we will work through the cue combination model in this chapter using numerical methods. We assume that the experimenter introduces a cue conflict between the auditory and the visual stimulus: $s_1 = 5$ and $s_2 = 10$. The standard deviations of the auditory and of the visual noise are $\sigma_1 = 2$ and $\sigma_2 = 1$, respectively. We assume a flat prior over s.

(a) Randomly draw an auditory measurement $x_{\text{obs},1}$ and a visual measurement $x_{\text{obs},2}$ from their respective distributions.

(b) Plot the corresponding elementary likelihood functions, $p(x_{\text{obs},1}|s)$ and $p(x_{\text{obs},2}|s)$, in one figure.

(c) Calculate the combined likelihood function, $p(x_{\text{obs},1}, x_{\text{obs},2}|s)$, by numerically multiplying the elementary likelihood functions. Plot this function.

(d) Calculate the posterior distribution by numerically normalizing the combined likelihood function. Plot this distribution in the same figure as the likelihood functions.

(e) Numerically find the PME of s, that is, the value of s at which the posterior distribution is maximal.

(f) Compare with the PME of s computed from equation (5.14) using the measurements drawn in (a).

(g) In the above, we simulated a single trial and computed the observer's PME of s, given the noisy measurements on that trial. If an analytical solution does not exist for the distribution of the PMEs, we can repeat the above procedure many times to approximate this distribution. Here, we practice this method even though an analytical solution is available in this case. Draw 100 pairs $(x_{\text{obs},1}, x_{\text{obs},2})$ and numerically compute the observer's PME for each pair, as in (e).

(h) Compute the mean of the PMEs obtained in (g) and compare with the mean estimate predicted using equation (5.19).

(i) Make a histogram of the PME.

(j) *Relative (auditory) bias* is defined as the mean PME minus the true auditory stimulus, divided by the true visual stimulus minus the true auditory stimulus. Compute relative auditory bias for your estimates.

Problem 5.8 A major assumption in the derivation of our cue combination model was that the measurements x_1 and x_2 were conditionally independent, equation (5.1). Here, we consider a generalization in which they are not. We replace equation (5.19) by a *bivariate normal distribution* with the same mean s for both measurements, standard deviations σ_1 and σ_2, and correlation ρ:

$$p(x_1, x_2|s) = \frac{1}{2\pi\sigma_1\sigma_2\sqrt{1-\rho^2}} e^{-\frac{1}{1-\rho^2}\left(\frac{(x_1-s)^2}{2\sigma_1^2}+\frac{(x_2-s)^2}{2\sigma_2^2}-\frac{\rho(x_1-s)(x_2-s)}{\sigma_1\sigma_2}\right)}. \tag{5.35}$$

Assume that the prior is flat.

(a) Step 2: Derive equations for the mean and variance of the posterior over s. Hint: follow the steps of problem 3.4 (completing the square).

(b) Perform a sanity check by setting $\rho = 0$, which corresponds to conditionally independent measurements. You should obtain equations (5.27) and (5.28).

(c) Plot the variance of the posterior as a function of ρ for $\sigma_1 = 1$, $\sigma_2 = 2$.

(d) Interpret the plot. Specifically, explain how the dependencies of the variables make sense.

Problem 5.9 Combining survey data from multiple people to get at the population mean can also be considered cue combination. Suppose a pollster asks respondents to what extent they agree or disagree with a particular statement. For simplicity, we represent amount of agreement by a real-valued variable x. As individuals in the population differ, we model x as following a normal distribution with mean s and variance σ^2. The variance will generally depend on how and when the data are collected. To the pollster, the variable of interest is s. The pollster collects responses $x_{\text{obs},1}, x_{\text{obs},2}, \ldots, x_{\text{obs},N}$ from N people; these responses are assumed to be conditionally independent.

(a) Show mathematically that the (normalized) likelihood function of s is a normal distribution with mean equal to the sample mean of the responses, denoted by $\bar{x}$, and variance $\frac{\sigma^2}{N}$. Thus, the MLE of s is simply $\bar{x}$.

(b) A poll aggregator is trying to do "cue combination" of one poll with N_1 respondents and individual variance σ_1^2, another with N_2 respondents and individual variance σ_2^2 (subscripts now refer to different polls, not to individuals within the poll). We denote the respective MLEs by $\hat{s}_1$ and $\hat{s}_2$. Show that the combined MLE of s is equal to

$$\hat{s}_{\text{ML, combined}} = \frac{w_1\hat{s}_1 + w_2\hat{s}_2}{w_1 + w_2}, \tag{5.36}$$

where

$$w_i = \frac{N}{\sigma_i^2} \tag{5.37}$$

for $i = 1, 2$.

(c) Interpret this result.

6

Learning as Inference

How do we learn from observations?

In most of this book, we assume that the generative model and its parameters are already known to the decision maker. In experiments, this can be made plausible by some combination of giving clear instructions, exposing the participant to a large number of example stimuli, testing the participant on properties of the relevant distributions, and providing a block of training trials (the data of which do not get analyzed). However, the process of learning generative models is a rich area of study by itself and the topic of this chapter.

Plan of the Chapter

We begin by considering how an observer can learn the probability of a binary event from repeated observations. We show that the posterior probability can be written in recursive form, updating after each time step. We see that, for a given number of observations of each outcome, the order of the observations affects the evolution of the posterior, but not the final posterior. We discuss nonuniform priors and conjugate priors and the link between Bayesian learning and reinforcement learning. Subsequently, we discuss learning the precision of Gaussian distribution from observations, learning the parameter of a relationship between two variables, and the learning of categorically distinct causal models of the world. In section 6.3 we describe a link between Bayesian learning and reinforcement learning.

6.1 The Many Forms of Learning

Learning is ubiquitous. All animals learn, and learning occurs in all domains of behavior, from learning to categorize objects as an infant, to learning how to control one's muscles while picking up an object, to learning a new language. Learning allows the observer or agent to perform a task better or to be better adapted to an environment. Some forms of learning are perceptual; for example, an experienced soccer player will be able to predict much better than a novice player where a ball is going to land. Others are more cognitive, such as you currently learning to do Bayesian modeling. Some forms of learning involve instruction or imitation—as when a young songbird learns their song from a parent—while others rely on feedback about the state of the world. Much learning even takes place in the absence of any feedback; for example, infants learn to parse spoken language in part by keeping track of co-occurrence frequencies of syllables. Learning happens over many timescales. We can learn the layout of an unfamiliar space in seconds. Learning about someone's personality traits sometimes takes a single behavior and sometimes years of interaction. Learning the statistical properties of the environment happens during the development of an individual, but continues throughout that individual's lifetime, across generations, and even across evolutionary timescales.

Within a Bayesian context, the primary form of learning is learning a parameter of a generative model. For example, a subject does not typically start an experiment knowing exactly the distribution of stimuli they will experience, and instructions are limited in what they can convey. The key concept of this chapter is that learning such a parameter is itself an inference process: the parameter takes the role of the world state in previous chapters. Thus, we use the same machinery that we already introduced in a different context. A complication is that learning and regular inference often have to take place at the same time: while the observer is not yet sure about the parameters of the generative model, they still have to make inference decisions based on that generative model. We will address this complication.

There is a close connection between learning the value of a fixed world state and evidence accumulation over time, as discussed in section 5.5. In fact, the basic form of the generative model is identical: a single world state, generating conditionally independent observations. In what is commonly called evidence accumulation, integration plays out on a shorter time scale (tens of milliseconds to seconds), with all observations being made within a single experimental. The world state still changes across trials on a slower timescale. In an experimental setting, no judgments are typically required while the evidence accumulation occurs. In what is commonly called learning, each trial corresponds to one observation, and the world state may be fixed across a much longer time scale, anywhere from seconds to years. In an experimental setting, the progress of learning might be probed by asking the subject for a judgment on each trial.

Note on notation: In chapter 5, we dropped the subscript "hyp" from "s_{hyp}" in step 2 of Bayesian modeling. From now on, we will also drop the subscript "obs" to indicate a specific value of the observation in step 2. It has served its didactic purpose, but it causes clutter and is not common in the literature. You will have to keep in mind that in step 1, the observation or measurement is always a random variable, whereas in step 2, it is always a specific value of that variable. (And to make matters more confusing, in step 3, that specific value becomes a variable.) In addition, we will mostly drop the subscript indicating the random variable from probability distribution notation; for example, we will write $p(s)$ instead of $p_s(s)$. For details on this convention, see appendix A.

6.2 Learning the Probability of a Binary Event

Imagine you are shipwrecked and stranded on a desert island. The good news is that it rains frequently. The bad news is that the interval between days with rain is unpredictable. To ration the rainwater that you save, you would like to estimate the probability that it will rain on a given day. After three days, you have observed:

dry—rain—dry

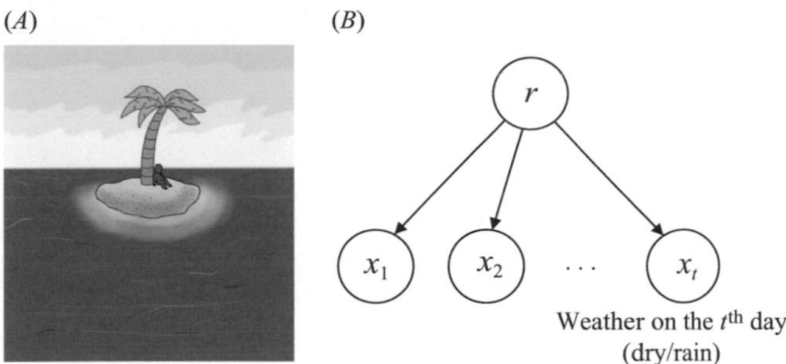

Figure 6.1
Learning a probability of a binary event. (*A*) Picture to make this chapter look better. (*B*) Generative model of the task of learning a probability r based on t conditionally independent binary observations.

Your best guess of the probability at this point may be $\frac{1}{3}$. However, given the small number of days, you are pretty uncertain about that guess. As a devoted Bayesian with a lot of free time, you would like to quantify that uncertainty, or even better, calculate a posterior probability distribution over the probability that it will rain. In other words, what is the probability that the probability of rain is 0.1? 0.3? 0.5? 0.9? And so on. So far in this book, we have not yet seen a probability as the world state variable of interest, but it is perfectly legitimate, very common in the real world, and it does not really change the math. The only thing to get used to is the slightly awkward expression "probability distribution of a probability."

Step 1 is to formulate the generative model. The graphical model is shown in figure 6.1B. It contains a top-level variable representing the probability of rain on a given day (i.e., the rainfall rate), which we will denote by r. As mentioned in the introductory section, we assume that this world state does not change over a very long time; in our example, that could mean that over the period of a month, the probability of rain is constant. Obviously, this is a simplification, and it is also possible to learn a slowly changing world state; we consider this situation later. The generative model is very similar to that for cue combination (figure 5.4, but the nature of the variables is different: a probability instead of a real-valued world state, binary instead of real-valued measurements, and time points instead of cues.

We have to assume a prior distribution over r. You may realize that across the world, there are few places where it rains almost every day (say, $r > 0.9$) or almost never (say, $r < 0.1$). Perhaps, you believe in the absence of any observations, that r is most probable to lie in the interval $[0.4, 0.6]$. These are examples of aspects of the prior distribution over r. For the moment, we assume a completely uniform distribution over r:

$$p(r) = 1. \tag{6.1}$$

Exercise 6.1 Why is this a properly normalized distribution?

At the bottom of the generative model, we find the sequence of binary observations corresponding to the observed weather (rain/dry) on days 1 to t, which we denote by $x_1, x_2, \ldots, x_t$. Each x_i can be 0 (dry) or 1 (rain). On a given day,

$$p(x_i = 1) = r \tag{6.2}$$

$$p(x_i = 0) = 1 - r. \tag{6.3}$$

This is an example of a *Bernoulli process*: a random variable with two possible outcomes, with each outcome having a fixed probability. The best-known example of a Bernoulli process is a coin flip and by analogy with this we model rain versus dry as a weighted coin flip. We do not have a single observation but a sequence of them, one on each day. For simplicity, you assume that the weather is independent across days, conditioned on r. (In reality, weather across days is very far from conditionally independent, and places can have long spells of dry versus rain days.) Then, we can write for the probability distribution associated with a sequence:

$$p(x_1, x_2, \ldots, x_t|r) = p(x_1|r)p(x_2|r)\cdots p(x_N|r), \tag{6.4}$$

which can be written in product notation as

$$p(x_1, x_2, \ldots, x_t|r) = \prod_{i=1}^{t} p(x_i|r). \tag{6.5}$$

This concludes the specification of the generative model. We are now ready to do inference!

In step 2, we compute a posterior distribution over r given your observations so far (dry—rain—dry), which we will denote by $\mathbf{x}$. We start with Bayes' rule,

$$p(r|\mathbf{x}) \propto p(r)\, p(\mathbf{x}|r). \tag{6.6}$$

We now use equation (6.1) for the prior and equation (6.4) for the likelihood:

$$p(r|\mathbf{x}) \propto p(x_1|r)\, p(x_2|r)\cdots p(x_t|r). \tag{6.7}$$

Now, we realize that every factor in this product is equal to either r or $1-r$, as those are the only possible values. Thus, we can simplify the expression to

$$p(r|\mathbf{x}) \propto r^{n_{\text{rain}}}(1-r)^{n_{\text{dry}}}, \tag{6.8}$$

where n_{rain} and n_{dry} are the numbers of rainy and dry days observed so far (with $n_{\text{rain}} + n_{\text{dry}} = t$). This type of distribution is called a *beta distribution* (see box 6.1). The normalization is provided by the so-called *beta function*, but this is not essential to our understanding.

What is important about equation (6.8) for the posterior is that every time you observe a rain day, it gets multiplied with an increasing function of r, namely $f(r) = r$, and every time you observe a dry day, it gets multiplied with a decreasing function, namely $f(r) = 1 - r$. We could call these individual factors *instantaneous* or *momentary likelihood functions*, as they are tied to an additional observation at a single time point. Figure 6.3 shows the evolution of the posterior when the observations are dry—rain—dry. The integration of the new data every day is an application of Bayes' rule using the day's likelihood function.

6.2.1 Prediction

Having computed the posterior distribution, $p(r|x_1, x_2, \ldots, x_t) = p(r|\mathbf{x})$, you may wonder, as you fall asleep thirsty after spending several days on the desert island, what is the probability that it will rain the following day? This is an instance of the problem we alluded to in section 6.1: observers have to do inference using a generative model while still being uncertain about the parameter of that model. The solution is to average (integrate) over the unknown parameter, here r: we compute the probability of any future outcome by integrating its probability under each world state, multiplied by the posterior probability of that world state:

Box 6.1
Beta Distribution

The beta distribution is defined over a random variable Y that takes values between 0 and 1. Most often, that random variable is a probability itself, so the beta distribution is a probability distribution over a probability variable. The probability density of the beta distribution is

$$p(y) = \frac{1}{\mathrm{B}(\alpha, \beta)} y^{\alpha-1}(1-y)^{\beta-1}. \tag{6.9}$$

Here, α and β are the parameters of the distribution, both restricted to be positive; $\mathrm{B}(\cdot, \cdot)$ denotes the *beta function*, a special function whose role (and in fact, definition) is to normalize the beta distribution. For the purposes of this book (and for most of Bayesian modeling), we do not need to know anything else about the beta function. In case you need it, all common numerical computation packages have the beta function preprogrammed. We show several examples of the beta distribution in figure 6.2. The mean of a beta-distributed random variable y is

$$\mathbb{E}[y] = \frac{\alpha}{\alpha + \beta} \tag{6.10}$$

and its variance is

$$\mathrm{Var}[y] = \frac{\alpha\beta}{(\alpha+\beta)^2(\alpha+\beta+1)}. \tag{6.11}$$

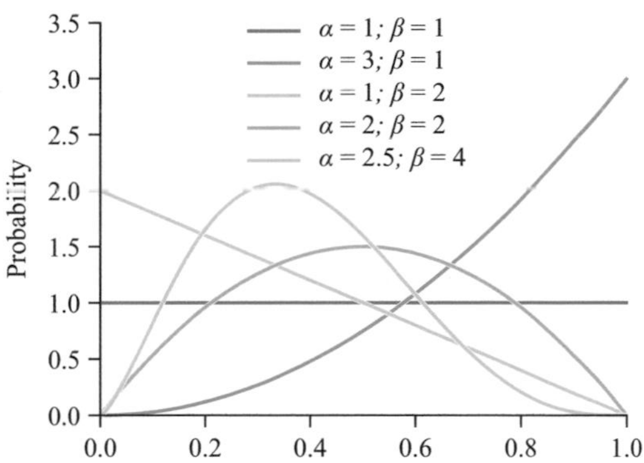

Figure 6.2
Examples of the beta distribution for different parameter combinations.

$$p(x_{t+1} = 1|\mathbf{x}) = \int p(x_{t+1} = 1|r)p(r|\mathbf{x})dr. \tag{6.12}$$

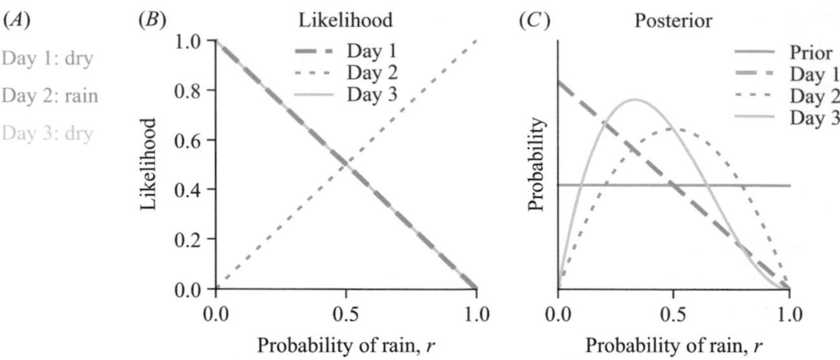

Figure 6.3
Evolution of the posterior over a Bernoulli probability. (*A*) Observations on days 1, 2, and 3. (*B*) Instantaneous likelihoods on days 1, 2, and 3. (*C*) Posterior distributions on days 1, 2, and 3.

In essence, we are stating that our belief it will rain tomorrow is our belief that r has a given value *and* it will rain if r has that value, OR that r has another value *and* it will rain if r has that other value, and so on. The result is called a *posterior predictive distribution*, and the average is an example of *marginalization*, which we will discuss in greater detail in chapter 8 (see also appendix section B.11.2).

Because we are dealing here with a binary outcome whose probability, given r, is simply r itself, our integral reduces to:

$$p(x_{t+1}=1|\mathbf{x}) = \int rp(r|\mathbf{x})dr. \tag{6.13}$$

The reader will recognize that this is the formula for the mean of the posterior distribution. Thus, the probability with which we expect rain to occur the next day is the posterior mean estimate (PME) of the rate parameter r.

Combining the fact that equation (6.8) is a beta distribution, equation (6.10) for the mean of a beta-distributed variable, and the fact that $n_{\text{rain}} + n_{\text{dry}} = t$, we find that the posterior mean estimate of r is equal to

$$\hat{r}_{\text{PM}} = \frac{n_{\text{rain}}+1}{t+2}. \tag{6.14}$$

This expression was derived in the eighteenth century by Pierre-Simon Laplace and is known as Laplace's Rule of Succession. This posterior mean estimate is different from the mode of the posterior (the MAP estimate), which is

$$\hat{r}_{\text{MAP}} = \frac{n_{\text{rain}}}{t}. \tag{6.15}$$

Step 3 does not apply, since we are considering neither internal noise nor decision noise. In other words, conditioned on the stimuli, there is no variability in the estimate.

6.2.2 Update Equations

Just like when we discussed evidence accumulation in section 5.5, we can write the posterior in recursive form, updating it after each time step. Specifically, the analogue of equation (5.31) is

$$p(r|x_1,\dots,x_{t+1}) \propto p(r|x_1,\dots,x_t)p(x_{t+1}|r), \tag{6.16}$$

where it is understood that $p(r|x_1,\dots,x_t)$ for $t=0$ is the prior distribution. This is possible thanks to the conditional-independence assumption, equation (6.4), which embodies that we believe that days only depend on the overall rain probability, that all days are conditionally independent given this probability. However, writing the posterior using an update equation is not so useful, since the explicit solution, equation (6.8) is straightforward.

6.2.3 Uncertainty

As in section 3.4.1, the standard deviation of the posterior can be used as a measure of uncertainty. In our case, the standard deviation is

$$\text{Std}\,(r|\mathbf{x}) = \frac{1}{t+2}\sqrt{\frac{(n_{\text{rain}}+1)(n_{\text{dry}}+1)}{t+3}}. \tag{6.17}$$

Exercise 6.2 Show this using equation (6.11) for the variance of a beta-distributed random variable.

6.2.4 Binomial Distribution

The order of the observation (rain days and dry days) affects the evolution of the posterior, but not the final posterior. For example, the posterior would be identical for dry-rain-dry as for rain—dry—dry or dry—dry—rain. Stated differently, we could in step 1 alternatively have summarized the sequence of binary observations as a single count. Then, the generative model would have been

$$p(n_{\text{rain}}|r) = \binom{t}{n_{\text{rain}}} n_{\text{rain}}^{r}(t-n_{\text{rain}})^{1-r}. \tag{6.18}$$

This is an example of a *binomial distribution*. It is defined over counts (integers), starting at 0. If we had used this generative model, it would not have changed any part of our inference in step 2.

Exercise 6.3 Why not?

6.2.5 Nonuniform Prior

So far, we have assumed a uniform distribution over r (see equation (6.1)). To generalize this, it is common to choose the prior to be a beta distribution:

$$p(r) \propto r^{\alpha_0-1}(1-r)^{\beta_0-1}, \tag{6.19}$$

where α_0 and β_0 are the parameters (both must be positive) and we left out the normalization. When $\alpha_0=\beta_0=1$, the prior is uniform. Using equation (6.19), the posterior distribution over r is again beta-distributed, but now with parameters

$$\alpha = \alpha_0 + n_{\text{rain}} \tag{6.20}$$

$$\beta = \beta_0 + n_{\text{dry}}. \tag{6.21}$$

Exercise 6.4 Show mathematically why this is the case.

Let us take stock of what we did. We showed earlier that the likelihood corresponding to a Bernoulli or binomial generative model is proportional to a beta distribution. We have now shown that multiplying a beta prior with a beta likelihood produces a beta posterior. This is elegant, because going from prior to posterior, we need to update only the parameters of the distribution, just as we did in chapter 3 when going from a normal prior to a normal posterior.

For a given generative model, a prior distribution that is such that the posterior is a member of the same parametric family of distributions as the prior is called a *conjugate prior* for that generative model. In other words, the beta distribution is a conjugate prior for the Bernoulli (or binomial) distribution.

But how can we justify choosing a prior merely based on elegance? Should a prior not reflect the statistics of the world? Yes and no. In practice, it is difficult to know or control the prior distribution that an observer has over a probability. The best we can often do is choose a sufficiently flexible distribution (usually that means having at least two parameters) and fit the parameters to an individual subject's data (and, if applicable, to a particular experimental condition). From this perspective, a beta prior is as good as any alternative, in which case the elegance can be a tiebreaker.

Finally, whereas being stranded on an island may not be particularly relevant as an experimental paradigm, many problems that are relevant have the same mathematical structure. For example, in an iterated trust game, the subject might be figuring out how trustworthy their partner from binary observations. There, the variable trustworthiness would take the place of the probability of rain in our island example.

6.3 Linking Bayesian Learning to Reinforcement Learning

There is an interesting connection between Bayesian learning and reinforcement learning. In section 6.2.2, we wrote down an update equation for the posterior when the prior is uniform, equation (6.16). We could similarly write down a recursive update equation for the posterior mean estimate at time t (which is the expected value of whether it will rain at time $t+1$):

$$\hat{r}_t = \hat{r}_{t-1} + \frac{1}{t+2}(x_t - \hat{r}_{t-1}). \tag{6.22}$$

(Here, we left out the "PM" subscript from $\hat{r}$ to avoid clutter.) This equation resembles the Rescorla-Wagner rule in reinforcement learning (box 6.2), in which the value of a state gets updated based on the difference between a received reward and the predicted reward—the *prediction error*. In equation (6.22), the expectation of rain after day $t-1$ acts as the value of the state at t, and whether it actually rained on day t acts as the received reward. The factor $\frac{1}{t+2}$ acts as the *learning rate*, which represents to what extent the prediction gets updated. It makes intuitive sense that the learning rate decreases as t increases: as the prediction is based on more observations, any one observation has less power to change the prediction. This is true beyond this specific example: a Bayesian learning rule typically has a learning rate that decreases in time, in a specific way that is fully dictated by the assumptions of the model. This time dependence contrasts with the standard

Rescorla-Wagner rule, in which the learning rate is by default a constant (although that rule can be generalized).

We will prove equation (6.22) in problem 6.2, but we work through a numerical example here. Since we start with a uniform prior, the prior expectation of rain is $\hat{r}_0 = \frac{1}{2}$. Let us say that it rains on your first day on the island. Then $x_1 = 1$, and the "prediction error" is $x_1 - \hat{r}_0 = 1 - \frac{1}{2} = \frac{1}{2}$. The posterior expectation of rain after day 1 is $\hat{r}_1 = \frac{1}{2} + \frac{1}{3} \cdot \frac{1}{2} = \frac{2}{3}$, which is indeed the expected value of a posterior $p(r|s_1) \propto r$.

Although we have linked Bayesian learning and reinforcement learning here, several caveats apply. First, the link was only at the level of the *readout* of the posterior, not of the full posterior distribution. The Rescorla-Wagner rule and many other reinforcement learning models do not treat uncertainty explicitly, although attempts to remedy this are ongoing. Second, even though in our case the expectation of rain was naturally interpreted as being valuable, not all Bayesian learning can be interpreted as the updating of a value-like quantity; the next section will illustrate this. Third, there is no guarantee in Bayesian learning in general that the difference between two successive updates is proportional to a "prediction error."

Box 6.2
Rescorla-Wagner Model

One of the simplest learning models is the Rescorla-Wagner model, which describes how the value of state is updated over time based on rewards. We denote by $V_t(s)$ the value of state s at time t, which can be thought of as the expected reward of s in the long run. The agent receives a reward R_t at time t while being in state s (actions are only implicit in this model). Then the value $V_t(s)$ of that state can be updated in the next time step, to

$$V_{t+1}(s) = V_t(s) + \lambda(R_t - V_t(s)). \tag{6.23}$$

Here, $R_t - V_t$ is the difference between the received reward and the expected reward; this difference is also called the *reward prediction error*. The parameter λ is called the *learning rate*. (A more common notation for the learning rate is α, but we already used that symbol in section 6.2 for one of the parameters of the beta distribution.) It is a number between 0 and 1 that describes how responsive the agent's value function is to the reward prediction error. If $\lambda = 0$, the value function is wholly unresponsive. If $\lambda = 1$, we have $V_{t+1}(s) = R_t$, which means that the old value is irrelevant and the new value is simply the received reward; this is typically much *too* responsive. In practice, λ is small but depends on the individual and on the experimental condition.

The basic structure of the Rescorla-Wagner model, in which value gets modified by a scaled version of a prediction error, is common to much of *reinforcement learning* and its application to neuroscience. Reinforcement learning is a branch of machine learning that studies how entities interacting with the world can figure out which actions are rewarding in which world states. These entities are typically called *agents* instead of observers, because their actions are part of a causal chain that continues after the action. A reinforcement learning model with the same basic structure but that also takes into account future rewards is *Q-learning*. Machine learning nowadays uses sophisticated neural networks to solve complex reinforcement learning problems (deep reinforcement learning).

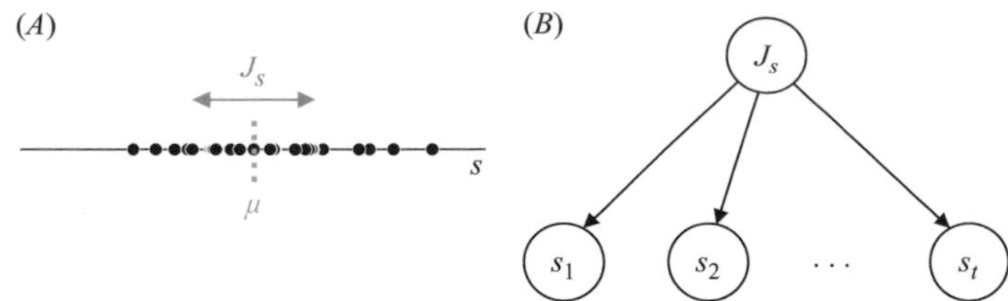

Figure 6.4
Precision learning task. (*A*) Samples of the stimulus s, drawn from a Gaussian distribution with mean μ and precision J_s. The length of the double arrow is twice the corresponding standard deviation. (*B*) Generative model. J_s is the parameter of interest, and $s_1, \ldots, s_t$ are the observations.

6.4 Learning the Precision of a Normal Distribution

In this section, we consider a very simple form of unsupervised learning, namely how an observer might learn the precision of a distribution from samples (figure 6.4A). We consider the Gaussian stimulus distribution considered in chapters 3 and 4:

$$p(s) = \frac{1}{\sqrt{2\pi\sigma_s^2}} e^{-\frac{(s-\mu)^2}{2\sigma_s^2}}. \tag{6.24}$$

For later convenience, we reparametrize this in terms of stimulus precision, $J_s \equiv \frac{1}{\sigma_s^2}$. We also make the dependence of the distribution on stimulus mean μ and stimulus precision J_s explicit:

$$p(s|\mu, J_s) = \sqrt{\frac{J_s}{2\pi}} e^{-\frac{J_s}{2}(s-\mu)^2}. \tag{6.25}$$

In chapter 3, we assumed that the observer knows this distribution—specifically, that they know μ and J_s. In reality, this information needs to be learned. For simplicity, we consider the case that μ is known but J_s needs to be learned; thus, from now on, we do not mention μ as a variable to be conditioned on.

Step 1: Generative model The generative model (figure 6.4B) contains the world state variable of interest, J_s, at the top. The observations—at the bottom—are a set of stimuli $s_1, \ldots, s_t$, which we will collectively denote by a vector $\mathbf{s}$. We assume that the stimuli are independently drawn conditioned on μ and J_s, so that

$$p(\mathbf{s}|J_s) \equiv p(s_1, s_2, \ldots, s_t|J_s) \tag{6.26}$$

$$= p(s_1|J_s)p(s_2|\mu, J_s) \cdots p(s_t|J_s). \tag{6.27}$$

In product notation, this can be written as $\prod_{i=1}^{t} p(s_i|\mu, J_s)$. Each individual s_i follows equation (6.25), so that

$$p(s_i|J_s) = \sqrt{\frac{J_s}{2\pi}} e^{-\frac{J_s}{2}(s_i-\mu)^2}. \tag{6.28}$$

Step 2: Inference Inference consists of learning the parameter J_s from specific observed samples s_1, s_2, s_t. Thus, the posterior of interest is $p(J_s|\mathbf{s})$. To calculate this posterior, we apply Bayes' rule and assume a uniform prior:

$$p(J_s|\mathbf{s}) \propto p(\mathbf{s}|J_s)p(J_s) \tag{6.29}$$

$$\propto p(\mathbf{s}|J_s). \tag{6.30}$$

We now make use of equations (6.27) to write the posterior as proportional to a product of instantaneous likelihood functions, each of which is based on an individual observed stimulus:

$$p(J_s|\mathbf{s}) \propto \prod_{i=1}^{t} p(s_i|\mu, J_s) \tag{6.31}$$

$$= \prod_{i=1}^{t} \left(\sqrt{\frac{J_s}{2\pi}} e^{-\frac{J_s}{2}(s_i-\mu)^2} \right) \tag{6.32}$$

$$\propto J_s^{\frac{t}{2}} e^{-\frac{J_s}{2}\sum_{i=1}^{t}(s_i-\mu)^2}. \tag{6.33}$$

This expression has several interesting aspects. First, the observed stimuli come only in a specific combination, namely $\sum_{i=1}^{t}(s_i - \mu)^2$. This is the sum of the squares of the observed stimuli to the known mean μ. This combination makes sense, since the larger its value is, the lower precision tends to be. Second, the dependence of J_s is one we have not seen before. This type of distribution is called a *gamma distribution* (see box 6.3). It is a common distribution for variables that cannot take negative values (such as J_s). Thus, in our example, each individual likelihood as well as the posterior has a gamma distribution form. If we had chosen a nonuniform prior, a convenient choice would have been a gamma distribution as well. Thus, the gamma distribution is a conjugate prior for inferring the precision of a normal distribution.

Just like in sections 5.5 and 6.2.2, we can write the posterior in the form of a recursive update equation. In figure 6.6, we show snapshots from the learning process.

As the last part of step 2, we might want to commit to a readout. If our objective is to minimize the expected squared error in estimating J_s, we should use the PME. (This objective can be challenged, as the squared error is mostly meaningful if a variable can take values across the entire real line.) We work this out in problem 6.5. Again, step 3 does not apply, since we are considering neither internal noise nor decision noise.

6.4.1 Why Not Infer Variance?

We formulated the case study in terms of inferring the precision parameter of a normal distribution. However, we could have alternatively formulated it as an inference of the variance or the standard deviation of a normal distribution. This would not have been equivalent, because a uniform prior over precision is not the same as a uniform prior over variance. (If this is not clear, read appendix section B.12.1 on transformations of

Box 6.3
Gamma Distribution

The gamma distribution is defined over a random variable Y that takes values on the positive real axis, such as precision in our case study. The probability density of the gamma distribution is

$$p(y) = \frac{1}{\Gamma(k)\theta^k} y^{k-1} e^{-\frac{y}{\theta}}. \tag{6.34}$$

Here, k and θ are the parameters of the distribution, both restricted to be positive. k is called the shape parameter and θ the scale parameter. $\Gamma(\cdot)$ denotes the *gamma function*, a special function that is also preprogrammed in all numerical computation packages. We show several examples of the gamma distribution in figure 6.5. The mean of a gamma-distributed random variable y is

$$\mathbb{E}[y] = k\theta. \tag{6.35}$$

Its variance is

$$\text{Var}[y] = k\theta^2. \tag{6.36}$$

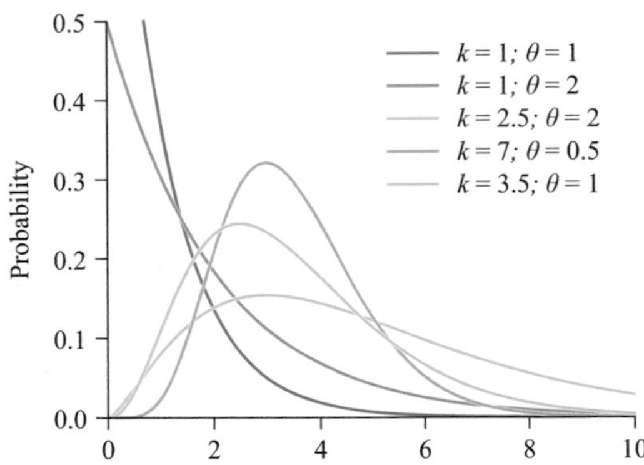

Figure 6.5
Examples of the gamma distribution for different parameter combinations.

random variables.) If the prior over variance were chosen uniform, then the posterior would have been an *inverse gamma distribution*. It is generally slightly simpler to work with gamma distributions than with inverse gamma distributions.

6.4.2 Prediction

As in section 6.2.1, the observer might be asked to predict the next observation, s_{t+1} while learning of the precision parameter J_s is still ongoing (through trial-to-trial feedback on the value of s_i). The posterior predictive distribution over s_{t+1} is obtained by marginalization over J_s:

$$p(s_{t+1}|\mathbf{s}) = \int p(s_{t+1}|J_s)p(J_s|\mathbf{s})dJ_s. \tag{6.37}$$

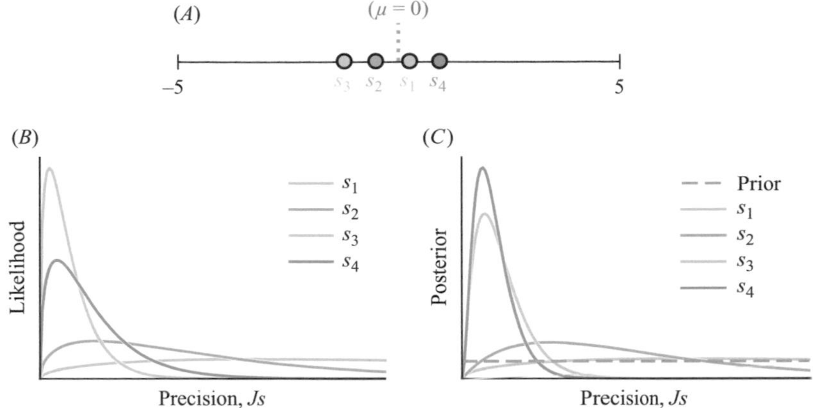

Figure 6.6
Evolution of the posterior over J_s. (*A*) Observations: $s_1 = -1.2, s_2 = -0.5, s_3 = 0.25$, and $s_4 = 0.9$. (*B*) Normalized likelihood functions associated with the most recent observation. (*C*) Posterior distributions given all observations up to the current time. The normalized likelihood functions and the posterior distributions are all gamma distributions. The prior distribution is improper (not normalizable; see section 3.5.2), which we indicate with a dashed line.

Equation (6.37) can be considered the prior over s_{t+1} that the observer has after time t. This "prior" can subsequently be combined with additional information, such as noisy measurements. We will evaluate equation (6.37) in problem 6.9.

6.5 Learning the Slope of a Linear Relationship

So far, we have studied how a Bayesian observer learns a parameter of a probability distribution either over a binary outcome or over a real-valued variable. A slightly more complicated situation arises when the parameter that has to be learned is one that defines the relation between two variables. Yet, the Bayesian approach will not require much modification.

To exemplify this, we consider a toddler who is learning how to control her limbs. To first approximation, the command signal (e.g., firing rate of motor cortex neurons) sent to the spinal cord adjusts muscle force linearly. But what is the slope of this relationship between motor command and force output for a particular muscle? Without this knowledge, the toddler will be clumsy; as she acquires this knowledge, her motor control will improve. We consider the toddler's first attempts to learn this relationship. For simplicity, we assume that the toddler already understands that the relationship between motor command and force output is linear (though, clearly, this too must be learned), but she doesn't know the slope. Her goal is to estimate the slope, k, relating command signal s to force output F:

$$F(s) = ks. \tag{6.38}$$

Figure 6.7 shows the results of ten iterations in which the toddler uses different command signal magnitudes to push against the floor with her arms, while she judges the force she produces. She is able to judge the force based on feedback from her proprioceptors.

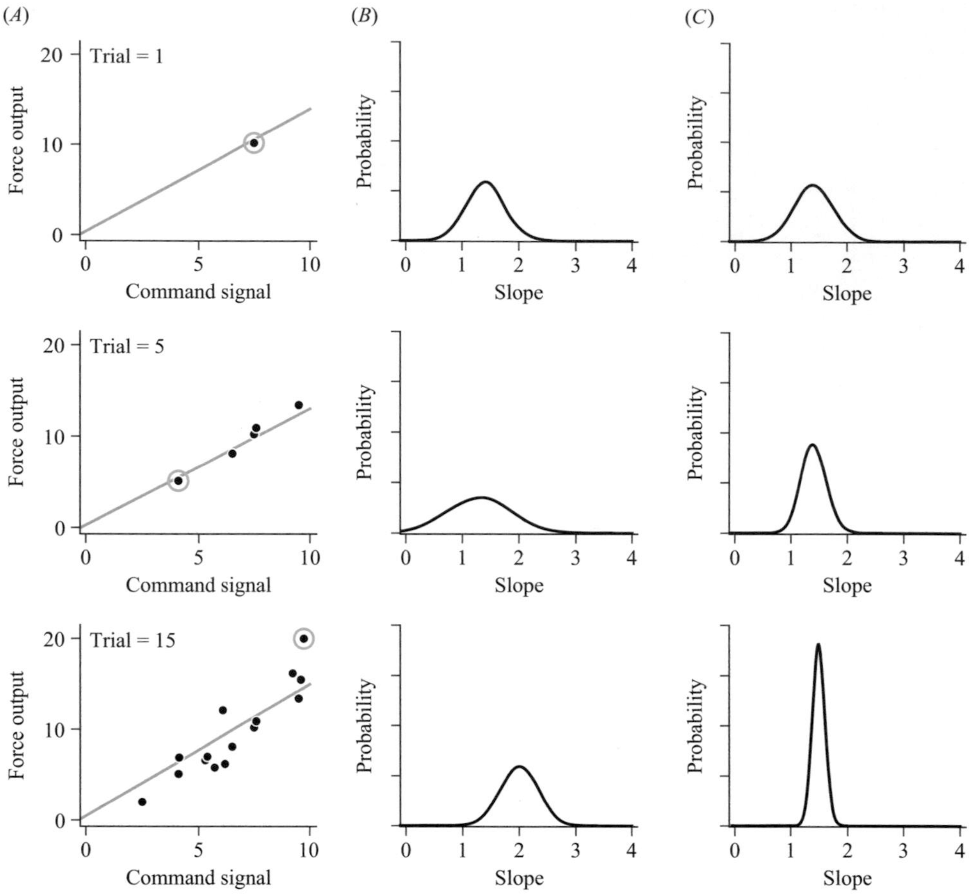

Figure 6.7
An observer learns the slope relating motor command signal magnitude (x axis) to muscle force production (y axis). (A) Scatter plots showing the data accumulating from trial 1 to 5 to 15. The line in each plot shows the posterior mean slope estimate based on all trials up to and including the one shown (data point circled in green). (B) Single-measurement likelihood functions from the corresponding trials. (C) Posterior distribution over slope. The actual slope value used to generate the data was $k = 1.5$, with $\sigma = 2$. We assumed that the prior $p(k)$ is flat.

These sensory signals are noisy, so we model her force measurement, f_t, on the tth trial as a sample drawn from a Gaussian distribution around the actual force produced:

$$p(f_t|k, s_t) = \mathcal{N}(f_t; ks_t, \sigma^2). \tag{6.39}$$

The observations are then command-measurement pairs (s_t, f_t), where we assume s_t to be known exactly. Unlike in section 6.4, we also assume that the toddler knows the measurement noise level σ; if she did not, then the inference described here would have to be combined with the inference in section 6.4.

After each push, with her knowledge of the noise distribution, the toddler can construct a likelihood function reflecting the probability of the measurement given the slope. We assume a prior $p(k)$. She can then calculate her posterior over slope; it can be written as a variation of equations (5.30) and (6.8),

$$p(k|s_1,f_1,s_2,f_2,\ldots,s_T,f_T)=p(k)\prod_{t=1}^{T}p(f_t|k,s_t), \tag{6.40}$$

or as a variation of the update equations, equations (5.31) and (6.16):

$$p(k|s_1,f_1,s_2,f_2,\ldots,s_{T+1},f_{T+1})\propto p(k|s_1,f_1,s_2,f_2,\ldots,s_T,f_T)p(f_{T+1}|k,s_{T+1}). \tag{6.41}$$

The difference with the case studies in previous sections is that now the commands $s_1,\ldots,s_T$ are needed in the likelihood, but since we assume these commands are known to the toddler, we can simply condition every probability on them.

6.6 Learning the Structure of a Causal Model

An important form of learning is learning about the causal structure of the world: clouds may cause rain (but not the other way round), a button may cause a device to turn on, and smoking causes cancer. Causal learning is the basis of classical conditioning and associative learning. Much of science is about understanding the causal structures underlying complex systems—molecular pathways, neural circuits, the global economy, and the climate. In some cases, intervention is possible.

In our case study, we consider an observer who tries to determine the causal relationships among three nodes, A, B, and C (figure 6.8). For simplicity, we assume that only three structures are possible and that the observer knows this.

Step 1: Generative model The three structures, which we denote by H_1, H_2, and H_3, have a priori the same probability:

$$p(H_1)=p(H_2)=p(H_3)=\frac{1}{3}. \tag{6.42}$$

We next assume that all nodes start out off. If a node in the structure gets turned on, then as a result, only the nodes directly connected by an arrow may also get turned on. However, there is a 0.2 probability of "failure," when, despite the presence of a connecting arrow, the node at the end of the arrow fails to turn on. If multiple arrows emanate from the same node, then the causal effects along those arrows are independent. Nodes do not spontaneously turn on. Taking H_1 as an example, the causal rules of that structure include (but are not limited to) the following:

$$p(\text{A on}\rightarrow\text{B on}\,|\,H_1)=0.8;$$

$$p(\text{B on}\rightarrow\text{A on}\,|\,H_1)=0;$$

$$p(\text{C on}\rightarrow\text{B on}\,|\,H_1)=0;$$

$$p(\text{B on}\rightarrow\text{C on}\,|\,H_1)=0.8;$$

$$p(\text{A on}\rightarrow\text{C on}\,|\,H_1)=0.8\cdot 0.8=0.64.$$

Step 2: Inference Let us consider a scenario in which an observer sees B being turned on and, as a result, C also gets turned on but A does not. Thus, the observation is "B on $\rightarrow$

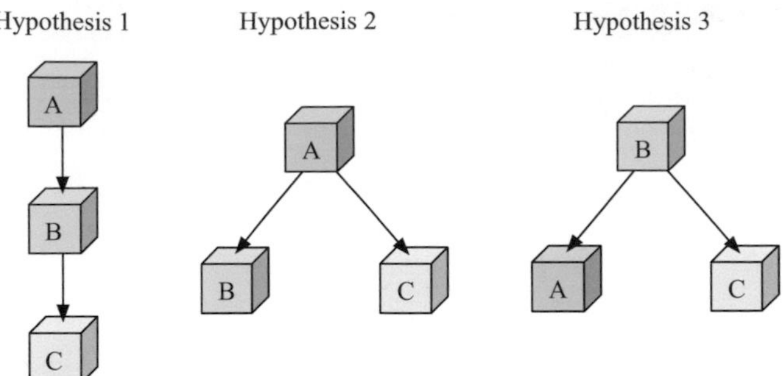

Figure 6.8
Three possible causal structures among three nodes. Although these diagrams look like generative models, they have a different meaning; each diagram is itself a world state in a generative model.

A off, C on." The likelihood of a hypothesis is the probabilities of this observation under the hypothesis. Using the causal structures, we find

$$\mathcal{L}(H_1; \text{B on} \rightarrow \text{A off, C on}) = p(\text{B on} \rightarrow \text{A off, C on}|H_1) = 1 \cdot 0.8 = 0.8;$$

$$\mathcal{L}(H_2; \text{B on} \rightarrow \text{A off, C on}) = p(\text{B on} \rightarrow \text{A off, C on}|H_2) = 1 \cdot 0 = 0;$$

$$\mathcal{L}(H_3; \text{B on} \rightarrow \text{A off, C on}) = p(\text{B on} \rightarrow \text{A off, C on}|H_3) = 0.2 \cdot 0.8 = 0.16.$$

Combining the priors with the likelihoods, we can now calculate the posterior probabilities of the causal structures to be roughly 0.833, 0, and 0.167, respectively.

Exercise 6.5 Verify numerically that this is true.

We have learned something about the causal structure of the world from our observations. More observations will allow us to learn more and perhaps even determine the causal structure with high certainty. Obviously, this was a very simple example and real-world causal inference is more complicated in many ways. First, causal structures might have more than three nodes. Second, we assumed that the causal rules and the associated probabilities were known, which is often not the case. Third, we restricted the inference problem to three possible structures; often, there is a combinatorial explosion of the number of possible structures. On the bright side, in the real world, we might not depend on observations that are given to us, but we might be able to intervene on the system.

Box 6.4
Non-Bayesian Learning in Artificial Neural Networks

Popular models of learning are artificial neural networks. In their simplest form, multilayer perceptrons, they accept an input x and calculate an output $\hat{y}$ which is meant to be similar to the true output y (say in the mean squared sense). The simplest function would be a linear function: $\hat{y} = \sum_i W_i x_i$. We may call this a single-layer neural network. We could then build a two-layer neural network with one matrix of weights ($\mathbf{W}_{12}$) from the first to the second layer and a vector of vectors ($\mathbf{W}_{23}$) from the second to the third as $\hat{y} = \sum_i W_{23,i} \sum_j W_{12,ij} x_j$. However, stacking two layers of linear transformations is just another linear transformation, written in a more complicated way. However, dependencies in the real world are usually nonlinear. Consequently, neural networks usually stack linear transformations and nonlinearities. For example, they may use the *rectified linear unit* function: $\phi(z) = \max(0, z)$ and say two layers (this class of functions can approximate most meaningful functions in human behavior). So they may use

$$\hat{y} = \sum_i W_{23,i}\phi\left(\sum_j W_{12,ij}x_j\right). \tag{6.43}$$

In this nonlinear function, artificial neural networks (ANNs) then tend to implement gradient descent with respect to the vector of all weights W on a loss function, for example, squared error $C(\hat{y}, y) = (\hat{y} - y)^2$.

$$\Delta \mathbf{W}_i = -\eta \frac{\partial C}{\partial W_i}, \tag{6.44}$$

where η is a (usually small) constant. This learning rule allows performance to progressively get better.

This update structure resembles the one that we saw in the Rescorla-Wagner model; that model can be interpreted as gradient descent on the squared difference between the received reward and the expected reward. We will discuss later how artificial neural network learning can be used to obtain a likelihood in domains where that is hard (box 14.4) and how artificial neural networks can be seen as a competing model to Bayesian ones to describe modeling of behavior (box 15.2).[1]

6.7 More Learning

There are many forms of learning that we did not cover here:

- In the examples we considered, the variable to be inferred does not change. This stands in contrast to forms of inference in which the variable of interest changes from trial to trial. We will treat this in chapter 12.
- Providing observations to allow for learning is called training or teaching. Optimal teaching can also be treated computationally but we do not do so in this book.
- We also did not discuss *active learning*, in which the learner has some control over the acquisition of observations.
- We provide a brief introduction to learning in neural networks in box 6.4.

6.8 Summary

In this chapter, we described learning from observations as a Bayesian computation. We learned:

- Learning parameters of a distribution can be treated as a Bayesian inference problem, where the parameters are treated as the world state.

1. Within this book we avoid the use of matrix notation, but in the domain of neural networks, it is shockingly useful.

- A posterior distribution can be formulated over the rate parameter of a Bernoulli (or a binomial) distribution, and over the precision parameter of a Gaussian distribution.
- This computation involves the multiplication of likelihood functions, similar to evidence accumulation. Conjugate priors can facilitate the integration of priors with new data.
- The posterior predictive distribution incorporates the learning of parameters into subsequent prediction or inference.
- The Bayesian rule for learning the probability of a reward has similarities and differences with the Rescorla-Wagner rule in reinforcement learning.
- Sometimes, the causal structure of the world is not known and has to be learned from observations.

6.9 Suggested Readings

- Daniel E. Acuña and Paul Schrater. "Structure Learning in Human Sequential Decision-Making." *PLoS Computational Biology* 6, no. 12 (2010): e1001003.
- Patricia W. Cheng. "From Covariation to Causation: A Causal Power Theory." *Psychological Review* 104, no. 2 (1997): 367–405.
- Anna Coenen, Bob Rehder, and Todd M. Gureckis. "Strategies to Intervene on Causal Systems Are Adaptively Selected." *Cognitive Psychology* 79 (2015): 102–133.
- Alison Gopnik, Clark Glymour, David M. Sobel, Laura E. Schulz, Tamar Kushnir, and David Danks. "A Theory of Causal Learning in Children: Causal Maps and Bayes Nets." *Psychological Review* 111, no. 1 (2004): 3–32.
- Charles Kemp, Andrew Perfors, and Joshua B. Tenenbaum. "Learning Overhypotheses with Hierarchical Bayesian Models." *Developmental Science* 10, no. 3 (2007): 307–321.
- Pierre-Simon Laplace. "Memoir on the Probability of the Causes of Events." *Statistical Science* 1, no. 3 (1986): 364–378.
- Tamas J.Madarasz, Lorenzo Diaz-Mataix, Omar Akhand, Edgar A. Ycu, Joseph E. LeDoux, and Joshua P. Johansen. "Evaluation of Ambiguous Associations in the Amygdala by Learning the Structure of the Environment." *Nature Neuroscience* 19, no. 7 (2016): 965–972.
- Jenny R. Saffran, Richard N. Aslin, and Elissa L. Newport. "Statistical Learning by 8-Month-Old Infants." *Science* 274, no. 5294 (1996): 1926–1928.
- Joshua B. Tenenbaum. "Bayesian Modeling of Human Concept Learning." *Advances in Neural Information Processing Systems 11*, edited by Michael S. Kearns, Sara A. Solla, and David A. Cohn: 59–68. Cambridge, MA: MIT Press, 1998.
- Ting Xiang, Terry Lohrenz, and P. Read Montague. "Computational Substrates of Norms and Their Violations during Social Exchange." *Journal of Neuroscience* 33, no. 3 (2013): 1099–1108.
- Fei Xu and Joshua B. Tenenbaum. "Word Learning as Bayesian Inference." *Psychological Review* 114, no. 2 (2007): 245–272.

6.10 Problems

Problem 6.1 We build on the case study of inferring a Bernoulli probability in section 6.2. We noted that the probability of rain the next day was given by the posterior mean estimate, and was different from the most probable value of the rainfall rate, r, on the island, which is the MAP estimate.

(a) Suppose you have been on the island for four days and it has rained only once. Show that the MAP estimate for r is 0.25 and, according to Laplace's rule of succession (equation (6.14)), the PME is 0.33.

(b) Explain intuitively why it makes sense that your estimate for the probability of rain the next day is different from (and, in this case, greater than) your estimate of the most probable rainfall rate.

(c) Verify equation (6.14) numerically, by discretizing r into several hundred or thousand values equally spaced between 0 and 1. Given one rainy day out of four, calculate the likelihood for each r value, and enter these into Bayes' formula with a uniform prior. Calculate the posterior mean as $\sum rp(r|\mathbf{x})$.

Problem 6.2 We build on the case study of inferring a Bernoulli probability in Section 6.2.

(a) Prove equation (6.22) for the PME, starting from equation (6.14).

(b) Generalize equation (6.22) to the case of a beta prior with parameters α_0 and β_0.

Problem 6.3 We start with equation (6.8) for the posterior over r in the island case study.

(a) Write down an expression for the posterior over r if you observe only rain days.

(b) Plot this posterior for $n_{\text{rain}} = 1, 2, 5, 10$ (four curves in the same plot).

(c) For how many days would it have to rain for the posterior mean to be greater than 0.9?

Problem 6.4 This problem builds on sections 5.5 and 6.3. Evidence accumulation can be formulated as a learning rule.

(a) Show through a mathematical derivation that the posterior mean estimate can be written recursively as $\hat{s}_{t+1} = \hat{s} + \lambda_{t+1}(x_{t+1} - \hat{s}_t)$, and find an expression for the "learning rate" λ_t in terms of σ and t.

(b) Does the learning rate increase or decrease as time goes by? Explain intuitively why this makes sense.

(c) Generalize (a) and (b) to the situation where each measurement x_t has its own variance σ_t^2.

Problem 6.5 We build on section 6.4 for learning the precision parameter of a Gaussian distribution.

(a) Using the posterior in equation (6.33), find an expression for the posterior mean estimate.

(b) Interpret this expression.

(c) Is it possible to write the updating of the PME of J_s in Rescorla-Wagner form, similar to section 6.2.2? If not, why not?

(d) Find an expression for the posterior standard deviation.

(e) Modify equation (6.33) if the learner has a prior $p(J_s)$ that is a gamma distribution with scale parameter k_0 and shape parameter θ_0.

(f) Modify the answers to parts (a), (b), and (d) accordingly.

Problem 6.6 Create a movie in which each frame corresponds to a trial in learning the slope parameter k in the relation $F = ks$ (see section 6.5). The toddler performs twenty trials. On each trial, she sends a command s that is drawn from a uniform distribution on [2, 10], and her force measurement is drawn from a Gaussian distribution with mean $1.5s$ and standard deviation 2. The prior over k is flat. Each frame should look like a row in figure 6.7: (*center*) the likelihood function over k computed from the measurement on the tth trial; (*right*) the posterior based on the measurements made up to and including the tth trial; (*left*) a graphical representation of the data, with the line corresponding to the posterior mean estimate of k based on the measurements made up to and including the tth trial. Make sure that the axes do not change from frame to frame. Choose the ranges on both axes large enough. Make sure that the numbers on the axes are easily legible and that the lines in your plots are sufficiently thick. Save your movie.

Problem 6.7 In the island case study of section 6.2, what is the distribution of the number of days elapsed between rain days? Explain.

Problem 6.8 In section 6.2, we assumed that the binary outcome was independent across days. Assume instead that the probability of rain on a day after a rain day is r_r and after a dry day r_d.

(a) Derive the posterior probability that it will rain on day $t+1$ after a series of observations $x_1, \ldots, x_t$.

(b) Give an intuition for the resulting expression.

Problem 6.9 In section 6.4.2, we specified how the posterior distribution over the precision parameter J_s is used in prediction.

(a) Evaluate the integral for the posterior predictive distribution of s_{t+1}, Eq. (6.37).

(b) How does this distribution compare to a situation where J_s is known?

Problem 6.10 This problem combines problem 5.9 with the present chapter. We encountered a new distribution, the beta distribution, which is generally useful when inferring from counts of occurrence of two categories the underlying probabilities of those categories. We consider a pollster who is interested in the *proportion* of a population supporting a candidate. Thus, the world state of interest, denoted by s, is a probability (between 0 and 1). The individual's poll response x_i is binary: whether or not they support the candidate. The pollster's prior is flat, and individual responses are conditionally independent given s. Suppose that the pollster polls N people and receives n positive responses.

(a) Show mathematically that the posterior distribution over s is a beta distribution with parameters $n+1$ and $N-n+1$. As a result, the MLE is $\hat{s}_{\text{ML}} = \frac{n}{N}$ and the posterior mean estimate is $\hat{s}_{\text{PM}} = \frac{n+1}{N+2}$.

(b) A poll aggregator is trying to do "cue combination" of one poll with N_1 respondents and another with N_2 respondents. The respective numbers of positive responses in both polls are n_1 and n_2, and the respective MLEs are $\hat{s}_{\text{ML},1}$ and $\hat{s}_{\text{ML},2}$. Show that the combined MLE of s is

$$\hat{s}_{\text{ML, combined}} = \frac{N_1\hat{s}_{\text{ML},1} + N_2\hat{s}_{\text{ML},2}}{N_1 + N_2}. \tag{6.45}$$

In other words, the combined MLE is a weighted average of the individual MLEs. This parallels equation (5.14) for combining two normally distributed measurements.

(c) Things are less intuitive with the PME. We denote the individual PMEs by $\hat{s}_{\text{PM},1}$ and $\hat{s}_{\text{PM},2}$. Show that the combined PME of s is

$$\hat{s}_{\text{PM, combined}} = \frac{(N_1+2)\hat{s}_{\text{PM},1} + (N_2+2)\hat{s}_{\text{PM},2} - 1}{N_1 + N_2 + 2}. \tag{6.46}$$

(d) Show that this expression makes sense in the special case of $N_2 = 0$.

Problem 6.11 Prospect theory, a popular (and Nobel Prize–winning) theory in behavioral economics, posits that people overestimate small probabilities, such as p(I will die of Ebola), and underestimate large probabilities, such as p(I will die of cancer). In section 6.2, we wrote about thinking about probabilities of probabilities. Use that concept to describe why people should, indeed, overestimate small and underestimate large probabilities.

7

Discrimination and Detection

How do we determine which of two stimuli occurred?

In previous chapters, we discussed the basics of Bayesian modeling, often using the example of a spatial localization task. In such a task, the variable of interest—location on a line—is continuous, that is, it takes on a continuum of values. The task was to estimate the location along this continuum—so that in principle, the subject has an infinite number of possible responses. However, many if not most tasks in the lab ask for a choice between only two alternatives. This is called a binary choice or binary decision. Binary decisions are like other discrete inference problems in that MAP estimation maximizes accuracy. But unique to binary decisions is that the prior, the posterior, and the estimate distributions are each characterized by a single number. This allows us to characterize behavior using a set of specialized tools, such as receiver operating characteristics.

Plan of the Chapter

This chapter is structured around the same three-step process as previous chapters: generative model, inference process, and estimate distribution. We will use the same basic task as in chapters 3 and 4, combining a measurement with a prior. This chapter will link Bayesian models to signal detection theory.

7.1 Example Tasks

Binary choices are common in the real world, for example: Will it rain today? Can I trust this person? Can I make it to the bus stop in time when I run? Is this email spam or not? Each of these questions has a yes/no answer, and the corresponding random variable (whether it will rain today, etc.) is therefore binary. Many examples we encountered in previous chapters also featured binary decisions.

Two types of binary decisions are particularly important, if only because they correspond to popular psychophysical paradigms. Imagine you are a radiologist trying to determine whether a tumor is present on a noisy X-ray. Such tasks, in which the observer decides whether a stimulus is present or absent, are *detection* tasks. Now imagine you are standing by the side of the road and see the silhouette of a moving car in the distance. You are

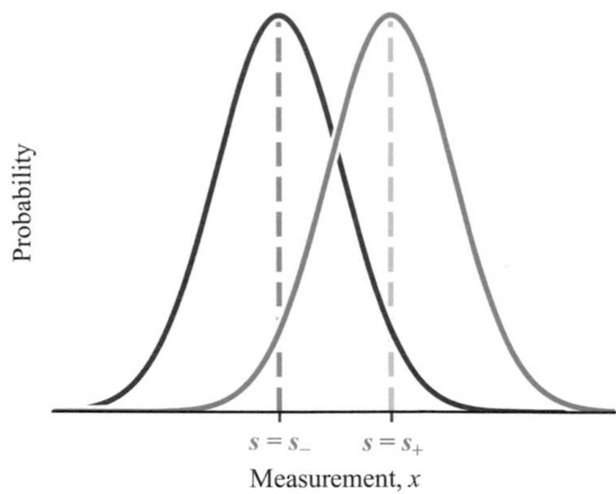

Figure 7.1
Noise distributions for the yes/no task of discriminating between s_+ and s_-.

trying to determine whether the car is coming towards or away from you. Tasks in which the observer decides between two nonzero values or categories of the stimulus variable (motion direction) are called *discrimination* tasks. Detection and discrimination tasks appear in many laboratory experiments: Was the motion to the left or to the right (discrimination)? Was a vertical line present in the display (detection)? Did you feel a stimulus on your finger (detection)? And so on. Even when the underlying stimulus variable is continuous (e.g., duration), it is common to phrase the task in terms of a choice between two options (e.g., which stimulus lasted longer), while continuously manipulating the stimuli from trial to trial.

7.2 Discrimination

In this chapter, the stimulus s can take just two values, which we call s_+ and s_-, and the observer chooses between them. For example, the observer reports whether an oriented pattern is tilted 1° to the right or 1° to the left of vertical. Such a task is called a *two-alternative forced choice task* by some, and a *yes/no task* by others.

7.2.1 Step 1: Generative Model

The generative model is $s \to x$, as in chapter 3. The difference with that chapter is that the stimulus s takes only two possible values, also called *alternatives* and denoted by s_+ and s_-. These two are simply numbers, such as -1 and 1, or 0 and 3. They are *not* random variables. We will consider in chapter 8 what happens when choosing between two *classes* of stimuli. The stimulus distribution is a discrete probability distribution, with values $p(s = s_+)$ and $p(s = s_-)$, which sum to 1 and reflect the frequencies with which the stimulus values are presented. The measurement x follows the usual Gaussian noise distribution $p(x|s)$. The noise distribution is shown in figure 7.1 for both possible values of s.

7.2.2 Step 2: Inference

Suppose $s_+ = 1°$ and $s_- = -1°$, and that on a given trial, your measurement is $0.1°$. Would you report that the stimulus was s_+ or s_-? You probably would say s_+ simply because the

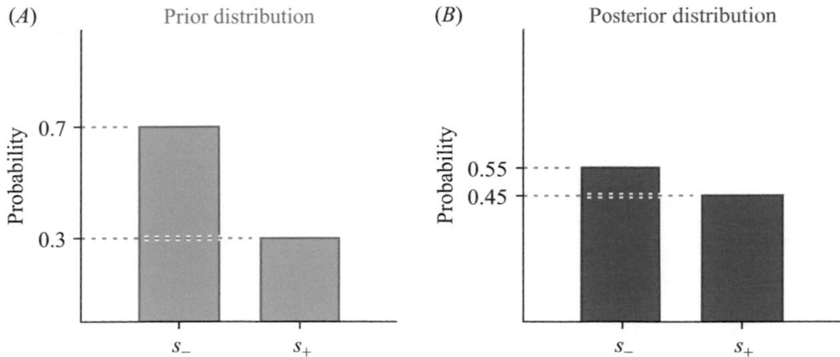

Figure 7.2
Example prior and posterior distribution over a binary variable.

measurement is closer to s_+ than to s_-. But that is not what a Bayesian observer would necessarily do. To see that, imagine that you knew that s_- was far more common in the world (or in the experiment) than s_+. In that case, a measurement that is only slightly closer to s_+ than to s_- would probably have been produced by s_-. In this subsection, we will work out how the Bayesian observer would decide.

Exercise 7.1 Describe a real-world discrimination task where one stimulus, say s_+ is far more probable than another stimulus s_-.

Describing the Bayesian observer requires calculating the posterior over s, $p(s|x)$. Since s takes on two values, the posterior distribution is a discrete probability distribution, with values $p(s=s_+|x)$ and $p(s=s_-|x)$, which have to sum to 1 (figure 7.2). Bayes' rule tells us that

$$p(s|x) = \frac{p(x|s)p(s)}{p(x)}. \tag{7.1}$$

For a binary variable, the posterior distribution is uniquely determined by the ratio of the posterior probabilities of the two alternatives. We can calculate this ratio using Bayes' rule:

$$\frac{p(s=s_+|x)}{p(s=s_-|x)} = \frac{\frac{p(x|s=s_+)p(s=s_+)}{p(x)}}{\frac{p(x|s=s_-)p(s=s_-)}{p(x)}} \tag{7.2}$$

$$= \frac{p(x|s=s_+)p(s=s_+)}{p(x|s=s_-)p(s=s_-)} \tag{7.3}$$

This ratio is called the *posterior ratio* or *posterior odds*. Its interpretation is that of the probability of one alternative *relative to* that of the other. We see that the normalization $p(x)$ drops out and is thus irrelevant to this ratio. For example, if your posterior probability that the stimulus was s_+ is 80 percent, then your posterior ratio is $\frac{0.80}{0.20} = 4$. The posterior ratio is always positive but can grow arbitrarily large. For example, if the posterior probability of s_+ is 99 percent, then the posterior ratio is $\frac{0.99}{0.01} = 99$. Knowing the posterior ratio, one can calculate the probability of each of the alternatives, and vice versa.

In Bayesian calculations, you will often see equation (7.3) with the natural logarithm taken of both sides. We denote this log ratio by d:

$$d \equiv \log \frac{p(s=s_+|x)}{p(s=s_-|x)} = \log \frac{p(x|s=s_+)p(s=s_+)}{p(x|s=s_-)p(s=s_-)} \tag{7.4}$$

Taking the logarithm simplifies many mathematical derivations, as we will see below. The quantity d is called the *log posterior ratio* (LPR; also *log posterior odds*). If the posterior probability of s_+ is 0.80, then the LPR is $\log \frac{0.80}{0.20} = \log 4 = 1.39$. The LPR contains the same information as the posterior distribution itself; after all, we can exponentiate it and calculate the probability of each alternative from it as we did above.

The posterior probabilities of the two alternatives can be recovered from d as follows:

$$p(s=s_+|x) = \frac{1}{1+e^{-d}}; \tag{7.5}$$

$$p(s=s_-|x) = \frac{1}{1+e^{d}}. \tag{7.6}$$

The former is called the *logistic function* of d, the latter is 1 minus that.

Exercise 7.2 Show analytically that this relation of their difference being 1 is correct.

The LPR takes values between $-\infty$ and ∞. The LPR also has a symmetry property: flipping the posterior probabilities of the two alternatives is equivalent to flipping the sign of the LPR. For example, if the posterior probability of s_+ is 20 percent and that of s_- 80 percent, then the LPR is $\log \frac{0.20}{0.80} = \log 0.25 = -1.39$. When two alternatives have the same posterior probability, the LPR is equal to 0. If the LPR is positive, then $p(s=s_+|x)$ is greater than $p(s=s_-|x)$. Therefore, the MAP estimate is

$$\hat{s}_{\text{MAP}} = \begin{cases} s_+ & \text{if } d > 0; \\ s_- & \text{if } d < 0. \end{cases} \tag{7.7}$$

Thus, Bayesian binary decision making is concisely described in terms of LPR.

We now point out some common terminology. The inequality used to determine the MAP estimate, $d > 0$, is also called the *decision rule* of the Bayesian MAP observer, and d is called the *decision variable* (hence our notation d). The decision rule can be thought of as the binary equivalent of the mapping from x to s in estimation tasks (chapters 3 and 4). The scalar value to which the decision variable is compared in order to make a decision, here 0, is also called the *decision criterion*, or simply the *criterion*. The terminology of decision rule, decision variable, and decision criterion is not specific to Bayesian models. Any inequality of the form $f(x) > k$, with f any function and k any scalar, can serve as a model for how the observer turns a measurement into a decision. In general, a *criterion* is any fixed value to which a variable is compared to produce a decision.

LPRs also simplify other aspects of the derivations. Since the logarithm of a product is the sum of the logarithms, the right-hand side of equation (7.4) can be rewritten as a sum:

$$d = \log \frac{p(s=s_+)}{p(s=s_-)} + \log \frac{p(x|s=s_+)}{p(x|s=s_-)}. \tag{7.8}$$

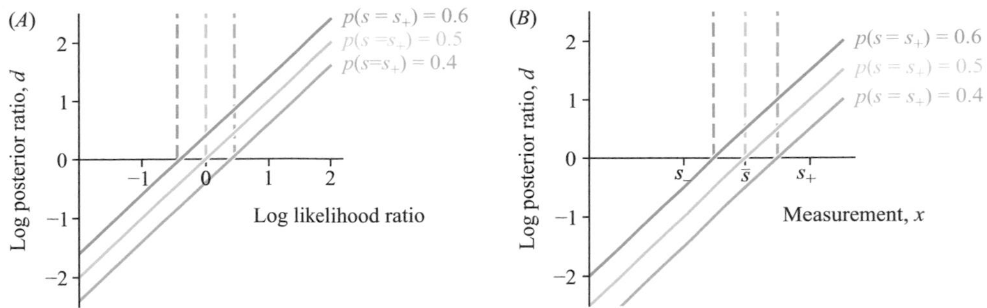

Figure 7.3
(*A*) Relation between LPR and log likelihood ratio. This relation is general and does not only apply to the Gaussian model. Going from red to purple to blue the prior more strongly favors s_+, the criterion on the log likelihood ratio shifts toward smaller values (dashed lines), indicating that the subject has a stronger tendency to report s_+. (*B*) Relation between LPR and measurement *in the Gaussian model.* The relation is linear with slope $\frac{\Delta s}{\sigma^2}$. The purple line also represents the log *likelihood* ratio, and the corresponding "neutral" criterion is the midpoint of s_+ and s_-.

Each of these terms has an intuitive meaning. The term on the right-hand side is called the *log likelihood ratio* (LLR) and reflects the amount of evidence provided by the measurement x. The term on the left-hand side is the log prior ratio and reflects the observer's relative prior beliefs in the two alternatives. The sum of these terms is the LPR (figure 7.3).

Whenever d is greater than zero, s_+ is most probable. The MAP decision rule is thus to report s_+ when the sum of the log prior ratio and the LLR is positive:

$$\text{MAP rule: Report } s_+ \text{ when } \log\frac{p(s=s_+)}{p(s=s_-)}+\log\frac{p(x|s=s_+)}{p(x|s=s_-)}>0. \tag{7.9}$$

When s_- occurs with higher probability than s_+ (in other words, when the the log prior ratio is negative), the optimal decision rule is to report s_+ only when the measurement x provides sufficiently strong evidence in favor of s_+ to overcome the prior bias in favor of s_-.

MAP estimation maximizes the probability of the response being correct. To understand why the MAP rule in this case, equation (7.9), is a good strategy, we consider a world in which a person has either brown eyes or blue eyes. Imagine you are standing in front of a classroom of people and you are asked about one person's eye color. The quality of information is affected by your distance from the person. Suppose that in your region of the world, brown eyes are more common than blue eyes. If you are asked about someone nearby, your sensory information will be of high quality and you will be able to base your decision predominantly on this sensory information. If you are asked about someone further away, the quality of the sensory information is worse or even uninformative. The lower the quality of the visual information, the more you will rely upon your knowledge of the prevalence of brown eyes in the general population. When no information is available at all, your best bet is to always respond that the person has brown eyes. This increasing effect of the prior as the quality of sensory information decreases is exactly expressed by equation (7.9). The LLR will tend to be smaller in magnitude (either positive or negative) when the sensory information is of lower quality ("tend to" because this term is a random variable that inherits its distribution from the distribution of x). The relative quality of prior and likelihood information determines which information dominates. In summary, in a yes/no discrimination task, the prior has the effect of *shifting the decision criterion.*

7.2.3 Gaussian Model

If the measurement x follows a Gaussian distribution when conditioned on s, we can further evaluate the LLR by substituting the expression for $p(x|s)$ This produces a particularly simple expression for the LPR,

$$d = \log \frac{p(s=s_+)}{p(s=s_-)} + \frac{\Delta s}{\sigma^2}(x - \bar{s}), \tag{7.10}$$

where we introduced the notation

$$\bar{s} \equiv \frac{s_+ + s_-}{2} \tag{7.11}$$

for the midpoint between s_+ and s_-, and

$$\Delta s = s_+ - s_- \tag{7.12}$$

for their difference. We will derive equation (7.10) in problem 7.5. We plotted the LPR d as a function of the measurement x in figure 7.3. It is a linear function of the measurement. Importantly, it also depends on the sensory uncertainty level σ.

To sum up, in an equal-variance Gaussian measurement model, the LPR is linear in the measurement. Although very convenient, this property is specific to the equal-variance Gaussian model and rarely true otherwise (see problems).

The LLR is positive whenever x is greater than the midpoint of s_+ and s_-, and negative otherwise. This is intuitive: the measurement provides evidence for s_+ if it lies closer to s_+ than to s_-. The factor $\frac{\Delta s}{\sigma^2}$ helps determine the magnitude of the LLR. This factor tells us that for the same x, the strength of the evidence is larger when the two stimuli to be discriminated are further apart (i.e., $s_+ - s_-$ is larger) or when the noise in the measurement (σ) is smaller.

7.2.4 Decision Rule in Terms of the Measurement

Substituting equation (7.10) for the LLR into equation (7.9) for the decision rule and solving for x, we arrive at the optimal decision rule for our yes/no discrimination task with Gaussian measurement noise: to report that the stimulus is s_+ when

$$x > k_{\text{MAP}}, \tag{7.13}$$

where the *MAP criterion on the measurement* is

$$k_{\text{MAP}} = \bar{s} - \frac{\sigma^2}{\Delta s} \log \frac{p(s=s_+)}{p(s=s_-)}. \tag{7.14}$$

While the MAP criterion on the LPR is always 0, the MAP criterion in the measurement depends on the stimuli to be discriminated, the uncertainty level (or noise level), and the prior probabilities. The observer must in general have knowledge of all these variables in order to be optimal.

In the special case that the prior is flat, $p(s=s_+)=p(s=s_-)=0.5$ (purple lines in figure 7.3), the log prior ratio is zero, and the observer would report s_+ simply when

$$x > \bar{s}. \tag{7.15}$$

This makes sense: x gets compared to the midpoint of s_- and s_+, which we could call the "neutral criterion". By comparing equations (7.14) and (7.15), we can identify the term $-\frac{\sigma^2}{\Delta s}\log\frac{p(s=s_+)}{p(s=s_-)}$ as the *criterion shift* due to having a non-flat prior. If the prior favors s_+, for example $p_s(s+)=0.6$ (blue line in figure 7.3B), then the negative log prior ratio would be a negative number (-0.405). As a consequence, the measurement x could be closer to s_- than to s_+ and yet the observer would respond $\hat{s}=s_+$. The opposite happens when the prior favors s_- (red line). In other words, the prior *biases* the observer toward reporting the alternative with the highest prior probability. This phenomenon is similar to how a Gaussian prior biases the observer toward its mean in the continuous estimation task discussed in chapters 3 and 4.

7.2.5 Multiple Tasks Can Have the Same Bayesian Decision Rule

Each binary decision has only one Bayesian decision rule, aside from the fact that the same rule can be written in mathematically equivalent forms, for example $x > 0$ would be equivalent to $e^x > 1$. However, different tasks can have the same decision rule. As a simple example, consider the decision rule in equation (7.15). There are many ways of choosing pairs of stimuli (s_+, s_-) that have the same mean and therefore the same decision rule. Therefore, it is not possible to reconstruct the task from the decision rule.

7.2.6 Step 3: Response Distribution

In this section, we are interested in the distribution of the response over many trials in which the experimental condition is held fixed. In our task, the experimental condition is completely specified by s, which can take two values. Therefore, the distribution of the stimulus estimate is given by the probability of reporting either $\hat{s}=s_+$ or $\hat{s}=s_-$, when x is drawn from either $p(x|s=s_+)$ or $p(x|s=s_-)$, for a total of four possibilities. These four numbers can be reduced to two, since the probability of estimating the stimulus as s_+ is 1 minus that of estimating it as s_-. Thus, the distribution of the estimate is determined by the following two probabilities, corresponding to the observer's correct and incorrect reports of s_+:

$$p(\hat{s}=s_+|s=s_+)=p(d>0|s=s_+); \tag{7.16}$$

$$p(\hat{s}=s_+|s=s_-)=p(d>0|s=s_-). \tag{7.17}$$

For convenience, we assume in this subsection that the prior is flat. Evaluating equation (7.17) allows us to calculate predictions for how we expect an observer to behave across multiple trials.

As the third step of Bayesian modeling, we need to compute the probability that equation (7.15) is satisfied when x is drawn from either $p(x|s=s_+)$ or $p(x|s=s_-)$. We can think of this in two equivalent spaces: d space (figure 7.4A) and x space (figure 7.4B). Here, we focus on the latter (in problem 7.9, on the former). In figure 7.4B, we copied the plot of both distributions from figure 7.1. Equation (7.15) is satisfied whenever the measurement falls to the right of the vertical line at k_{MAP}. Thus, graphically, the probability that equation (7.15)

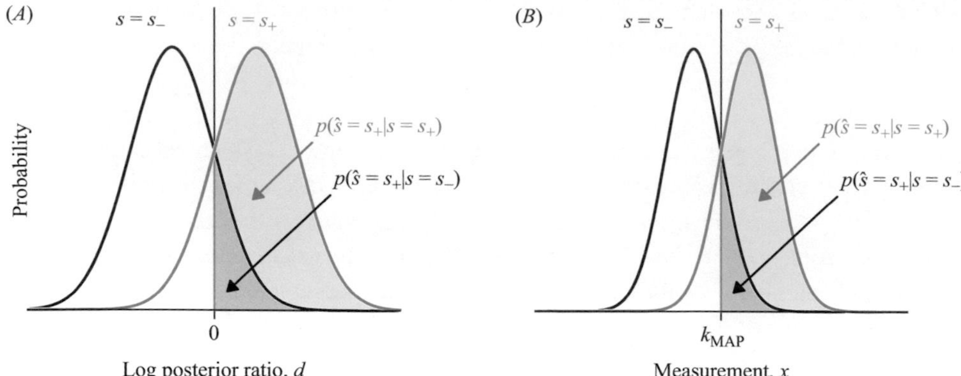

Figure 7.4
Response probabilities in discrimination. The following are two equivalent ways of visualizing step 3. (*A*). In LPR space (*d* space). Distributions of the decision variable (the LPR) conditioned on the true world state, s_+ or s_-. The probability that the MAP estimate is s_+ is equal to the shaded area when the true stimulus is s_- (gray) or s_+ (teal). The means of the distributions are $-\frac{(\Delta s)^2}{2\sigma^2}$ and $\frac{(\Delta s)^2}{2\sigma^2}$, and the standard deviation of each is $\frac{\Delta s}{\sigma}$ (see problem 7.9). The optimal criterion is 0. (*B*) In measurement space (*x* space). The means of the distributions are s_- and s_+, and the standard deviation of each is σ. The optimal criterion on the measurement is given by equation (7.14). The shaded areas have the same meaning as in (*A*).

is satisfied is the area under the probability density function to the right of the line. Mathematically, calculating this area corresponds to integrating the density function from k_{MAP} to infinity:

$$p(\hat{s}=s_+|s=s_+)=\int_{k_{\text{MAP}}}^{\infty}\mathcal{N}(x;s_+,\sigma^2)dx=\frac{1}{\sqrt{2\pi\sigma^2}}\int_{k_{\text{MAP}}}^{\infty}e^{-\frac{(x-s_+)^2}{2\sigma^2}}dx; \tag{7.18}$$

$$p(\hat{s}=s_+|s=s_-)=\int_{k_{\text{MAP}}}^{\infty}\mathcal{N}(x;s_-,\sigma^2)dx=\frac{1}{\sqrt{2\pi\sigma^2}}\int_{k_{\text{MAP}}}^{\infty}e^{-\frac{(x-s_-)^2}{2\sigma^2}}dx. \tag{7.19}$$

In one of the great tragedies of mathematics, these Gaussian integrals cannot be evaluated in closed form, that is, in terms of already known functions. Therefore, we have to just live with the integral expressions. However, since these integrals are so common, special notation is typically introduced for them. We will denote the cumulative distribution of a normal distribution by Φ and the cumulative distribution of a *standard* normal distribution by Φ_{standard}. (Some other texts use Φ for the cumulative standard normal distribution.) Then, one can show that

$$p(\hat{s}=s_+|s=s_+)=\Phi\left(s_+;k_{\text{MAP}},\sigma^2\right)=\Phi_{\text{standard}}\left(\frac{s_+-k_{\text{MAP}}}{\sigma}\right); \tag{7.20}$$

$$p(\hat{s}=s_+|s=s_-)=\Phi\left(s_-;k_{\text{MAP}},\sigma^2\right)=\Phi_{\text{standard}}\left(\frac{s_--k_{\text{MAP}}}{\sigma}\right). \tag{7.21}$$

How we obtain these equations is explained in more detail in box 7.1. Equations (7.20) and (7.21) represent predictions for how often subjects would estimate the stimulus s_+ when it is in reality s_+, or when it is in reality s_-. These predictions can be compared against experimental results, in parallel to what we found in section 4.2 for the case of continuous estimation.

Finally, we can substitute equation (7.14) into equations (7.20) and (7.21) to find

$$p(\hat{s}=s_+|s=s_+) = \Phi_{\text{standard}}\left(\frac{\Delta s}{2\sigma} + \frac{\sigma}{\Delta s}\log\frac{p(s=s_+)}{p(s=s_-)}\right); \tag{7.22}$$

$$p(\hat{s}=s_+|s=s_-) = \Phi_{\text{standard}}\left(-\frac{\Delta s}{2\sigma} + \frac{\sigma}{\Delta s}\log\frac{p(s=s_+)}{p(s=s_-)}\right). \tag{7.23}$$

Signal detection theory Signal detection theory would postulate an observer who applies some criterion k to x. This criterion may or may not be equal to the Bayesian criterion k_{MAP}. Then, equations (7.20) and (7.21) would still hold, but with k_{MAP} replaced by k. Within signal detection theory, the difference between k and k_{MAP} is also called *bias*, but we do not use that term, in order to avoid confusion with estimation bias as defined in section 4.5. We will discuss the relationship between Bayesian inference and signal detection theory a bit more in section 7.6.

Box 7.1
Cumulative Normal Distribution

A cumulative distribution is obtained from a regular probability distribution by summing the values up from left to right, keeping a running tally. For example, the cumulative normal distribution belonging to a normal distribution with mean μ and variance σ^2 is defined as

$$\Phi\left(y;\mu,\sigma^2\right) \equiv \frac{1}{\sqrt{2\pi\sigma^2}}\int_{-\infty}^{y} e^{-\frac{(x-\mu)^2}{2\sigma^2}}\,dx. \tag{7.24}$$

Here, y is the argument of the cumulative distribution, whereas x is just an integration variable. The *standard normal distribution* is a normal distribution with mean 0 and variance 1. We give its cumulative distribution a special notation,

$$\Phi_{\text{standard}}(y) \equiv \Phi(y;0,1). \tag{7.25}$$

The following properties hold:

$$\Phi(y;\mu,\sigma^2) = \Phi(y-\mu;0,\sigma^2) \tag{7.26}$$

$$\Phi(y;\mu,\sigma^2) = \Phi_{\text{standard}}\left(\frac{y-\mu}{\sigma}\right) \tag{7.27}$$

$$\Phi(y;\mu,\sigma^2) + \Phi(-y;-\mu,\sigma^2) = 1 \tag{7.28}$$

$$\Phi_{\text{standard}}(y) + \Phi_{\text{standard}}(-y) = 1. \tag{7.29}$$

Exercise 7.3

(a) Derive these properties.

(b) Show how we can use some of these properties to derive equation (7.20) from equation (7.18) (or equation (7.21) from equation (7.19)).

7.3 Detection

In section 7.1, we introduced the example of a radiologist determining whether a patient has a tumor based on an X-ray. This is an example of a *detection task*: Is the tumor present or not? There are many other daily-life examples. As we are showering, we have to determine whether or not our phone rang. On the road, we have to determine whether there is a bump ahead or not. If we have a gas stove, detecting the smell of gas can keep us out of danger. In general, the task is to determine whether a signal is present in noise.

Mathematically, detection is closely related to the discrimination task discussed so far. In the discrimination task, the observer had to discriminate between two stimulus values, s_+ and s_-. In its simplest form, detection is the special case in which s_+ is positive and $s_- = 0$, that is, the observer is discriminating between a certain nonzero value and zero. In many cases, the variable s is somewhat abstract; for instance, s could be a composite of different image features that a radiologist uses to judge tumor presence. For simplicity, however, we still conceptualize s as a one-dimensional variable.

The Bayesian model we described for discrimination can therefore also be used for detection. For example, the decision variable is obtained by substituting $s_- = 0$ into equation (7.10) to obtain

$$d = \log \frac{p(s = s_+)}{p(s = 0)} + \frac{s_+}{\sigma^2}\left(x - \frac{s_+}{2}\right). \tag{7.30}$$

Similarly, from equation (7.20), the probability of correctly detecting the stimulus is

$$p(\hat{s} = s_+ | s = s_+) = \Phi_{\text{standard}}\left(\frac{s_+}{2\sigma} + \frac{\sigma}{s_+}\log\frac{p(s = s_+)}{p(s = 0)}\right). \tag{7.31}$$

In a detection task, the probability of reporting "present" when the signal is present is called the *hit rate*, *detection rate*, *sensitivity*, or *true positive rate*, whereas the probability of reporting "present" when the signal is absent is called the *false-alarm rate* or *false-positive rate*. These probabilities can all be recognized as areas under the curves in figure 7.4. This terminology stems from *signal detection theory*. 1 minus the hit rate is the *miss rate* or *false-negative rate*, and 1 minus the false-alarm rate is called the *correct-rejection rate*, *specificity*, or *true-negative rate* (see figure 7.5). These four terms can also be applied to a discrimination task, such as discriminating a $-1°$ from a $1°$ orientation, but in such tasks, it is arbitrary which stimulus is regarded as the "signal."

In our example, hit and correct-rejection rates are equal, as are the false-alarm and miss rates. Consequently, hit and false-alarm rates sum to 1. In general, however, hit and false-alarm rates do not need to sum to 1.

7.4 Confidence

In a binary decision, the sign of the LPR determines the MAP decision. However, the LPR also has a magnitude or absolute value. A decision made with an LPR of 0.1 is made less confidently than one with an LPR of 1: after all, a lower absolute value means that the posterior probabilities of the two alternatives are closer to each other. Therefore, a natural measure of confidence in a binary decision is the magnitude of the LPR:

		Reported world state: Present	Reported world state: Absent	Total
True world state	Present	Hits (true positives)	Misses (false negatives)	1
True world state	Absent	False alarms (false positives)	Correct rejections (true negatives)	1

Figure 7.5
Terminology for the four types of response frequencies in a detection task.

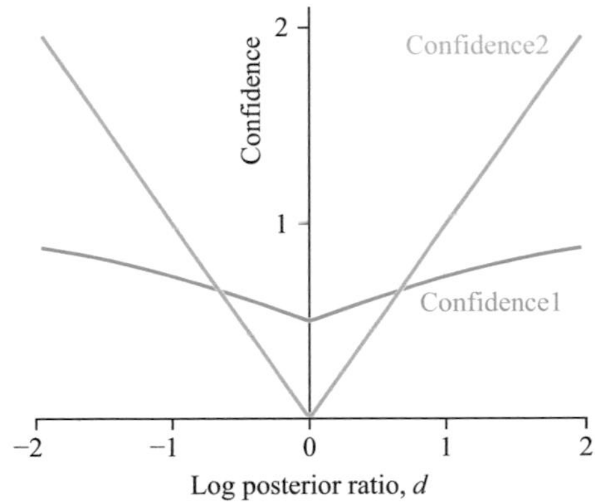

Figure 7.6
Two measures of confidence as a function of the LPR. Red: the absolute value of the LPR. Blue: the posterior probability of the MAP choice.

$$\text{confidence2} = \left| \log \frac{p(s=s_+|x)}{p(s=s_-|x)} \right|. \tag{7.32}$$

This is plotted in figure 7.6. Confidence can decrease due to a nonflat prior. For example, when the LLR is 0.3, and the log prior ratio is -0.4, then confidence decreases from 0.3 to 0.1 due to the introduction of the nonflat prior.

In chapter 3, we introduced a different measure of confidence, namely the posterior probability of the response (estimate). We will here call that measure confidence1:

$$\text{confidence1} = p(s=\hat{s}|x). \tag{7.33}$$

In the current task, the estimate is the MAP estimate, so confidence1 is equal to $p(s=s_+|x)$ if $p(s=s_+|x) > 0.5$, and equal to $1-p(s=s_+|x)$ if $p(s=s_+|x) < 0.5$. The two measures, confidence1 and confidence2, are related through:

$$\text{confidence1} = \frac{1}{1+e^{-\text{confidence2}}}. \tag{7.34}$$

We will derive this in problem 7.8.

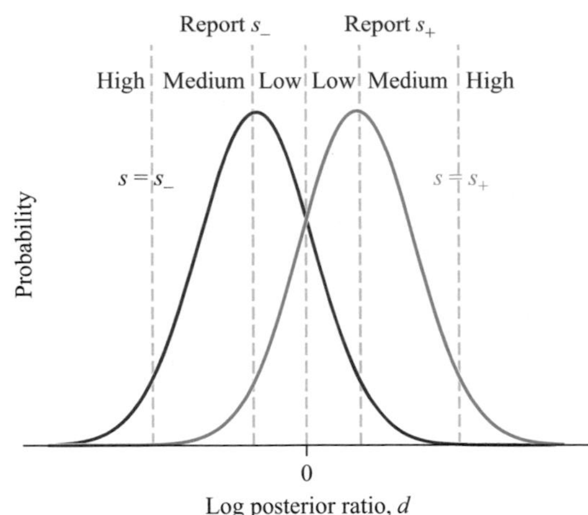

Figure 7.7
Confidence ratings (here low, medium, high) subdivide the regions $d < 0$ (report s_-) and $d > 0$ (report s_+). In this hypothetical experiment, the observer has a total of six response categories, ordered as indicated. When d is high in absolute value, the confidence rating is higher.

In equation (7.34), the logistic function, which we first encountered in equation (7.5), makes an appearance again—and for much the same reason. Since the logistic function is monotonically increasing, the two measures of confidence are in one-to-one correspondence, and both are legitimate measures of confidence. In figure 7.6, both measures are plotted as a function of the LPR.

Exercise 7.4 Open question: Does one of the two measures make more sense to you than the other? Why or why not?

Having established that confidence (either measure) corresponds to distance from the origin on the decision variable axis, we can use the Bayesian model to predict not only the observer's responses on the discrimination (or detection) task, but also how often this decision is made with high or with low confidence. The dividing line between low and high confidence is a parameter that an experimenter could fit to a human subject's data. This is the idea behind a *confidence rating* experiment: the subject is asked not only for a binary judgment on the discrimination task, but afterwards also to rate confidence, let us say as low, medium, or high. Thus, there are now six possible responses: two estimates times three confidence ratings. We saw before that the Bayesian observer makes a binary judgment by determining in which of two regions in the decision space (positive and negative) the LPR falls. Similarly, the Bayesian observer now chooses one of the six possible responses by determining in which of six decision regions the LPR falls (figure 7.7). Three of these regions together form the negative axis, and three the positive axis. From left to right, these regions would correspond to estimating the stimulus as s_- with high, medium, and low confidence, and estimating the stimulus as s_+ with low, medium, and high confidence. A total of five decision criteria separate these regions. When there are M confidence ratings, the number of criteria is $2M - 1$.

7.5 Further Characterizing the Response Distribution

In the binary tasks that are the topic of this chapter, we can further characterize the response distribution. The following applies both to discrimination and to detection, although the terminology (hit rate, false-alarm rate, etc.) is mostly associated with detection.

7.5.1 Receiver Operating Characteristic

In section 7.3, we defined hit and false-alarm rates with respect to one particular decision criterion. In a task with confidence ratings, we can associate a hit and a false-alarm rate with any criterion dividing two adjoining decision regions. In the example with three criteria, the highest criterion would separate s_+ estimates made with medium confidence from those made with high confidence. The generalized hit rate of the Bayesian model is equal to the area under the distribution of the decision variable when $s=s_+$, $p(d|s=s_+)$, to the right of a particular criterion k (which was previously always zero). Similarly, the generalized false-alarm rate is equal to the area under the $s=s_-$ distribution of the decision variable, $p(d|s=s_-)$, to the right of the same criterion k. In equations:

$$H(k)=p(d>k|s=s_+); \tag{7.35}$$

$$F(k)=p(d>k|s=s_-). \tag{7.36}$$

If there are three confidence ratings, this leads to five pairs of hit and false-alarm rates, one for each criterion. Plotting hit rate $H(k)$ against false-alarm rate $F(k)$ gives us five points in a plot with horizontal and vertical axes both ranging from 0 to 1. For example, the second point in this plot would have as y coordinate the proportion of s_+ responses made with either medium or high confidence under the $p(d|s=s_+)$ distribution, and as x coordinate the same proportion under the $p(d|s=s_-)$ distribution. In the limit of having a very large number of confidence ratings, the plot would contain a smooth curve passing through the origin and through (1,1). This would correspond to the decision criterion k moving continuously along the decision axis from right to left, at each value producing a hit and a false-alarm rate (figure 7.8). This curve is called the *receiver operating characteristic* (ROC). It characterizes the distributions of the decision variable given either stimulus in a more complete manner than the original hit and false-alarm rates can; the latter are essentially only one point on the ROC. The ROC is *parameterized by* the criterion. The ROC is one of the most important concepts in signal detection theory.

In the main case under study in this chapter, the hit rate is equal to the correct rejection rate and the false-alarm rate is equal to the miss rate. As a consequence, the ROC is symmetrical around the negative diagonal.

Exercise 7.5 Why is this the case?

However, this is not the case in general, and in a problem we will see an example of an ROC that is asymmetric around the negative diagonal.

In an actual experiment, an empirical ROC is obtained from the response frequencies in each of the $2M$ response categories, for each of the two stimuli. The way to do this is by creating a table of 2 rows and $2M$ columns (figure 7.9, rows I and II). The top row corresponds to the true stimulus being $s=s_+$, the bottom row to $s=s_-$. Each column corresponds to a

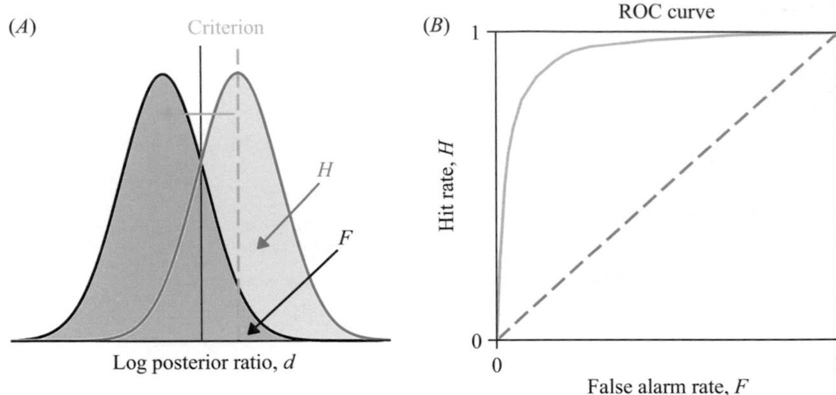

Figure 7.8
Theoretical receiver operating characteristic. (*A*) Distribution of the LPR d when the stimulus is s_+ (teal curve) or s_- (gray curve). The criterion (dashed gold line) defines a hit rate H (teal area), and a false-alarm rate F (gray area). (*B*) By sweeping the criterion from right to left and plotting H against F, we obtain the theoretical ROC.

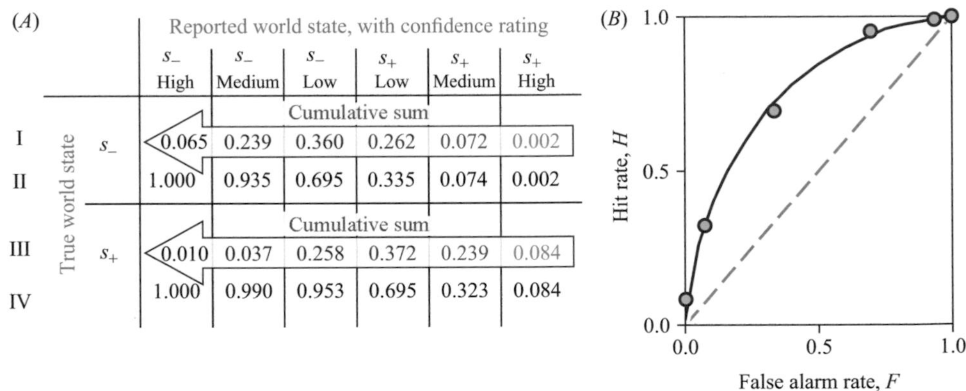

	True world state	s_- High	s_- Medium	s_- Low	s_+ Low	s_+ Medium	s_+ High
		Cumulative sum					
I	s_-	0.065	0.239	0.360	0.262	0.072	0.002
II		1.000	0.935	0.695	0.335	0.074	0.002
		Cumulative sum					
III	s_+	0.010	0.037	0.258	0.372	0.239	0.084
IV		1.000	0.990	0.953	0.695	0.323	0.084

Figure 7.9
Empirical receiver operating characteristic. (*A*) Rows I and III show the response proportions in a hypothetical experiment when the world state was s_+ (I) or s_- (III). The subject reported s_+ or s_- with low, medium, or high confidence rating (columns). We first take the cumulative sum of proportions from right to left; this produces rows II and IV. Then, we plot row II (hit rate) against row IV (false-alarm rate). This produces the green points in (*B*). The black curve represents the theoretical ROC underlying these data.

response category. The left M columns correspond to $s=s_-$ responses, in order of decreasing confidence. The right M columns correspond to $s=s_+$ responses, in order of increasing confidence. Each cell in the table contains the frequency of responses in each category, divided by the total number of responses across all categories for that stimulus. Thus, the sum of the numbers in each row equals 1. Next, create a new table in which each cell contains the sum of the number in the corresponding cell and all cells to the right of it in the same row in the original table. In other words, the new table (rows II and IV in figure 7.9) is built by cumulatively summing the numbers in the original table from right to left, for each row separately. In the new table, each column corresponds to a (hit, false-alarm) rate pair. The leftmost pair should, by construction, always be equal to (1,1). Finally, the hit rate (row II) is plotted against the false-alarm rate (row IV) (figure 7.9). When a model accurately describes an observer, the ROC obtained from that model should go through the points of the empirical ROC.

Note that it does not matter where the observer places their confidence criteria in d space—the ROC will be the same.

7.5.2 Discriminability

In previous sections, we used the decision rule $d > 0$, where d is the LPR. There are, however, scenarios in which the observer would use a criterion different from 0. One possibility is that the observer is suboptimal and makes a wrong assumption about the prior probabilities. In that case, the criterion will be replaced by an unknown number. A second scenario is that the observer may (rationally or not) attach greater importance to estimating one of the stimuli correctly than to estimating the other correctly (we will elaborate on this situation in a later chapter).

In these scenarios, the decision rule takes the form $d > k$. The hit and false-alarm rates with respect to this unknown criterion, equations (7.36), become

$$H(k) = \Phi_{\text{standard}}\left(\frac{s_+ - k}{\sigma}\right); \tag{7.37}$$

$$F(k) = \Phi_{\text{standard}}\left(\frac{s_- - k}{\sigma}\right). \tag{7.38}$$

Adopting a different k changes various measures of the observer's performance. For instance, the observer's proportion correct, which is $p(s = s_+)H + p(s = s_-)F$, now depends on k. However, the adoption of a different k does not change the observer's ROC. Rather, the effect of a change in k is simply to move the observer to a different point on the same ROC. The fact that the observer's ROC is unaffected by the criterion suggests that it should be possible to derive a numerical measure of performance that does not depend on k. One such criterion-invariant measure of performance is the *area under the ROC*, but the most common criterion-invariant measure is discriminability. Discriminability, denoted by d' (read "d prime"), is a way to quantify how well separated the distributions of the decision variable under the two alternatives are. This measure is defined as

$$d' = \Phi^{-1}_{\text{standard}}(H) - \Phi^{-1}_{\text{standard}}(F). \tag{7.39}$$

where $\Phi^{-1}_{\text{standard}}$ refers to the inverse function of Φ_{standard}. This means that $\Phi^{-1}_{\text{standard}}(y)$ is the value x for which $\Phi_{\text{standard}}(x) = y$. You can think of this as a "reverse lookup." In some texts, z (z-score) is used to denote $\Phi^{-1}_{\text{standard}}$.

Discriminability is sometimes called *sensitivity*, but we avoid that term as it can also refer to the hit rate.

Equation (7.39) becomes more intuitive when we choose $k = k_{\text{MAP}}$. Then, we find, after some calculation, that

$$d' = \frac{\Delta s}{\sigma}. \tag{7.40}$$

This remarkably simple expression does not depend on the criterion (or on the prior)! No matter how much or how little the observer may be biased, sensitivity only reflects the distributions of the decision variable conditioned on s, that is, the sensory evidence. The more these distributions overlap, the lower d' is. As the ratio of the difference between the two stimuli to be discriminated and the level of sensory noise, d' can be interpreted as the observer's signal-to-noise ratio for the task.

Note that some text use equation (7.40) instead of equation (7.39) as the definition of d'. However, this would be much less general.

The introduction of performance measures that are invariant under changes in criterion was one of the main accomplishments of signal detection theory. However, an important caveat is that if the noise model is not equal-variance Gaussian, then equation (7.40) ceases to hold and d', as defined in equation (7.39), does become criterion-dependent.

Box 7.2
Discriminability or Accuracy?

There might seem to be a tension between discriminability, d', and accuracy. In signal detection theory, it is common to regard discriminability as a better measure of performance than accuracy, because it is independent of the criterion. By contrast, accuracy is maximized by the Bayesian MAP observer, so it would make sense to use accuracy as a measure of performance. This apparent tension is resolved by noting that the Bayesian MAP observer does not use just any criterion but rather the optimal one (the one that maximizes the posterior). Thus, accuracy is an entirely valid measure of performance. However, it might still be useful to split up accuracy into hit rate H and 1 minus false-alarm rate F. Discriminability is equivalent to accuracy (i.e., perfectly correlated with accuracy) when the distributions of the decision variable conditioned on s are Gaussian distributions with the same variance. In other cases, discriminability is of limited use in Bayesian models.

7.6 The Relation between Bayesian Inference and Signal Detection Theory

Signal detection theory has been widely applied in many domains, ranging from detecting objects on radar (for which the theory was originally developed), to characterizing the performance of clinical diagnostic tests, to studying recall of words from memory. Each of these situations involves the observer using noisy information (a radar image, physiological measurements taken from a patient, or a sense of familiarity) to classify a stimulus into one of two categories (presence or absence of an object, presence or absence of the disease, having seen the word before or not).

Signal detection theory modeling is in some sense a subset of Bayesian modeling, and in other ways a superset. It is a subset because signal detection theory has mostly concerned itself with binary discrimination or detection tasks. Signal detection theory modeling is a superset because, within that realm, it has not limited itself to the optimal decision rule. Bayesian models typically emphasize that the optimal estimate or decision rule depends on uncertainty, as was the case in equation (7.14). Studies in which observers are optimal even when optimality requires knowledge of sensory uncertainty teach us something about the representation of uncertainty. This goal is usually not present in signal detection theory modeling studies.

7.7 Extensions

Binary and continuous variables are two ends on a spectrum. A stimulus variable that is discrete but has a large number of possible values comes close to being continuous. An example would be choosing in which of eight directions a cloud of dots is moving. All probability distributions in the Bayesian model would be probability mass functions rather than probability density functions. In that sense, all Bayesian inference on a discrete stimulus variable is very similar to binary decisions. However, many of the concepts introduced in this chapter, such as LPR, decision rules, and ROCs, are not natural concepts when there are more than two alternatives.

The type of binary decisions considered in this chapter have been rather limited, namely only those where the class C uniquely specifies the stimulus (whose values we denoted s_+ and s_-). Much more general is the case where each class C determines a *distribution* over the stimulus. For example, in a typical orientation discrimination task, the subject is not asked to discriminate between 2° to the right and 2° to the left of the vertical but between any leftward tilted and any rightward tilted stimulus. To treat this case properly, we need to introduce the concept of marginalization, which we will do in the next chapter.

7.8 Summary

In this chapter we have introduced the Bayesian framework for binary decision making, along with signal detection theory. We have learned the following:

- Bayesian binary decision making is based on the LPR.
- In discrimination or detection, the Bayesian MAP rule can be defined in terms of a criterion applied to the measurement. This criterion depends on the prior ratio and on the sensory uncertainty level.
- Proportion correct does not distinguish between the two types of correct responses, hits and correct rejections.
- Discriminability, d', is defined as $d' = \Phi^{-1}_{\text{standard}}(H) - \Phi^{-1}_{\text{standard}}(F)$.
- Within signal detection theory, hit rate and correct rejection rate depend on discriminability and criterion or bias. The ROC is a curve obtained by varying the criterion.
- When the conditional distributions of the decision variable are Gaussian with equal variance, we have $d' = \frac{\Delta s}{\sigma}$.
- Signal detection theory models are both a subset and a superset of Bayesian models.

7.9 Suggested Readings

- George A. Gescheider. *Psychophysics: The Fundamentals*. New York: Psychology Press, 2013.
- David M. Green and John A. Swets. *Signal Detection Theory and Psychophysics*. Vol. 1. New York: Wiley, 1966.

- Michael J Hautus, Neil A Macmillan, and C. Douglas Creelman. *Detection Theory: A User's Guide.* (2021).
- W. Wesley Peterson, Theodore G. Birdsall, and William C. Fox. "The Theory of Signal Detectability." *Transactions of the IRE Professional Group on Information Theory* 4 (1954): 171–212.
- Frederick A. A. Kingdom and Nicolaas Prins. *Psychophysics: A Practical Introduction.* 2nd ed. London: Academic Press, 2016.
- Felix A. Wichmann and N. Jeremy Hill. "The Psychometric Function: I. Fitting, Sampling, and Goodness of Fit." *Perception and Psychophysics* 63, no. 8 (2001): 1293–1313.
- Thomas D. Wickens. *Elementary Signal Detection Theory*. Oxford: Oxford University Press, 2001.

7.10 Problems

Problem 7.1 In medicine, it is common to encounter the terms *sensitivity* and *specificity* for a diagnostic test for a disease; these are synonyms for the true-positive rate and the true-negative rate, respectively. In addition, the (objectively correct) prior probability of a disease is called its *prevalence*. The *positive predictive value* (PPV) is the probability that someone has the disease given that they test positive. Use Bayes' rule to show that

$$\text{PPV} = \frac{\text{sensitivity} \cdot \text{prevalence}}{\text{sensitivity} \cdot \text{prevalence} + (1 - \text{specificity}) \cdot (1 - \text{prevalence})}. \tag{7.41}$$

Problem 7.2 Suppose the prior distribution and the posterior distribution are as in figure 7.2.

(a) Calculate the likelihood ratio.

(b) Does the sensory evidence alone (without the prior) indicate that the stimulus was s_+ or s_-?

(c) Do the prior and the likelihood favor the same alternative?

Problem 7.3 In this problem, we explore the relationship between posterior probabilities and LPR numerically.

(a) Create a vector of ninety-nine possible posterior probabilities of s_+, from 0.01 to 0.99 in steps of 0.01. For each value, calculate the LPR $d = \log \frac{p(s=s_+|x)}{p(s=s_-|x)}$. Then plot this ratio as a function of the posterior probability of s_+. This should show that every posterior probability corresponds to exactly one LPR and the other way round (we are dealing with monotonic functions). Knowing one is as good as knowing the other.

(b) Why did we not include the posterior probabilities 0 and 1?

(c) Suppose you know the LPR d. Express the posterior probability of $s = s_+$, $p(s = s_+|x)$, as a function of d only. Do the same for $p(s = s_-|x)$.

(d) If the LPR is 0.1, what are the posterior probabilities of s_+ and s_-? What if the LPR is 1?

Problem 7.4 We formulated the decision rule as reporting one alternative $d > 0$ and the other when $d < 0$. Why does the case $d(x) = 0$ usually not have to be considered? What would the observer do when $d(x) = 0$?

Problem 7.5 Prove equation (7.10) for the log likelihood in the Gaussian measurement model.

Problem 7.6 Suppose the stimulus s can take two values: $s_+ = 1°$ and $s_- = -1°$. Suppose that the measurement is normally distributed around s with standard deviation $0.5°$. On a given trial, the observer's measurement is $-0.1°$, and $s = s_+$ occurs on 80 percent of trials. Would an optimal observer report that the stimulus was s_+ or s_-? Provide all the steps in your reasoning.

Problem 7.7 We want to choose our criterion k for a decision-making task such that the probability of being correct is maximized. Starting from equation (7.20), derive an expression for the criterion k.

Problem 7.8 Above we introduced two confidence measures confidence1 and confidence2. Prove equation (7.34) for the relation between the two confidence measures.

Problem 7.9 In the context of our discrimination task, assume a flat prior, so that the Bayesian decision variable d becomes the LLR. We can think of d as a random variable that "inherits" its distribution from the distribution of x. Derive that the conditional distributions of the decision variable, as depicted in figure 7.4, are

$$p(d|s = s_+) = \mathcal{N}\left(d; \frac{(\Delta s)^2}{2\sigma^2}, \frac{(\Delta s)^2}{\sigma^2}\right) \tag{7.42}$$

$$p(d|s = s_-) = \mathcal{N}\left(d; -\frac{(\Delta s)^2}{2\sigma^2}, \frac{(\Delta s)^2}{\sigma^2}\right). \tag{7.43}$$

Problem 7.10 This problem was designed by Ronald van den Berg when he was a postdoc with Wei Ji Ma. In the first row of the figure below, each plot shows the probability distributions (probability density functions) of the log posterior ratio under each of the two alternatives in a binary decision task. The second row displays ROC curves. Indicate for each ROC to which plot in the top row it belongs.

Problem 7.11 We will simulate the ROC in a detection problem. An observer is trying to detect a signal of strength $s_+ = 3$ in noise ($s_- = 0$). The noise has a normal distribution with standard deviation $\sigma = 2$. On each trial, an experimenter presents noise (probability 0.4), or noise plus signal (probability 0.6). The task of the observer is to respond whether the signal is present or absent.

(a) Simulate the stimulus (signal or noise) on each of the 100,000 trials. Save as a column vector.

(b) Simulate the measurement on each trial.

(c) Based on the measurements in part (b), calculate two measurement histograms: one for the trials when the signal was present and one for the trials when the signal was absent. Use as basis for your histograms a set of fifty bins, linearly spaced between -10 and 10. Normalize both histograms. Plot both in the same plot as lines (not as bars).

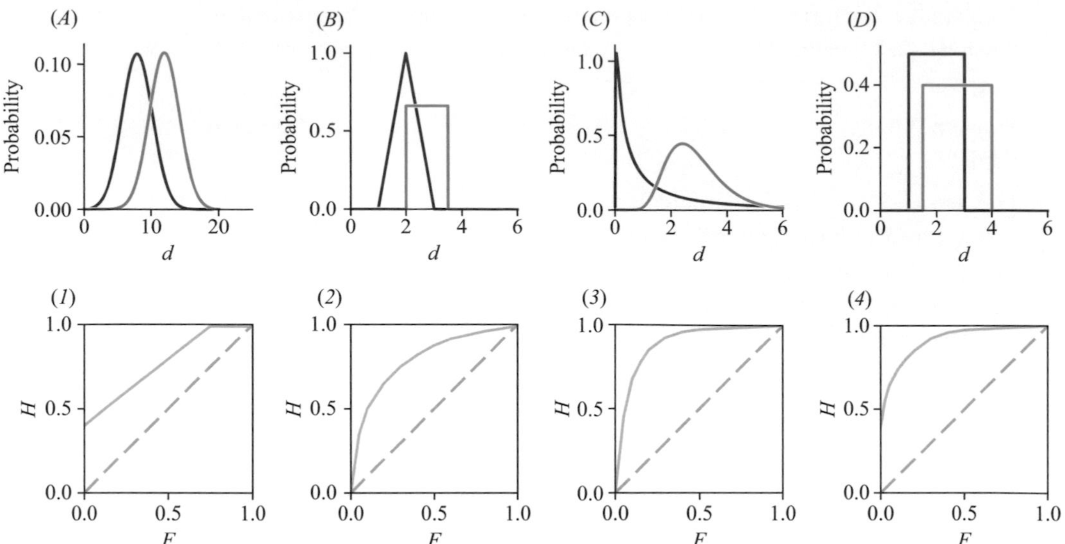

(d) Based on the measurements in part (b), calculate the LPR on each trial. Calculate and plot the histograms of the LPR analogous to the histograms of the measurement in part (c).

(e) Assume now that on each trial, the observer also provides a confidence rating by reporting "high confidence" when the absolute value of the LPR exceeds 2, "medium confidence" when it lies between 1 and 2, and "low confidence" when it is lies between 0 and 1. Create a 2×6 table of the two possible stimuli (signal present or absent) and the six possible responses. In each cell, put the frequency of the response (normalized by row).

(f) Calculate the empirical ROC by cumulatively summing the response frequencies.

(g) Plot the resulting points on top of the theoretical ROC based on equations (7.38).

(h) Simulate and describe what happens to the ROC when you reduce the signal strength to $s_+ = 2$?

(i) Interpret the change. Why does it make intuitive sense?

Problem 7.12 Here, we combine cue combination (chapter 5) with the discrimination task of the current chapter. A judge in a trial is trying to determine whether a suspect is guilty. The juror's prior is 0.5. The judge has three conditionally independent pieces of evidence. If they had had only one of these pieces (any one of them), the posterior probability that the suspect is guilty would have been 60 percent. Now that they have all three pieces, what is the posterior probability that the suspect is guilty? This problem uses only the conditional independence assumption equation (5.1) from chapter 5 not the Gaussian noise assumption equation (5.3). Please note that the use of Bayes' rule is not entirely unchallenged in the world of jurisdiction, but we would like you to answer this question as if this problem was already overcome.

Problem 7.13 Like the previous problem, this problem is about cue combination for a binary stimulus. However, we are now more specific about the measurement, and assume a Gaussian measurement model. Consider a stimulus s that takes two values, s_+ and s_-, with equal probability. The observer makes not one but two, conditionally independent measurements, x_1 and x_2, drawn from normal distributions with mean s and variance σ^2.

(a) Derive an expression for the log posterior ratio.

(b) What is the optimal decision rule in terms of the measurements?

(c) Repeat parts (a) and (b) for N instead of two conditionally independent measurements.

Problem 7.14 This is a mathematical problem that extends the formalism of this chapter to measurement noise with unequal (stimulus-dependent) variance. Consider our discrimination task with two possible stimulus values, s_+ and s_-. Assume equal probabilities (flat prior). In the chapter, we assumed that the distributions of the measurement, $p(x|s=s_+)$ and $p(x|s=s_-)$, were normal with equal variance. Now assume instead that their variances are different and have values σ_+^2 and σ_-^2, respectively.

(a) Show that the log posterior ratio is given by

$$d=\log\frac{\sigma_-}{\sigma_+}-\frac{1}{2}\left(\frac{(x-s_+)^2}{2\sigma_+^2}-\frac{(x-s_-)^2}{2\sigma_-^2}\right). \tag{7.44}$$

(b) Now assume $s_+=3$, $s_-=0$, $\sigma_+=3$, and $\sigma_-=1$. Plot the LPR as a function of x.

(c) Interpret the shape of this function. Compare and contrast with figure 7.3.

(d) Without assuming specific values, simplify the Bayesian decision rule $d>0$ to a set of inequalities for x. Why do you get two inequalities rather than one?

(e) Derive expressions for the hit and false-alarm rates in terms of the standard cumulative normal distribution Φ_{standard}.

(f) Numerically calculate the hit and false-alarm rates for the values in part (b).

Problem 7.15 We continue our extension to unequal variances but now treat the problem through simulations. Assume $s_+=3$, $s_-=0$, $\sigma_+=3$, and $\sigma_-=1$.

(a) Simulate the stimulus (signal or noise) on each of the 100,000 trials. Save as a column vector.

(b) Simulate the measurement on each trial.

(c) Based on the measurements in part (b), calculate two measurement histograms: one for the trials when the signal was present and one for the trials when the signal was absent. Use as basis for your histograms a set of fifty bins, linearly spaced between -10 and 10. Normalize both histograms. Plot both in the same plot as lines (not as bars).

(d) Based on the measurements in part (b) and the answer to problem 7.14a, calculate the LPR on each trial. Calculate and plot the histograms of the LPR analogous to the histograms of the measurement in part (c).

(e) Based on the measurements in part (b), calculate the hit and false-alarm rates. Compare to what you found in problem 7.14f.

8

Binary Classification

How do we determine to which of two categories a stimulus belongs?

In chapter 7, we introduced discrimination and detection tasks, tasks that require the observer to decide between two specific stimulus values, which we called s_+ and s_-. We concluded the chapter by noting that these tasks far from cover all binary decision tasks: in the real world as well as in the laboratory, the choice in a binary decision is often not between two specific stimulus values, but between two *categories* or *classes*, each of which comprises multiple, in some case infinitely many, stimulus values.

Plan of the Chapter

In the present chapter, our focus is on such *binary classification* tasks. In these tasks, the observer is asked to report not the stimulus, only the stimulus category. Yet, the stimulus value is still unknown to the observer. Such tasks are richer than discrimination and detection, because the dependence of the observer's behavior on the stimulus can be studied. In terms of the mathematical machinery, binary classification requires a Bayesian observer to integrate over all possible values of this unknown stimulus, an operation known as *marginalization*. Marginalization is a central operation in Bayesian models of virtually all tasks, except the very simplest ones. Thus, this chapter is the gateway to a large domain of applications of Bayesian models.

8.1 Example Tasks

Binary classification tasks are common in the world. Three examples:

- Deciding whether a distant car on a country road is moving toward or away from you. You are categorizing the velocity vector by its sign (positive or negative).
- Deciding whether the person approaching you is the friend you are waiting for or not. You are categorizing the image as "friend" or "other."
- Deciding whether a cloud is a rain cloud or not.

Binary classification tasks are also very common in perception experiments in the laboratory. Three examples:

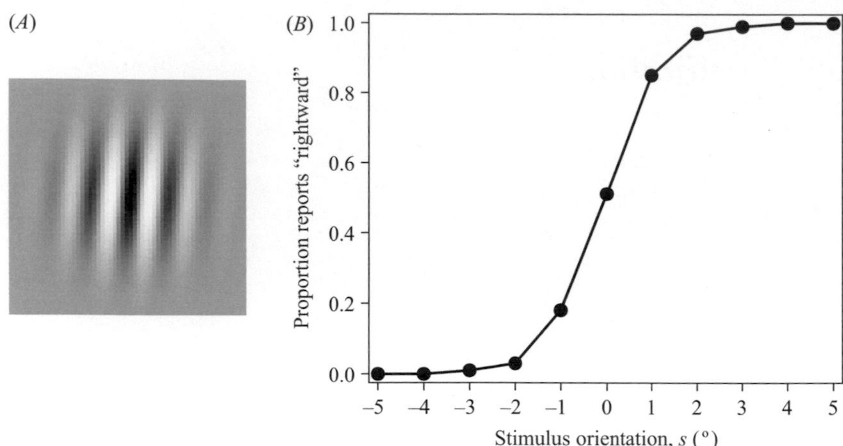

Figure 8.1
A binary classification experiment. (*A*) Example Gabor stimulus. (*B*) Psychometric curve.

- Deciding whether a noisy cloud of moving dots has a net motion direction to the right or left. You are categorizing the net motion vector by its sign (left or right).
- Deciding in a split second whether a natural scene contains an animal or not.
- Deciding if a stimulus that has both a size and an orientation belongs to category 1 or category 2, both of which are defined by the experimenter.

Although any of the examples above could be used to build a formal model, we will here use an extremely simple single-feature example, so that few additional assumptions are needed and we can focus on the essence of classification. Consider the following common visual task. You are briefly shown an oriented pattern like the one in figure 8.1A. The orientation of the pattern varies among many possible values, and your job is to report through a key press whether the pattern was tilted left (counterclockwise) or right (clockwise) of vertical. The data in this task consist of a stimulus and a category report on each trial. If the stimuli are discrete, for example from $-5°$ to $5°$ in steps of $1°$, then it is common to calculate for each presented stimulus the proportion of trials in which the subject reports one alternative, say "right." This proportion can be plotted as a function of the stimulus (figure 8.1B). The result is an example of a *psychometric curve*: a curve that has some summary of human behavior plotted against a physical quantity that is varied by the experimenter. It stands to reason that the more rightward the stimulus is, the greater will be the proportion of "rightward" responses. One outcome of the Bayesian modeling in this section will be a model for the psychometric curve. Let us first consider two caveats about the psychometric curve:

- If the stimulus can take a large number of values or is continuous, then to plot a visually useful psychometric curve, the stimulus values have to be binned or otherwise grouped together. While useful for visualization, such grouping loses information. Therefore, any quantitative data analysis is ideally based on the raw data.
- Even if the stimulus can take only a small number of values, as in figure 8.1B, then the psychometric curve does not capture the entirety of the raw data: what is missing is the number of trials on which each stimulus is presented. In many experiments, this

Table 8.1
Task types.

Task type	Number of distinct stimuli	Possible responses
Discrimination	2	2 stimulus identifiers
Detection	2 (absent and present)	2 (absent and present)
Identification	$n > 2$	n stimulus identifiers
Search (detection)	n	2 (absent and present)
Search (localization)	n	n locations
Classification / categorization	n	$< n$ categories

number is the same for all possible stimuli and for all subjects, so it only needs to be reported once.

Even though a task like this is often called a "discrimination" task, that terminology is inaccurate (table 8.1); it is not discrimination, but falls under the broader umbrella of *classification* or *categorization*. What is the difference? In discrimination, the number of values the stimulus can take and the number of possible responses are both two. In classification, the stimulus may take more than two values, but the number of possible responses (usually two) is smaller than the number of stimulus values. Thus, discrimination and the restricted form of detection in chapter 7 are special cases of classification.

For our classification experiment, we will follow the same recipe as in earlier chapters: generative model, inference, and distribution of estimates. However, the generative model will now have an interesting extra ingredient, namely *class-conditioned stimulus distributions*.

8.2 Generative Model

The generative model diagram is shown graphically in figure 8.4A. It has three nodes: class C, stimulus s, and measurement x. The observer is asked to report C, the world state of interest. As in earlier chapters, each node is associated with its own probability distribution. We will now discuss these one by one.

Class We designate the two possible values of C by 1 (in the example: rightward) and -1 (in the example: leftward). We could have chosen any two values here, but in the example of figure 8.1, C is naturally equal to the sign of s. Associated with C is a distribution $p(C)$, which is specified by two values, $p(C=1)$ and $p(C=-1)$, which have to sum to 1. These probabilities reflect the believed prevalence of rightward- and leftward-tilted stimuli in the experiment. In many experiments, a class is chosen randomly with probability 0.5, so that, if the observer knows this, $p(C=1)$ and $p(C=-1)$ equal 0.5; however, we will not restrict ourselves to that case.

Stimulus When the class C equals -1, the experimenter draws the stimulus randomly from one set of values; when $C=1$, the experimenter draws the stimulus randomly from the other set of values. We denote the corresponding stimulus distributions by $p(s|C=-1)$ and $p(s|C=1)$, respectively; these are *class-conditioned stimulus distributions* (CCSDs). Here, we assume that the observer's believed CCSDs are the same as the experimental CCSDs.

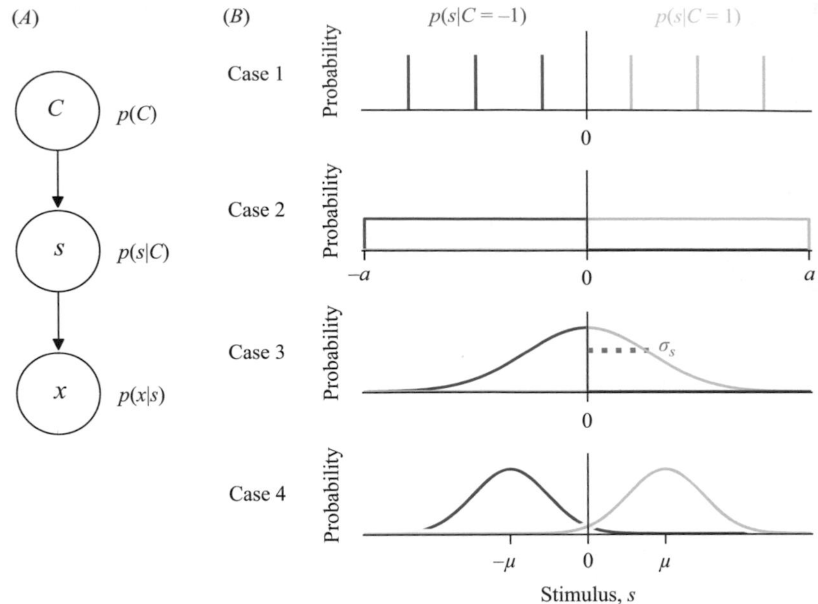

Figure 8.2
(*A*) Generative model diagram. (*B*) Examples of mirror-imaged class-conditioned stimulus distributions.

Measurement The final step of the generative model is the same as in previous chapters: we assume that the observer's measurement x is drawn from a normal distribution centered at the stimulus, with standard deviation σ:

$$p(x|s) = \frac{1}{\sqrt{2\pi\sigma^2}} e^{-\frac{(x-s)^2}{2\sigma^2}}. \tag{8.1}$$

8.2.1 Mirror-Symmetric Class-Conditioned Stimulus Distributions

In binary classification, importantly some CCSDs are pairs that are mirror images *of each other*. Four specific examples of this type of CCSD are (figure 8.2B):

- Case 1: The stimulus is discrete and chosen with equal probability from n possible values: for $C = 1$, they are $s_1, s_2, \ldots, s_n$ (all positive numbers), and for $C = -1$, they are mirror-symmetric: $-s_1, -s_2, \ldots, -s_n$ (all negative numbers). This procedure is called the *method of constant stimuli*.
- Case 2: The stimulus is continuous and drawn from a uniform distribution on an interval $[-a, 0]$ when $C = -1$, and from a uniform distribution on $[0, a]$ when $C = 1$, where a is a positive number.
- Case 3: The stimulus is continuous and drawn from a Gaussian distribution with mean 0, but then designated to be class -1 or class 1 based on its sign.
- Case 4: The stimulus is continuous and drawn from a Gaussian distribution (variance σ_s^2) with mean $-\mu$ when $C = -1$ and mean μ when $C = 1$.

8.3 Marginalization

So far, we have specified the distribution of the measurement conditioned on the stimulus, $p(x|s)$, and the distribution of the stimulus conditioned on the class, $p(s|C)$. We do not, however, have the distribution of the measurement conditioned on the class, $p(x|C)$, which is the distribution needed to do the inference. To obtain an expression for $p(x|C)$, we will first introduce an important general mathematical identity called *marginalization*. Marginalization is common in Bayesian models and inevitable in all but the simplest problems. The combination of Bayes' rule and marginalization powers virtually all of Bayesian modeling.

8.3.1 The Sum of Two Die Rolls

In probability theory, marginalization is the operation of turning a probability distribution over multiple variables into a distribution over one of them. For example, if a and b are discrete random variables, and $p(a, b)$ is their joint distribution, then summing the joint distribution over b produces the distribution over a:

$$p(a) = \sum_b p(a, b). \tag{8.2}$$

By using the definition of conditional probability, we can also write this as

$$p(a) = \sum_b p(a|b)p(b). \tag{8.3}$$

As an example, suppose we roll two fair dice, one at a time. The game we are playing rewards us if the total score from the two rolls is 10. What is the probability that this will occur? To find out, we can consider the probability of every value resulting from the first roll, and the probability of a total of 10 given that first value:

$$p(\text{total} = 10) = \sum_{i=1}^{6} p(\text{total} = 10|\text{first roll} = i)p(\text{first roll} = i). \tag{8.4}$$

We are *marginalizing* over the value of the first roll. To express the marginalization formula in words, we replace each product with "and" and each addition with "or." We are stating that the probability of a total of 10 is the probability that the first die lands 1 AND the total will be 10 given that the first lands 1, OR that the first lands 2 AND the total will be 10 given that the first lands 2, and so on. To compute the marginalization sum, we note that if the first die lands 1, 2, or 3, it is impossible for the total of the two dice to reach 10; if the first die lands 4, 5, or 6, then the second die would need to land 6, 5, or 4, respectively, and each of these events occurs with probability $\frac{1}{6}$. Thus, we have:

$$p(\text{total} = 10) = 0 \cdot \frac{1}{6} + 0 \cdot \frac{1}{6} + 0 \cdot \frac{1}{6} + \frac{1}{6} \cdot \frac{1}{6} + \frac{1}{6} \cdot \frac{1}{6} + \frac{1}{6} \cdot \frac{1}{6} = \frac{3}{36} = \frac{1}{12}. \tag{8.5}$$

Note that we care only about the probability of the total value, but in order to calculate this, we must nevertheless consider all possible values of the first roll. Because we must take the first roll, which we do not really care about, into consideration, the value of the first roll is called a nuisance variable.

For a second example, we use a variant of the example introduced in section 4.5.2. Suppose we want to know the probability that a randomly selected citizen of a particular country is a farmer. The country has twelve geographic regions. Suppose we find an almanac that reports the proportion of the population that lives within each region and also the proportion of farmers within each region. To obtain the desired probability, we multiply those two proportions for every region and then sum over all regions. Here, the region of residence is the nuisance variable:

$$p(\text{farmer}) = \sum_{i=1}^{12} p(\text{farmer}|\text{region}_i)p(\text{region}_i). \tag{8.6}$$

We are stating that the probability of randomly selecting a farmer is the probability that we will randomly select a person from region 1 AND that a randomly selected person from region 1 is a farmer, OR that we will randomly select a person from region 2 AND that a randomly selected person from region 2 is a farmer, and so on. Equivalently, we can conceptualize our procedure as first randomly selecting a region, with a probability proportional to the population of the region, and then randomly selecting a person from within that region.

Box 8.1
Etymology

Where does the name "marginalization" come from? In its most basic form, marginalization is captured by equation (8.2), $p(a) = \Sigma_b p(a, b)$. Thus, we can think of marginalization as the sum over one dimension—the dimension we are not interested in estimating—of a joint probability distribution. When the sum is repeated for each value of the relevant dimension (e.g., not just for dice totals of 10, but for all totals; not just for farmers, but for all occupations), then marginalization reduces the joint distribution to a distribution over just the dimension of interest. Represented graphically, the summation occurs toward the "margin" of the joint distribution, giving marginalization its name (figure 8.3).

8.3.2 Continuous Variables

In many cases we have to deal with latent variables that we need to marginalize that are continuous. If b is a continuous variable, marginalization consists of an integral:

$$p(a) = \int p(a, b)db \tag{8.7}$$

$$= \int p(a|b)p(b)db. \tag{8.8}$$

In such cases, everything about marginalization stays the same, the only difference is that sums are replaced with integrals.

One example of such a continuous marginalization is if we want to calculate the probability distribution of the sum of two continuous variables, a and b. This can be thought of as a continuous analog of the "sum of two dice" example. We assume that both a and b have

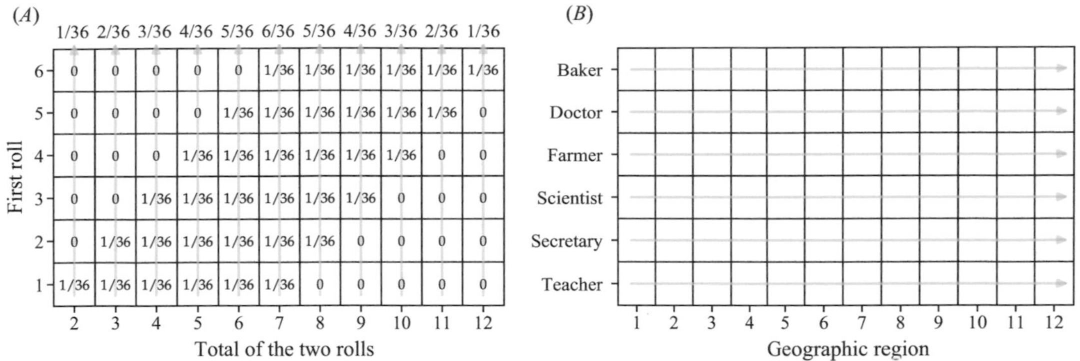

Figure 8.3
Marginalization. Each panel shows a joint probability distribution over two variables. The brown lines represent marginalization over the nuisance variable, a procedure that reduces the two-dimensional distribution to a one-dimensional distribution over the variable of interest. (*A*) Dice example. The value in each square is the probability of that particular (first roll value, total value) pair. Marginalization over the first roll value results in a probability distribution over the total value (top numbers). (*B*) Farmer example. Values within each square (not shown) would represent the proportion of citizens characterized by the corresponding (occupation, geographic region) pair. Marginalization over region results in a probability distribution over occupation. (Only a small subset of occupations is shown.)

Gaussian distributions:

$$p(a) = \mathcal{N}(a; \mu_a, \sigma_a^2); \tag{8.9}$$

$$p(b) = \mathcal{N}(b; \mu_b, \sigma_b^2). \tag{8.10}$$

Denoting the sum variable by c, we have

$$p(c) = \mathcal{N}(c; \mu_a + \mu_b, \sigma_a^2 + \sigma_a^2). \tag{8.11}$$

Exercise 8.1 Show this, either by solving the integral or by referring to an equation derived earlier.

8.3.3 Conditioned Marginalization

The basic form of marginalization, equations (8.2) and (8.3), still holds if all probabilities are already conditioned on other variables, for example, c:

$$p(a|c) = \sum_b p(a, b|c) = \sum_b p(a|b, c)p(b|c). \tag{8.12}$$

For example, suppose that we want to know the probability that a randomly selected middle-aged citizen (e.g., a person between forty-five and sixty-five years old) from the country mentioned above is a farmer. If we define $C = 1$ to be young adult, $C = 2$ to be middle-aged, and $C = 3$ to be older, then conditioning on $C = 2$, we could calculate:

$$p(\text{farmer}|C = 2) = \sum_{i=1}^{12} p(\text{farmer}|\text{region}_i, C = 2)p(\text{region}_i|C = 2). \tag{8.13}$$

The probability that a randomly selected middle-aged person is a farmer is the probability that a randomly selected middle-aged person is from region 1 AND that a randomly selected middle-aged person from region 1 is a farmer, OR that a randomly selected middle-aged person is from region 2 AND that a randomly selected middle-aged person from region 2 is a farmer, and so on.

As another example of conditioned marginalization, we consider the transmission of a particular virus. Suppose we are interested in the probability $p(x=1|C=1)$ that an unvaccinated person exposed to an infected unvaccinated person (we denote this exposure by $C=1$) themselves gets infected by a virus ($x=1$). A complication is that the virus comes in many variants, which we will denote by s_i, with $i=1,\ldots,n$ (here, n is the number of variants); the variable s_i is the nuisance variable in the problem. Each variant has its own transmission rate for unvaccinated people, $p(x=1|s_i,C=1)$, and its own prevalence among infected unvaccinated people, $p(s_i|C=1)$. To obtain the answer to our question, we first multiply those two proportions for every variant, yielding $p(x=1|s_i,C=1)p(s_i|C=1)$ for every i. This is interpreted as the probability that an infected unvaccinated person has variant s_i *and* transmits it. Finally, we sum over all variants. This calculation implements the equation

$$p(x=1|C=1)=\sum_{i=1}^{n}p(x=1|s_i,C=1)p(s_i|C=1). \tag{8.14}$$

Exercise 8.2 This book was finalized in its current version during the COVID-19 pandemic. Discuss how this phrasing of the problem would apply to pandemics and which of the modeling assumptions are justified. Specifically, what can you say about the interactions between people?

8.3.4 Using the Generative Model

Let us now return to the case study of our chapter. Recall that the generative model (figure 8.2A) specifies the distribution of the stimulus conditioned on the class, $p(s|C)$, which may or may not be mirror-symmetric, and the distribution of the measurement conditioned on the stimulus, $p(x|s)$. Despite its central role in the generative model, the stimulus s is a *nuisance variable*; it is neither the observation (measurement, x) nor the variable of primary interest (C). In order to make an inference, we must calculate the distribution of the measurement conditioned on the class, $p(x|C)$, which we obtain by marginalizing over s:

$$\text{when } s \text{ is discrete: } p(x|C)=\sum_{i=1}^{n}p(x|s_i,C)p(s_i|C); \tag{8.15}$$

$$\text{when } s \text{ is continuous: } p(x|C)=\int p(x|s,C)p(s|C)ds. \tag{8.16}$$

See appendix section B.11.3 for the derivation of these rules.

Equations (8.15)–(8.16) are completely general; they are mathematical identities that hold regardless of context. We now incorporate the structure of the generative model. In the generative model, the distribution of x depends only on s and not directly on C. In figure 8.2, this is graphically understood by the fact that the only arrow pointing to x comes from s; there is no arrow from C to x. In other words, when s is known, knowledge of C is redundant when one is interested in the distribution of x. Mathematically, this is expressed by the conditional distribution $p(x|s,C)$ being identical to $p(x|s)$. Substituting this into equations (8.15)–(8.16), we arrive at the following expressions for the class likelihood:

$$\text{when } s \text{ is discrete: } p(x|C) = \sum_{i=1}^{n} p(x|s_i)p(s_i|C); \tag{8.17}$$

$$\text{when } s \text{ is continuous: } p(x|C) = \int p(x|s)p(s|C)ds. \tag{8.18}$$

These equations act as a kind of chain rule to link the class, C, to the measurement, x, by way of the intermediate variable, the stimulus, s.

8.4 Inference

On a given trial, the observer makes a measurement x. Since the observer is interested in class, C, the posterior distribution we want to calculate is now $p(C|x)$, not $p(s|x)$. This is the first time in the book that the stimulus, s, does not appear in the posterior: the stimulus is not directly of interest, only class is. Nevertheless, the logic of inference is exactly the same as in previous chapters. Just as in chapter 7, the Bayesian observer decides based on the log posterior ratio, but now over class:

$$d \equiv \log \frac{p(C=1|x)}{p(C=-1|x)} \tag{8.19}$$

$$= \log \frac{p(C=1)}{p(C=-1)} + \log \frac{p(x|C=1)}{p(x|C=-1)}. \tag{8.20}$$

The likelihood of class 1, $p(x|C=1)$, and the likelihood of class -1, $p(x|C=-1)$, can both be obtained from the corresponding distributions in the generative model. Because s is continuous, these distributions are given by the marginalization equation (8.18). The resulting likelihoods are:

$$\mathcal{L}(C=1;x) \equiv p(x|C=1) = \int p(x|s)p(s|C=1)ds; \tag{8.21}$$

$$\mathcal{L}(C=-1;x) \equiv p(x|C=-1) = \int p(x|s)p(s|C=-1)ds. \tag{8.22}$$

From an inference perspective, these equations can be interpreted in terms of the "propagation" of uncertainty information: $p_{x|s}(x|s)$ as a function of s is the likelihood over the stimulus and represents sensory uncertainty. By contrast, $p_{x|C}(x|C)$ as a function of C is the likelihood over the class, which represents class uncertainty:

$$\underbrace{\mathcal{L}(C;x)}_{\text{likelihood of class } C} \equiv p(x|C) = \int \underbrace{p(x|s)}_{\substack{\text{likelihood of } s,\\ \mathcal{L}(s;x)}} \underbrace{p(s|C)}_{\substack{\text{CCSD}\\ \text{(task-dependent)}}} ds \tag{8.23}$$

(and similarly for discrete s). Thus, the uncertainty about the "lower-level" variable of s gets transformed into or propagated to uncertainty about the "higher-level" variable of class. This transformation is mediated by the learned, "top-down" knowledge of the CCSDs. Unlike

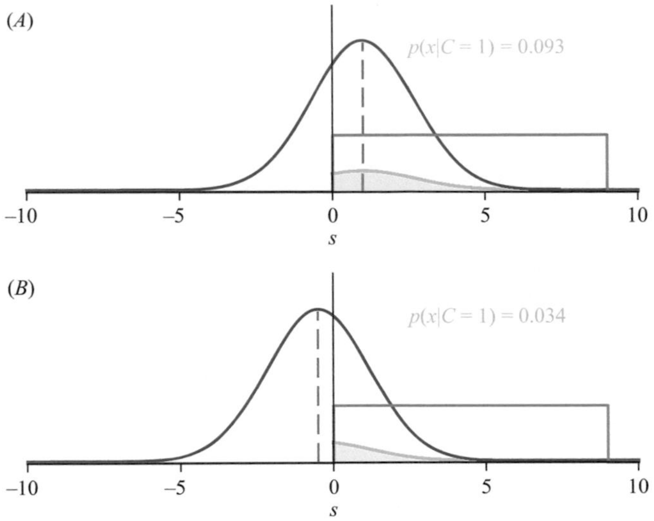

Figure 8.4
Graphical explanation of the calculation of the likelihood of a class.

the likelihoods, the CCSDs depend only on the task (to be precise, on the observer's beliefs about the task) and do not change from trial to trial.

Graphically, the likelihood of class 1 is the "overlap" between the likelihood over s and the CCSD for class 1: first multiply, then take the area (figure 8.4). If the resulting area is small, then it means that there was not much overlap.

Substituting equations (8.17)–(8.18) back into equation (8.20), we find for the LPR:

$$\text{when } s \text{ is discrete: } d = \log \frac{p(C=1)}{p(C=-1)} + \log \frac{\sum_{i=1}^{n} p_{x|s}(x|s_i)p(s_i|C=1)}{\sum_{i=1}^{n} p_{x|s}(x|s_i)p(s_i|C=-1))}; \tag{8.24}$$

$$\text{when } s \text{ is continuous: } d = \log \frac{p(C=1)}{p(C=-1)} + \log \frac{\int p_{x|s}(x|s)p(s|C=1)ds}{\int p_{x|s}(x|s)p(s|C=-1)ds}. \tag{8.25}$$

The optimal decision rule is

$$\text{"report } \hat{C} = 1 \text{ if } d > 0." \tag{8.26}$$

Like in chapter 7, the prior over C biases the observer's decision, and its effect is stronger when the sensory evidence, as expressed by the log likelihood ratio, is weaker.

Outside of a few special cases, it is impossible to obtain an analytical decision rule out of equation (8.26). The best strategy is typically to numerically restate the inequality $d > 0$ in terms of x.

Flat prior over class Analytical progress *can* be made in an important special case, namely that the observer uses a flat prior over class, $p(C=1) = p(C=-1) = 0.5$, and the CCSDs are mirror-symmetric. The prior conditions means that the observer—correctly or incorrectly—believes that the classes are equally frequent (see the discussion of suboptimal Bayesian observers in section 3.5). Since the prior is flat, MAP estimation is equivalent to reporting $C = 1$ when the likelihood $p(x|C=1)$ exceeds the likelihood $p(x|C=-1)$. This makes the problem completely symmetric, and the only sensible candidate for the optimal decision rule is $x > 0$. Proving this, however, is not easy and we leave it to problem 8.7.

We can, however, draw a lesson from this special case. The decision rule $x > 0$ does not depend on the CCSDs $p(s|C)$, even though the general rule, $d > 0$, does (see equation (8.25)). That means that when the prior is flat, the observer might have a completely wrong belief about the shape of the CCSDs, yet make the optimal decision, simply because the wrong belief is irrelevant to the decision rule. Thus, wrong beliefs about the generative model do not always cause suboptimal behavior. (Keep in mind that we are still assuming $p(C=1)=0.5$ and the CCSDs are mirror-symmetric. If either condition is violated, then the observer's beliefs about the CCSDs *do* matter both for the decision rule and for performance.) In short, in Bayesian modeling, an observer's wrong assumptions about the generative model do not always affect the decision rule.

Box 8.2
Problem with the Method of Constant Stimuli

While case 2 (discrete) is probably the most common CCSD in binary classification experiments, it is not ideal from the point of view of Bayesian modeling. The reason is that it is unlikely that the subjects learns exactly the experimental distribution, because it is so "spiky" and the locations of the "spikes" would have to be learned. Thus, the observer would most likely approximate the distribution by a continuous distribution, but it is not clear which one. This does not matter if the observer has a class prior of 0.5, but it will matter if you consider the possibility that the class prior is different from 0.5. It might still not matter very much, but you do not want to blindly count on that.

8.5 Response Distribution

When the stimulus is s, the response distribution is given by the probability of reporting either $\hat{C}=1$ or $\hat{C}=-1$ given s. We evaluate the former probability:

$$p(\hat{C}=1|s)=p(d>0|s) \tag{8.27}$$

$$=p(x>k|s), \tag{8.28}$$

where k is the criterion that we can numerically compute in step 2. Equation (8.28) is the probability that a measurement x drawn from $p(x|s)$ will be greater than k. We can further evaluate this probability using equation (8.1):

$$p(\hat{C}=1|s)=\int_k^\infty \mathcal{N}(x;s,\sigma^2)dx \tag{8.29}$$

$$=\Phi_{\text{standard}}\left(\frac{s-k}{\sigma}\right), \tag{8.30}$$

where we used the same steps that gave us equation (7.20), and Φ_{standard} is the cumulative standard normal distribution (see box 7.1). Equation (8.30) has sensible properties. First, it is a monotonic function: when s increases (e.g., the orientation becomes more rightward tilted), then the probability of reporting $C=1$ increases as well. The curve has the characteristic sigmoid shape seen in figure 8.1B). Second, when there is more noise or the stimulus is closer to the criterion, the quantity $\frac{s-k}{\sigma}$ is smaller in absolute value (closer to 0), and the probability of reporting class 1 will be closer to 0.5; this makes sense because in both scenarios the task will be harder.

Psychometric curve Now we are finally ready to plot the psychometric curve predicted by the Bayesian model. It is given by equation (8.30) as a function of s. The predicted psychometric curve is a cumulative normal distribution that crosses 0.5 when $s=k$. In psychophysics parlance, k is the *point of subjective equality* (PSE): the value of the stimulus for which the subject (in this case, the Bayesian observer) reports the two classes equally often. The *slope* of the psychometric curve is usually defined as the inverse of the standard deviation of the cumulative normal, that is, $\frac{1}{\sigma}$. Since in general, k depends on sensory noise level σ, the parameter(s) of the CCSDs, and the log prior ratio over class, the final psychometric curve also depends on all of those parameters.

Proportion correct In chapter 7, we computed hit rate H and false-alarm rate F in a detection task. Knowing H and F is equivalent to knowing H and the correct rejection rate, $1-F$. Here, we can similarly compute the probability of a correct report of class 1, denoted by PC_1, and the probability of a correct report of class -1, denoted by PC_{-1}. These probabilities are similar to the one in equation (8.30) for the response distribution, except that the conditioning is on a (true) class C instead of on the stimulus s. Let us examine PC_1:

$$\text{PC}_1 = p(\hat{C}=1|C=1) \tag{8.31}$$

$$= p(d>k|C=1) \tag{8.32}$$

$$= \frac{1}{n_{\text{trials},1}} \sum_{\substack{\text{trials } t: \\ C_t=1}} p(d>k|s_t), \tag{8.33}$$

where the sum is over all experimentally presented trials, labeled t, for which C_t was 1, $n_{\text{trials},1}$ is their number, and s_t is the stimulus on the tth trial. The sum is almost always calculated numerically. The expression for PC_{-1} is analogous.

Overall proportion correct according to the model is a weighted sum of PC_1 and PC_{-1}:

$$\text{PC} = p(C=1)\text{PC}_1 + p(C=-1)\text{PC}_{-1}, \tag{8.34}$$

where $p(C=1)$ and $p(C=-1)$ represent the true frequencies of $C=1$ and $C=-1$ trials.

Predictions for proportion correct, whether overall or split up by true C, are an impoverished description of predicted behavior. They involve a grand average over stimuli, whereas the psychometric curve predicts the probability of reporting class 1 for each value of the stimulus. More generally, when evaluating a model (after fitting its parameters) through its predictions for summary statistics of behavior, the summary statistics that are chosen are typically the result of a trade-off between ease of visualization (or the ease of numerical reporting) and the granularity of the statistic, with more granular meaning more informative. In this trade-off, trial-level predictions are usually hard to visualize but most informative, whereas a proportion correct is easiest to report but least informative. It is good practice to report the match between data and model using summary statistics of different levels of granularity.

8.6 Beyond Mirror-Image Class-Conditioned Stimulus Distributions

Although in many experiments that use binary classification, the CCSDs are mirror images of each other, this is unnecessarily restrictive. A world of new possibilities opens up if we consider other setups. For example, the friend and animal examples in section 8.1 do not feature symmetric classes. In both cases, one class is a restricted set of stimuli (images of friend/animal), whereas the other class is a broad, encompassing class (images of random people, or images not containing animals). To illustrate such cases, we simplify them to their essence: a narrow class "embedded in" a wide class. While images are complex stimuli, we can define such classes even for single-feature (one-dimensional) stimuli, and indeed experimenters have done so. The math does not fundamentally change from the previous section.

Step 1: Generative model We denote the classes by $C=1$ and $C=2$; since their CCSDs are not mirror images of each other, the notation $C=-1$ makes less sense for the second category. Class 1 stimuli are drawn from a Gaussian distribution with mean 0 and variance σ_1^2. Class 2 stimuli are drawn from a Gaussian distribution with mean 0 and variance σ_2^2, which is greater than σ_1^2. The distributions are illustrated in figure 8.4A, with example stimuli in panel *B*.

When the CCSDs overlap (regardless of whether they are mirror images of each other), the observer cannot reach perfect performance even in the absence of sensory noise. This situation produces ambiguity: the same stimulus could have come from more than one class, though usually not with the same probability. As we have noted previously (section 1.3), ambiguity is common in perception, one example being that when viewing a scene with one eye, the retinal image could have been produced by many three-dimensional scenes. Here, the CCSD is the three-dimensional scene-conditioned distribution of the retinal image.

Step 2: Inference The observer infers class C based on a measurement x. As in equations (8.17)–(8.18), the likelihood of class C is given by an integral over s:

$$\mathcal{L}(C;x)=p(x|C)=\int p(x|s)p(s|C)ds. \tag{8.35}$$

This integral has a closed-form solution, which we already encountered in equation (8.11):

$$\mathcal{L}(C;x)=\mathcal{N}(x;0,\sigma^2+\sigma_C^2), \tag{8.36}$$

where σ_C is either σ_1 or σ_2. The intuition is that the measurement x results from two independent noise processes: one external noise process with variance σ_C^2, and one internal noise process with variance σ^2. Per box 4.1, the overall variance is the sum of these variances.

If the prior were flat, the MAP decision would be the maximum-likelihood decision. The maximum-likelihood decision rule, in turn, one would be able to graphically deduce from a plot of the class-conditioned measurement distributions (CCMDs) $p(x|C)$ at a given noise

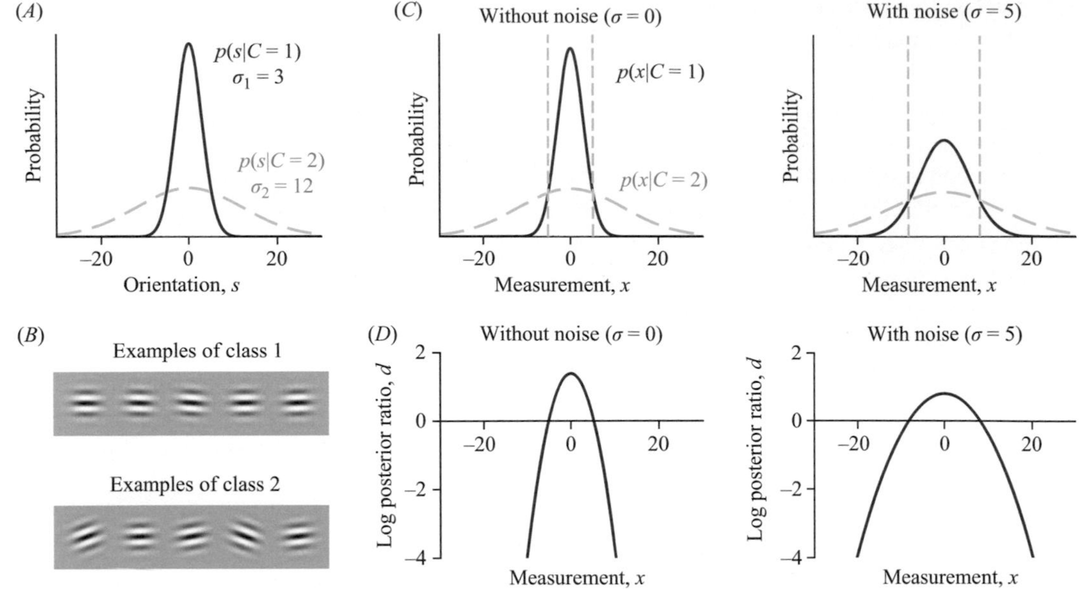

Figure 8.5
Embedded class task. (*A*) CCSDs over orientation (in degrees). The distributions have the same mean, but different standard deviations and are therefore not mirror images of each other. (*B*) Representative examples of stimuli in each class. (*C*) Class-conditioned *measurement* distributions at two different noise levels. The higher the noise level, the wider the CCMDs. The vertical dashed lines indicate the intersection points of the CCMDs of $C=1$ and $C=2$. (*D*) Corresponding LPR over C as a function of the measurement, assuming equal priors. When noise is higher, the evidence in favor of $C=1$ falls off more slowly. The horizontal dashed line is at $d=0$, the value that corresponds to the intersection points in (*C*). The higher the noise, the larger the region for which $d>0$ and the observer reports "$C=1$."

level σ (figure 8.5C): for any x, the maximum-likelihood decision is to pick the class for which the CCMD has a higher value. Thus, the decision switches at the intersection points of the two CCMDs.

If the prior is not flat, this is no longer the case. For a general prior, the LPR takes the form (see problem 8.8):

$$d=\log\frac{p(C=1)}{p(C=2)}+\frac{1}{2}\log\frac{\sigma^2+\sigma_2^2}{\sigma^2+\sigma_1^2}-\frac{x^2}{2}\left(\frac{1}{\sigma^2+\sigma_1^2}-\frac{1}{\sigma^2+\sigma_2^2}\right). \tag{8.37}$$

We plotted this expression in figure 8.5D.

This is the first time we have encountered an LPR that is *quadratic* in the measurement; previous LPRs were all linear in the measurement. This makes perfect sense, however. It is clear from figure 8.5D that the probability of x is higher under class 1 than under class 2 when x falls in a narrow region around 0. To make this more precise: the MAP observer reports class 1 when $d>0$. This inequality can be rewritten as one for the measurement, x. First, we observe that if $\log\frac{p(C=1)}{p(C=2)}+\frac{1}{2}\log\frac{\sigma^2+\sigma_2^2}{\sigma^2+\sigma_1^2}<0$, then d is negative and the observer reports class 2 regardless of the value of x.

Exercise 8.3 Why?

Second, if that condition is not satisfied, then the MAP decision rule becomes

$$|x| < \sqrt{\frac{2\log\frac{p(C=1)}{p(C=2)} + \log\frac{\sigma^2+\sigma_2^2}{\sigma^2+\sigma_1^2}}{\frac{1}{\sigma^2+\sigma_1^2} - \frac{1}{\sigma^2+\sigma_2^2}}}. \tag{8.38}$$

In other words, there are two decision criteria on the measurement, one on each side of 0. The observer reports class 1 when the measurement falls between those criteria. As in section 8.4, criteria depend on the sensory uncertainty (measurement noise) level, the CCSD parameters σ_1 and σ_2, and the log prior ratio. Note that the prior over C plays a relatively unimportant role. Again, we see that Bayesian inference is not all about priors.

Step 3: Response distribution. It is possible to express the probability that a Bayesian observer reports class 1 in terms of cumulative normal functions. We will do this in problem 8.5.

8.7 "Following the Arrows"

As we will be exploring generative models of increasing complexity, it is helpful to have a clear and straightforward recipe for deriving an expression for a posterior of our choice based on the information provided by the generative model. Recall that the generative model specifies exactly which distributions are given in the problem. Each variable that does not have any arrows pointing to it follows a regular probability distribution. Each variable that does have arrows pointing to it follows a conditional distribution, where the conditioning is on the variable(s) from which the arrows originate; its distribution does not depend on any other variables in the problem. This gives us the recipe we are looking for:

1. Evaluate the joint distribution over all variables in the generative model by "following the arrows." Start at the top with the variables that have no arrows pointing to them. Working your way down, write down the prior or conditional probability distribution associated with each node, and multiply all distributions obtained this way.
2. Compute any conditional distribution by writing out its definition and marginalizing the joint distribution (i.e., summing or integrating over the variables not in the conditional).

When a posterior over a state-of-the-world variable is computed, a marginalization is done over every variable in the generative model other than the observations and the state-of-the-world variable. Note that in this recipe, the joint distribution is central, not the likelihood or the prior. In fact, Bayes' rule, which expresses the joint in terms of the likelihood and the prior, is simply the first step in the recipe of evaluating the joint distribution!

In the present chapter, the joint distribution is $p(C,s,x)$. Following the arrows in figure 8.4A, we find $p(C,s,x) = p(C)p(s|C)p(x|s)$. The posterior we are interested in is $p(C|x)$, which we obtain from marginalization. For the continuous case:

$$p(C|x) \propto p(C,x) \tag{8.39}$$

$$= \int p(C,s,x)ds \quad (8.40)$$

$$= \int p(C)p(s|C)p(x|s)ds \quad (8.41)$$

$$= p(C) \int p(s|C)p(x|s)ds. \quad (8.42)$$

If the generative model looks like a "string" of variables, each of which receives an arrow only from the previous one, such as here $C \rightarrow s \rightarrow x$, it is called a *Markov chain*. We will encounter other Markov chains in chapter 12. As in section 3.3.3, the proportionality sign in the first line is saying, "we will calculate the *unnormalized* posterior (what we call the pro-toposterior); to get the posterior, normalize at the end." When the MAP estimate is the only quantity of interest, normalization is not even necessary.

8.8 Summary

In this chapter, we have analyzed how binary classification can be phrased in terms of Bayesian statistics. We learned:

- Psychometric curves characterize the dependency of the choice on the stimulus.
- In classification, the stimulus may take more than two values, but the number of possible responses (usually two) is smaller than the number of stimulus values.
- Even for flat priors, Bayesian classification can be complicated because of the need to marginalize over the stimulus.
- Class-conditioned stimulus distributions are often chosen to be mirror-symmetric, but breaking this convention achieves not only greater realism, but also makes it easier to study uncertainty-dependent decision rules.
- Ultimately, Bayesian models are about representing information in probability distributions—but these do not have to be priors.

8.9 Suggested Readings

- Rachel N. Denison, William T. Adler, Marisa Carrasco, and Wei Ji Ma. "Humans Incorporate Attention-Dependent Uncertainty into Perceptual Decisions and Confidence." *Proceedings of the National Academy of Sciences* 115, no. 43 (2018): 11090–11095.
- Thomas L. Griffiths and Joshua B. Tenenbaum. "Theory-Based Causal Induction." *Psychological Review* 116, no. 4 (2009): 661–716.
- Daniel Kersten, Pascal Mamassian, and Alan Yuille. "Object Perception as Bayesian Inference." *Annual Review of Psychology* 55 (2004): 271–304.
- Zili Liu, David C. Knill, and Daniel Kersten. "Object Classification for Human and Ideal Observers." *Vision Research* 35, no. 4 (1995): 549–568.

- Gregory L. Murphy, Stephanie Y. Chen, and Brian H. Ross. "Reasoning with Uncertain Categories." *Thinking and Reasoning* 18, no. 1 (2012): 81–117
- Ahmad T. Qamar, R. James Cotton, Ryan G. George, and Wei Ji Ma. "Trial-to-Trial, Uncertainty-Based Adjustment of Decision Boundaries in Visual Categorization." *Proceedings of the National Academy of Sciences* 110, no. 50 (2013): 20332–20337.

8.10 Problems

Problem 8.1 What would the psychometric curve of a noiseless observer look like?

Problem 8.2 In the context of section 8.5, an observer's *75 percent threshold* can be defined as the value of the stimulus for which the observer (who has observation noise of σ) reports "right" 75 percent of trials minus the value of the stimulus for which the observer reports "right" 50 percent of trials. How many standard deviations σ does the 75 percent-threshold correspond to? As a reminder:

$$p(\hat{C}=1|s)=\Phi_{\text{standard}}\left(\frac{s-k}{\sigma}\right).$$

Problem 8.3 Consider binary classification with a general class prior $p(C)$ and a Gaussian measurement distribution. Please refer to the CCSDs in section 8.2.

(a) Derive the decision rule of the Bayesian observer in case 2 (CCSDs uniform on an interval). The answer will involve more than one cumulative standard normal distribution (Φ_{standard}), and it is not a pretty rule.

(b) Repeat for case 4 (CCSDs are equivariant Gaussians).

Problem 8.4 Consider case 3 in section 8.2, where the stimulus is continuous and drawn from a Gaussian distribution with mean 0, where the class is -1 if the stimulus is below zero and 1 otherwise.

(a) Show that the log likelihood ratio (LLR) is equal to

$$\text{LLR}=\log\Phi_{\text{standard}}\left(\frac{x\sigma_s}{\sigma\sqrt{\sigma^2+\sigma_s^2}}\right)-\log\left(1-\Phi_{\text{standard}}\left(\frac{x\sigma_s}{\sigma\sqrt{\sigma^2+\sigma_s^2}}\right)\right). \quad (8.43)$$

(b) Set $\sigma=1$. Plot the LLR as a function of x in two cases: $\sigma_s=1$ and $\sigma_s=10$ (two curves in one plot). You should see that σ_s has a big effect.

(c) Set $\sigma=1$ and log prior ratio to 0.2. Numerically solve $d=0$ for x in two cases: $\sigma_s=1$ and $\sigma_s=10$. You may do this by choosing a fine grid for x, then finding the zero-crossing of d using interpolation.

(d) Use the two numbers found in (d) to plot the psychometric curves of the optimal observer for $\sigma_s=1$ and $\sigma_s=10$ (two curves in one plot). You should see that σ_s does not have much of an effect. Thus, CCSD parameters sometimes greatly affect confidence (log posterior ratio; LPR) while only minimally affecting choice (decisions).

Problem 8.5 Consider the binary classification (C_1 vs C_2) task of section 8.6 where both classes are equally probable ($p(C=1)=p(C=2)=0.5$.), both stimuli are drawn from Gaussian distributions with a mean of 0, and where the variance C_1 stimuli is $\sigma_1^2=9$ and that of C_2 stimuli is $\sigma_2^2=144$.

(a) Plot the distributions $p(s|C=1)$ and $p(s|C=2)$. Use the plot to explain why even an optimal observer cannot be 100 percent correct on this task.

(b) For general σ_1 and σ_2, derive equation (8.37), the equation for the LPR.

(c) Derive an expression for the probability that the optimal observer reports Class 1 when the true stimulus is s. Use the standard cumulative normal distribution, Φ_{standard}, defined in box 7.1.

(d) Plot this probability as a function of s (between -30 and 30) for $\sigma=10$. This is the psychometric curve of the optimal observer with $\sigma=10$. Do the same for $\sigma=1$. Plot both cases in the same plot.

(e) Interpret the differences between the two curves.

(f) Derive expressions for "hit rate" (probability of reporting class 1 when the true class was 1), "false-alarm rate" (probability of reporting class 1 when the true class was 2), and proportion correct.

(g) Plot all three expressions as a function of σ in the same plot.

(h) Interpret the plot.

Problem 8.6 In this problem, we discuss unmodeled errors. Suppose that we perform an experiment with binary responses ($r=0$ or $r=1$) and that $p(r|s)$ expresses the predicted probability of the observer's response—under an arbitrary model, Bayesian or non-Bayesian—when the stimulus is s.

(a) Suppose that the observer accidentally presses the wrong key on a proportion λ of all trials. How does this change the predicted probability of the observer's response? Remark, much of the human psychophysics literature makes such an assumption and calls the relevant effect lapse.

(b) Suppose that the observer makes a random guess on a proportion g of all trials (e.g., because he sometimes did not pay attention and didn't see the stimulus). How does this change the predicted probability of the observer's response?

Problem 8.7 Consider binary classification with a flat prior, mirror-image CCSDs (i.e., $p(s|C=1)=p(-s|C=-1)$), and a measurement distribution $p(x|s)$ that is symmetric around s (though not necessarily Gaussian). Show that the MAP observer has the decision rule "report $C=1$ if $x>0$."

Problem 8.8 Consider an observer performing binary classification with a class distribution $p(C)$ and mirror-image, nonoverlapping CCSDs $p(s|C=-1)$, nonzero for $s<0$, and $p(s|C=1)$, nonzero for $s>0$. In the chapter, we described the decision strategy of a Bayesian observer in this task. A student suggests an alternative decision strategy, namely that the observer first calculates the overall stimulus distribution,

$$p(s)=p(s|C=1)p(C=1)+p(s|C=-1)p(C=1), \tag{8.44}$$

then uses this as a prior to compute a posterior over s, then compares the mean of this posterior, which is an estimate of s, to 0 (the midpoint of our two CCSDs).

(a) Show using equations that the resulting decision rule is equivalent to the decision rule of a Bayesian observer who assumes incorrect CCSDs when doing inference, namely

$$q(s|C=-1) \propto -sp(s|C=-1) \tag{8.45}$$

$$q(s|C=1) \propto sp(s|C=1). \tag{8.46}$$

(b) Characterize for the CCSDs of cases 1 and 2 the differences between the student's strategy and the optimal strategy in terms of psychometric curve and proportion correct. You have to make your own choices for the parameters. Exploration includes examining whether those choices matter.

Problem 8.9 This problem is about model mismatch. Experimental psychologists often assume that the observer's beliefs about the CCSDs do not matter for behavior, as long as the CCSDs (whether true or assumed) are mirror images of each other. However, that is guaranteed only if $p(C=1)=0.5$. Here, we will show this concretely in case 3 from section 8.2. CCSDs are half-Gaussian with $\sigma_s=1$. For the prior probability of class 1, consider $p(C=1) \in \{0.1, 0.3, 0.5, 0.7, 0.9\}$. For the sensory noise, consider $\sigma \in \{0.5, 1, 2\}$.

(a) Consider a *suboptimal* observer who instead of the MAP decision rule uses the rule "report $C=1$ when $x>0$" (which as we know is only optimal if $p(C=1)=0.5$). Calculate proportion correct of this suboptimal observer for every combination of $p(C=1)$ and σ.

(b) Repeat for the optimal observer. You may use the expression for the LLR from equation (8.43).

(c) Plot all results in the same plot of proportion correct as a function of $p(C=1)$, with different lines (in different colors) corresponding to the different values of σ. Use solid lines for suboptimal, dashed lines for optimal.

There is a lesson for experimental design here: if there is any possibility that the observer has a nonflat prior (and usually this possibility exists), then it is good practice to perform simulations to determine whether the assumption that the observer makes about the CCSDs might matter.

Problem 8.10 This problem examines the interaction between cue combination (chapter 5) and binary classification (this chapter). We consider a task in which a category C takes values -1 (left) and 1 (right). On each trial, an experimenter chooses a value of C, with both values having equal probability. Then, the experimenter draws a stimulus value from a CCSD $p(s|C)$. Finally, the observer makes two measurements, x_1 and x_2, which we assume to be conditionally independent given s, that is, drawn independently from $p(x|s)$. A student claims that in this problem, the likelihood over C is

$$p(x_1, x_2|C) \propto p(x_1|C)p(x_2|C), \tag{8.47}$$

where

$$p(x_i|C) = \int p(x_i|s)p(x|C)ds \tag{8.48}$$

for $i = 1, 2$.

(a) Why is this incorrect?

(b) Write down the correct equation for $p(x_1, x_2|C)$ in terms of known probabilities.

(c) Work out an example MAP model in which the student's incorrect likelihood produces a psychometric curve $p(\hat{C} = 1|s)$ as a function of s that is different from the correct likelihood. You will need to make specific assumptions about $p(s|C)$ and $p(x|s)$.

Problem 8.11 We will fit a simple model of binary classification. Read appendix C if you are not familiar with model fitting and model comparison. The stimulus took values from -5 to 5 in steps of 1; we will denote the values by s_j, with $j = 1, 2, \ldots, 11$. On each trial, the subject responded whether the stimulus was positive (to the right; $\hat{C} = 1$) or negative (to the left; $\hat{C} = -1$). Download psychometric.csv from https://osf.io/84kpb/. The rows correspond to trials (500 trials in total). The first column contains the values of the presented stimulus, s, the second column the subject's classification responses, $\hat{C}$.

(a) Plot the psychometric curve, that is, the proportion of "right" responses as a function of the stimulus, without a connecting line.

(b) We will now fit the model in equation (8.30) with $k = 0$:

$$p(\hat{C} = 1|s, \sigma) = \Phi_{\text{standard}}\left(\frac{s}{\sigma}\right). \tag{8.49}$$

Explain why the log likelihood of σ takes the form

$$\log \mathcal{L}(\sigma; \text{data}) = \sum_{j=1}^{11} n_{j+} \log p(\hat{C} = 1|s = s_j, \sigma) + \sum_{j=1}^{11} n_{j-} \log p(\hat{C} = -1|s = s_j, \sigma), \tag{8.50}$$

where n_{j+} and n_{j-} are the numbers of trials on which the stimulus was s_j and the response was $\hat{C} = 1$ or $\hat{C} = -1$, respectively.

(c) Plot the log likelihood function as a function of σ. Use a grid for σ from 0.1 to 5 in 1000 steps.

(d) Plot the likelihood function (no log) over σ using the same grid.

(e) Explain why the likelihood values are extremely small.

(f) Find the MLE of σ on the grid.

(g) Instead of using a grid, find the MLE of σ using a built-in numerical optimization algorithm. Justify the algorithm that you chose.

(h) Plot the best model fit as a line in the plot obtained in part (a). Use the maximum-likelihood estimate (MLE) of σ from part (g), or if you did not complete that part, from part (f).

Problem 8.12 This is a follow-up on problem 8.11 and uses the same data file psychometric.csv. We will compare the simple model from problem 8.11 with a more general model in which the observer sometimes guesses randomly. The latter model takes the form

$$p(\hat{C}=1|s,\sigma,\lambda)=(1-\lambda)\Phi_{\text{standard}}\left(\frac{s}{\sigma}\right)+0.5\lambda, \tag{8.51}$$

where λ is an unknown guessing rate. This model has more flexibility due to the extra parameter.

(a) Plot the log likelihood landscape of this model as a heat map. For σ, use the same grid as in problem 8.11. For λ, use a grid from 0 to 0.3 in 1,000 steps.

(b) Find the MLEs of σ and λ on their respective grids.

(c) Now forget about the grids and instead find the MLEs of σ and λ using a built-in optimization algorithm. Justify the algorithm that you chose.

(d) Plot the psychometric curve (open circles) along with the best fits of both the simple model and the more complex model (solid lines in different colors). Use the MLEs from part (c) or, if you did not complete that part, from part (b).

(e) Calculate the AIC for both models. Calculate the AIC difference. Draw a conclusion in words.

(f) Calculate the BIC for both models. Calculate the BIC difference. Draw a conclusion in words.

(g) Calculate the ten fold cross-validated log likelihood for both models. Calculate the cross-validated log likelihood difference. Draw a conclusion in words.

9

Top-Level Nuisance Variables and Ambiguity

How can we deal with aspects of the world that affect our observations but are not directly related to the question we want to ask?

In chapter 8, we encountered the concept of *nuisance variables*, which are world state variables that are not of primary interest to the decision maker yet must be taken into account during inference because they affect the observations. There, the nuisance variable was intermediate between the variable of interest and the observations. The topic of the present chapter is nuisance variables that are themselves top-level variables.

Plan of the Chapter

We work through two classic examples of top-level nuisance variables: depth perception and color perception. These examples are mathematically similar, even though they come from two entirely different domains of perception. We discuss how top-level nuisance variables cause ambiguity, and show how the observer has to marginalize in order to compute the likelihood function over the world state variable of interest.

9.1 Example Tasks

We consider situations in which two world state variables together give rise to the observation or measurement. One of these variables is the one of interest, the other is a top-level nuisance variable. The distribution of a top-level variable does not depend on any other variables in the problem. The generative model can graphically be represented as in figure 9.1. It has a characteristic V-shape. Examples of top-level nuisance variables abound:

- When inferring the distance from an object, its size is a top-level nuisance variable, because the observation—the retinal size of the object—depends not only on one's distance from it but also on object size.
- Conversely, when inferring the size of an object, one's distance from it is a nuisance variable. This illustrates that what is a nuisance variable depends on the task.
- When inferring the color of a surface, the color of the incident light is a top-level nuisance variable.

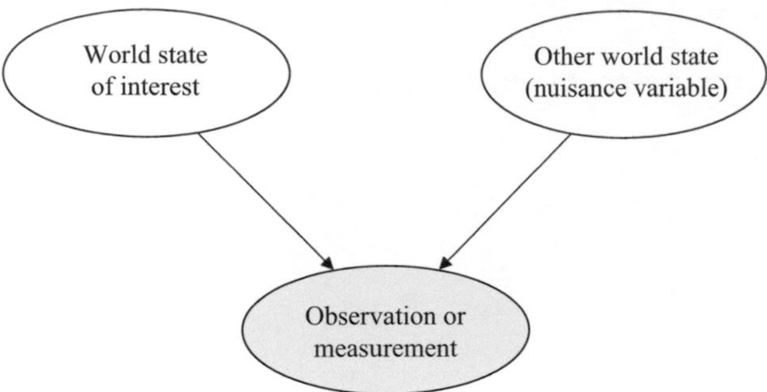

Figure 9.1
Generative model with a top-level nuisance variable. All case studies in this chapter have this structure.

- When classifying an object based on a visual image, the viewing angle is a top-level nuisance variable.
- When a pilot tries to infer the approach angle of their plane, the slope of the runway is a top-level nuisance variable.
- When inferring a person's intent from their spoken words, the amount of experience that the speaker has with the language is a nuisance variable, because people are less adept at expressing important nuances when they are early-stage learners of a language.

Top-level nuisance variables cause ambiguity even in the absence of noise, Namely, if the value of the nuisance variable were known, it might be possible to know the value of the variable of interest exactly. However, since the value of the nuisance variable is not known, the same observation is consistent with multiple (and often infinitely many) values of the variable of interest.

9.2 Size as a Top-Level Nuisance Variable in Depth Perception

A nuisance variable can introduce ambiguity where none would otherwise be present. A classic example is size-depth ambiguity. Consider a defensive driver who wants to maintain a safe distance between their car and the one ahead (figure 9.2A). To do so, they must accurately perceive the distance from the next car (distance away, in the three-dimensional world, is known as *depth*). Under good visual conditions, a driver has many cues to aid depth perception. However, when visual conditions are poor, as for instance in darkness or fog, the number of distance cues is diminished. One cue a driver can use provided simply that they can see the taillights of the car ahead, even if with just a single eye, is the size of the image of the car on their retina. The observer's task, then, is to estimate the distance from the next car, D, from the width of the retinal image of that car, x. For simplicity, we assume that the retinal image is flat rather than slightly curved.

Step 1: Generative model (figure 9.2B) The generative model contains three variables: width w, distance D, and measured width x. In view of the geometry of the problem, retinal width is completely determined by w and D,

$$x = \frac{lw}{D}, \tag{9.1}$$

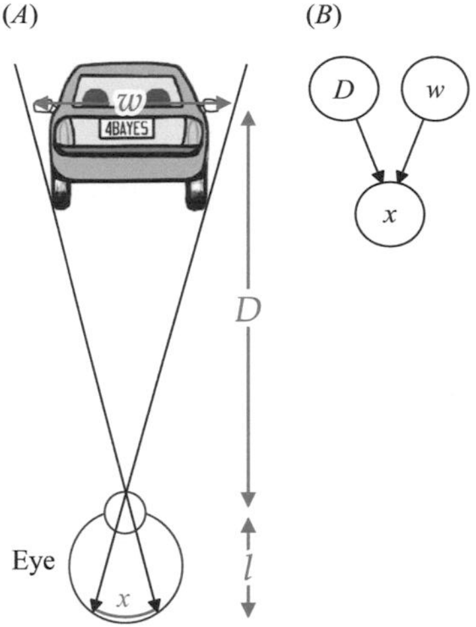

Figure 9.2
Depth perception from retinal size. (*A*) At distance D from the observer, a car of width w produces a retinal image of width x. In the absence of measurement noise, we know from trigonometry that $\frac{x}{l}=\frac{w}{D}$, where l is the distance from the observer's pupil to their retina; a smaller car, closer to the observer, would subtend the same visual angle and produce the same retinal image. The observer can therefore infer the distance from the car only if they have prior beliefs about the car's size. (*B*) The generative model. The nuisance variable, w, and the world state variable of interest, D, are both needed to generate the observation, x. We assume no noise in the observation.

where l is the diameter (length) of the eye (distance from the pupil to the fovea), assumed to be a fixed and known constant. To focus on the ambiguity induced by the top-level nuisance variable, we assume that the measurement is noiseless.

We have specified x in terms of w and D, but we still have to provide w and D with prior distributions. We first assume that these variables are independent:

$$p(D,w)=p(D)p(w). \tag{9.2}$$

Beyond that, we observe that w and D are restricted to the positive real line: none of them can take negative values. For such positive-valued variables (sometimes called "magnitude variables"), a normal distribution is not appropriate, because that choice would imply that the variable can take negative values. A common correct solution is to assume a *lognormal distribution* instead (see section 3.7). We assume a lognormal distribution over w:

$$p(w)=\text{Lognormal}(w;\mu_w,\sigma_w^2). \tag{9.3}$$

This is equivalent to saying that $\log w$ follows a normal distribution with mean μ_w and variance σ_w^2:

$$p(\log w)=\mathcal{N}(\log w;\mu_w,\sigma_w^2). \tag{9.4}$$

For $\log D$, we assume a flat (improper) prior for simplicity:

$$p(\log D)=\text{constant}. \tag{9.5}$$

(This prior would not be flat in D space. We will consider alternatives in problem 9.6.) It is helpful to rewrite equation (9.1) as

$$\log x=\log l+\log w-\log D. \tag{9.6}$$

Step 2: Inference The observer infers D from a given measured width x. The ambiguity in this problem consists of the fact that for a given x, there are infinitely many combinations of $\log w$ and $\log D$ that satisfy equation (9.1). To calculate the posterior distribution over $\log D$, we apply Bayes' rule:

$$p(\log D|\log x) \propto p(\log D)p(\log x|\log D) \tag{9.7}$$

$$\propto p(\log x|\log D), \tag{9.8}$$

where we have used the assumption from step 1 that the prior over $\log D$ is flat. We have so far not yet specified $p(\log x|\log D)$. To so so, we start with equation (9.6). Since we are conditioning on $\log D$ in $p(x|\log D)$, we treat $\log D$ as a constant. Moreover, $\log l$ is a constant. As a result, $\log x$ is equal to $\log w$ plus a constant. Since $\log w$ has a normal distribution with mean μ_w and variance σ_w^2 (equation (9.4), we can use the properties in box 4.1 to find that $\log x$ conditioned on $\log D$ has a normal distribution with a shifted mean:

$$p(\log x|\log D) = \mathcal{N}(\log x; \mu_w + \log l - \log D, \sigma_w^2). \tag{9.9}$$

Thus, equation (9.8) becomes

$$p(\log D|\log x) \propto \mathcal{N}\left(\log x; \mu_w + \log l - \log D, \sigma_w^2\right) \tag{9.10}$$

$$= \mathcal{N}\left(\log D; \mu_w + \log l - \log x, \sigma_w^2\right), \tag{9.11}$$

where we made use of the rule $\mathcal{N}(a; b, \sigma^2) = \mathcal{N}(a + c; b + c, \sigma^2)$ for any a, b, and c.

Exercise 9.1

(a) Why does this rule make intuitive sense?

(b) Show how we applied this rule to obtain equation (9.11).

Equation (9.11) implies that D itself (without the log) follows a lognormal distribution:

$$p(D|x) = \text{LogNormal}(D; \mu_w + \log l - \log x, \sigma_w^2). \tag{9.12}$$

(Conditioning on $\log x$ is the same as conditioning on x.) We have plotted several example posteriors in figure 9.3.

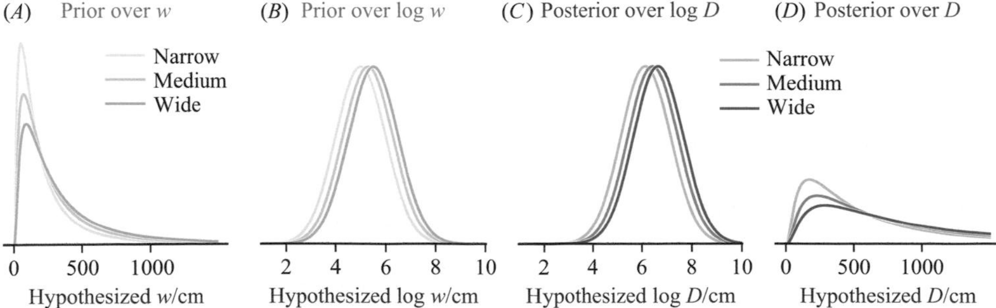

Figure 9.3
Priors and posteriors over w and D in the car example.

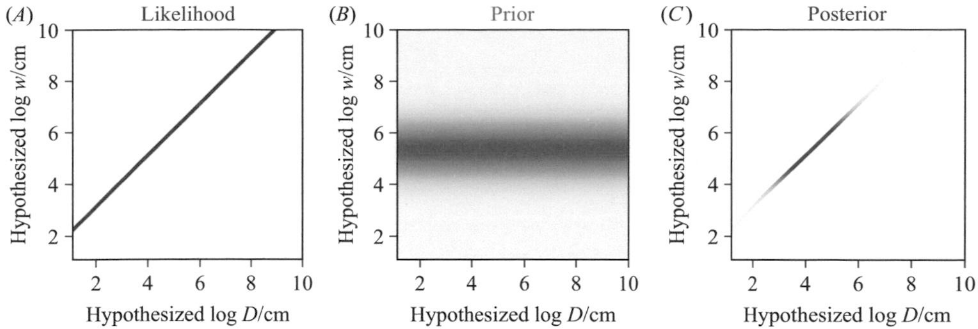

Figure 9.4
Two-dimensional prior, likelihood, and posterior in the car example.

The last part of step 2 is to read out the posterior. If the observer minimizes the expected squared error in the log domain (which makes more sense than in the domain of D itself), then they will report the mean of the posterior over $\log D$ which is

$$\log \hat{D} = \mu_w + \log l - \log x. \tag{9.13}$$

Transforming back to the original space, this means that

$$\hat{D} = \frac{l}{x} e^{\mu_w}. \tag{9.14}$$

This equation implies that the Bayesian observer in this problem uses prior knowledge of the width of the car, through the parameter μ_w, which is the mean of the logarithm of the width. At first, one might think that e^{μ_w} is simply $\mathbb{E}[w]$. However, this is not the case. The mean of a lognormally distributed variable is not the exponentiated first parameter. Instead, the exponentiated first parameter is the *median* of the variable (see section 3.7). Thus, e^{μ_w} is the median of w (computed using the prior over w). Finally, since we assumed that the observation is noiseless, $\frac{l}{x}$ is equal to the ratio of true distance to true width, $\frac{D}{w}$, which is known from the experimenter's point of view. As a result,

$$\hat{D} = D \frac{\text{Median}[w]}{w}. \tag{9.15}$$

Although this relation is simple, it demonstrates an interesting and deep point, namely that estimation of the world state of interest (here distance) can be biased by the prior over the nuisance variable (here width). The bias manifests as follows: when a car is wider than median, then the observer's estimate of distance will be lower than the true distance: you think the car is closer than it truly is, because it is bigger than you expect. Conversely, when a car is narrower than median, then the observer will overestimate its distance. This is a bias, but not one due to measurement noise combined a prior over the world state of interest, as it was, for example, in section 4.5. In the present case, there is no measurement noise, and the bias originates from marginalizing over the nuisance variable. The likelihood over the variable of interest (here D) is "inherited" from a prior over the nuisance variable (here w). This is typical for inference in generative models of the type in figure 9.1: the prior over the nuisance variable affects the likelihood over the variable of interest, as the two

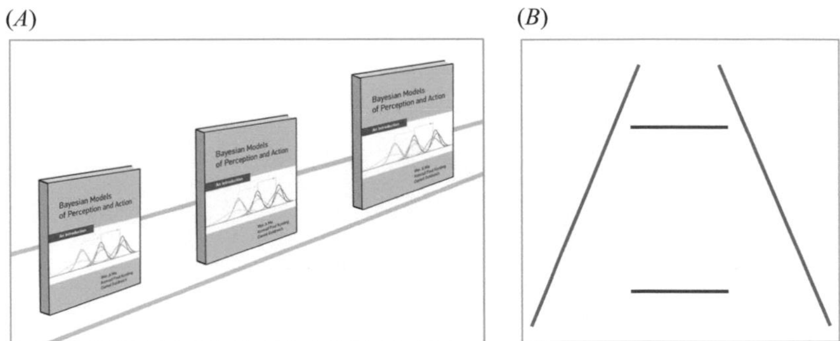

Figure 9.5
Two variants of the Ponzo illusion. In both cases, the topmost object may look taller (wider) even though it is physically equally tall/wide in the drawing.

variables are "coupled" through the observation. From the point of view of the generative model, a transformation of variables has occurred from w to x, turning the prior over w into a conditional distribution over x (given D); the latter is then used for the likelihood over D.

We have seen that the prior over the nuisance variable, w, narrows down the posterior over the world state of interest, D: the prior over car width resolves the ambiguity about distance. This prior has to be learned from experience. Most of us have a great deal of familiarity with cars, so we have considerable knowledge, gained from years of experience, regarding the distribution of the sizes and shapes of cars on the road. Indeed, priors over sizes of objects are sometimes referred to as a form of monocular cue about depth (distance).

So far, we have seen how our understanding of an object's size affects our perception of its distance. Conversely, our understanding of an object's distance also affects our perception of its size. Figure 9.5 illustrates two examples of the *Ponzo illusion*. Within each panel, the figures have the same size on the page (and therefore on the retina), but the topmost figure appears larger. This occurs because the brain interprets the two-dimensional picture as a three-dimensional scene, where the context suggests that the topmost figure is further away. The only way one object can have the same retinal image size as another and yet be farther away is if the further object is larger. Your thumb, viewed at arm's length, may occupy the same retinal area as a distant mountain (for this reason, you can use your thumb to block your view of the mountain). Size-depth ambiguity occurs because objects with an infinite set of possible physical sizes can produce the same retinal image size, depending on their distance from the observer. The more confident we are of the distance to an object, the more confident we can be of its size.

9.3 Marginalization Formulation

In section 8.3, we saw that the Bayesian decision maker deals with a nuisance variable through marginalization. In the previous section, we avoided explicit mention of marginalization. However, marginalization is still taking place in the background. In this section, we explain this; this explanation comes with a helpful two-dimensional visualization. As a first step, we write down a *two-dimensional* posterior distribution over both the variable of interest, D, and the nuisance variable, w:

$$p(\log D, \log w | \log x) \propto p(\log D, \log w) p(\log x | \log D, \log w). \tag{9.16}$$

Using the independence equation, equation (9.2), this becomes

$$p(\log D, \log w | \log x) \propto p(\log D) p(\log w) p(\log x | \log D, \log w). \tag{9.17}$$

In equation (9.17), $p(D)p(w)$ is a two-dimensional prior distribution. An example is shown in figure 9.4B. But what is the likelihood function over $\log D$ and $\log w$, $p(\log x | \log D, \log w)$? We know from equation (9.6) that x is a deterministic function of D and w. That means that a particular combination of D and w is either fully compatible with x or not at all; there is no in between. Pictorially, this means that the *two-dimensional likelihood function* is a sharp line (figure 9.4A): all combinations of D and w on this line have a nonzero likelihood (which we can think of as being infinite), and all combinations off the line have zero likelihood. Formally, we can write such a deterministic relationship as a *delta function*:

$$p(\log x | \log D, \log w) = \delta\left(\log x - (\log l + \log D - \log w)\right). \tag{9.18}$$

For background on the delta function, see appendix section B.8. Multiplying the two-dimensional prior with the two-dimensional likelihood produces a two-dimensional pro-toposterior, which we can numerically normalize to obtain a two-dimensional posterior (figure 9.4C). The effect of the sharp likelihood is to take a "slice" out of the two-dimensional prior. The final step is then to marginalize over the nuisance variable:

$$p(\log D | \log x) = \int p(\log D, \log w | \log x) d(\log w). \tag{9.19}$$

This means "collapsing" the two-dimensional posterior distribution into a one-dimensional distribution over D, by averaging over the second dimension, w. Recall from box 8.1 that the word "marginalization" in fact refers to "collapsing into the margin." We will show in problem 9.7 that the marginalization formulation gives rise to the same posterior over $\log D$ as in equation (9.11). For background on marginalization, see appendix section B.11.2.

Exercise 9.2 In equation (9.19), we integrate over $\log w$. Ask yourself under which circumstances we can do that analytically and under which circumstances we will need to solve this numerically.

9.4 In Color Perception

We now move to an entirely different domain of perception, namely color vision. Although color vision is not typically discussed alongside depth perception, we do so in this book because the observer's inference process is nearly identical. We will only sketch the ingredients of this example here, and further work it out in problem 9.9.

The essence of the computational problem does not require actual color. Therefore, to keep things simple, we consider surfaces that are a shade of gray (i.e., anywhere in between white and black). We see a surface when there is a light source. A light source emits photons

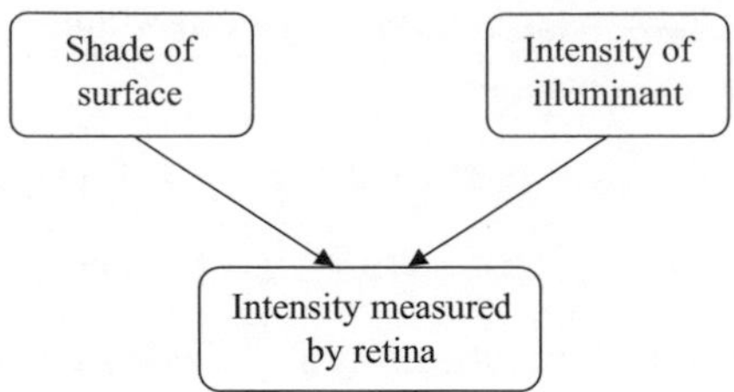

Figure 9.6
Generative model of color perception.

(light particles), each of which carries an amount of energy. The surface absorbs some proportion of photons and reflects the rest. Some of the reflected photons reach our eye and trigger the process of vision.

Step 1: Generative model (figure 9.6) The world states are surface shade and illuminant intensity. The *shade* of a surface is the grayscale in which a surface has been painted. It is a form of *color* of a surface. (Note: in this context, shade has nothing to do with shadow.) Technically, shade is reflectance: the proportion of incident light that is reflected. The shade of the paper determines its *reflectance* ρ: black paper might absorb 90 percent of the incident light and reflect only 10 percent, while white paper might absorb only 10 percent and reflect 90 percent. *The intensity of a light source* (illuminant), denoted by I, is the amount of light it emits; it is measured in "number of photons" or power. The intensity of the light source is a nuisance variable when inferring the shade of a surface. Both reflectance and intensity will have probability distributions associated with them. We will choose examples in a problem.

The measurement is the amount of light measured by the retina, which we will also refer to as *retinal intensity* and denote by x. Assuming no measurement noise, the retinal intensity is $x=\rho I$. In other words, if you make a surface twice as reflectant, it has the same effect on your retina as doubling the intensity of the light source. Since intensity and reflectance are magnitude variables, we immediately move to the log domain and write

$$\log x = \log \rho + \log I. \tag{9.20}$$

Although the variables have entirely different meanings, equation (9.20) parallels equation (9.6) in the car example. This illustrate that computation can connect disparate domains.

Step 2: Inference The observer needs to infer the surface shade (reflectance) ρ from retinal intensity x. From step 1, we understand that retinal intensity provides ambiguous information about the shade of a surface. A particular retinal light intensity is consistent with multiple (in fact infinitely many) possible combinations of true shade and illumination. For example, the same retinal intensity can be produced by dark paper in sunlight and by white paper in dim light. The main challenge of the inference process is to resolve the ambiguity. We use the term *lightness* for the perceived shade of a surface. It is the result of inference, not a physical state of the world.

So far, we have considered grayscale surfaces. However, analogous arguments apply to colored surfaces: the intensity of the illuminant is replaced by the set of intensities at different wavelengths, also called the *power spectrum* of the illuminant. The shade of the

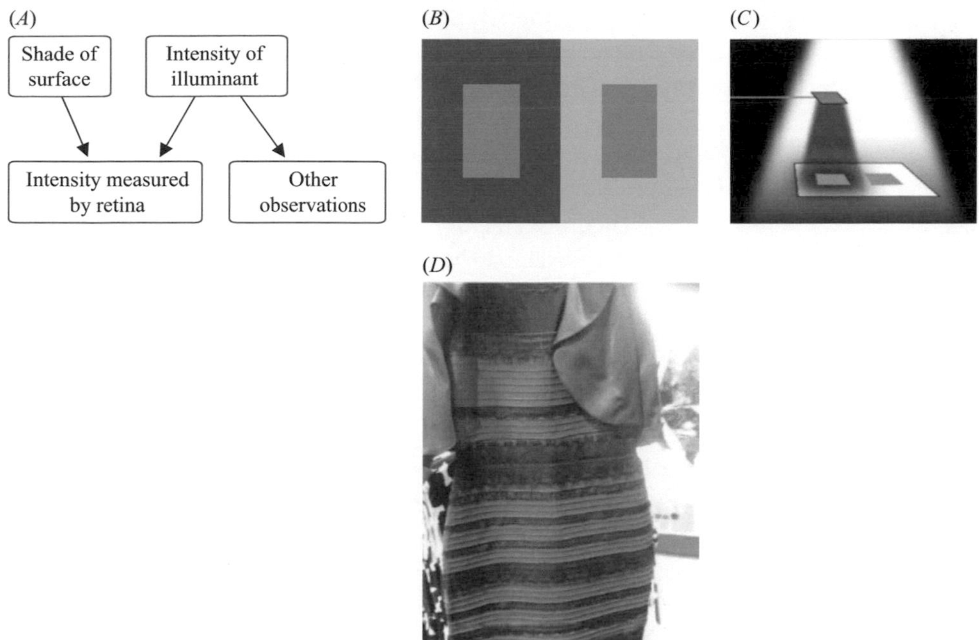

Figure 9.7
Context to help resolve ambiguity. (*A*) Generative model. (*B*) Simultaneous-contrast illusion. (*C*) Edward Adelson's explanation of the simultaneous-contrast illusion. (*D*) "The Dress": what colors do you perceive?

surface is replaced by the color of the surface, or technically the *reflectance curve*, which specifies what proportion of photons at each wavelength is reflected by the surface. Finally, the retinal intensity is replaced by the power spectrum of the light incident on the retina. The power spectrum of the illuminant is the nuisance variable when inferring the color of the surface.

Besides the prior, other ways of resolving ambiguity about surface shade/color are to remove the uncertainty about the intensity of the illumination, for example, by using a light source of known intensity, or—as in the Ponzo illusion—to use contextual cues: other clues in the visual scene that tell you the intensity or power spectrum of the light. Then, the generative model is as in figure 9.7A. Then, we can infer the intensity of the illumination from the other observations and, using that information, we can compute a posterior distribution over surface shade. This process is called *discounting the illuminant*. For example, if the other observations tell us that the surface is in the shadow, then we will believe that the shade of the surface is whiter than when we get the same retinal intensity but the other observations tell us that the surface is in sunlight.

This inference process implies that it is possible to perceive different shades (i.e., report different *lightness*) based on identical retinal intensities, as long as we are made to believe that the illumination differs. One illusion in which this seems to happen is the *simultaneous contrast illusion* (figure 9.7B). Although both central squares have the same shade, the right one looks darker. Edward Adelson has proposed that this happens because the surrounding big squares suggests different illumination for the two halves of the picture: for example, the right half might be in the light and the left half might be in the shadow (figure 9.7C). In this explanation, the brain uses other observations (here the surrounding big squares) to infer the illuminations, and based on the result of this inference, proceeds to infer the shades of the small squares.

Similarly, it is possible to perceive different colors based on identical retinal color, as long as we are made to believe that the illumination differs. A famous example of this phenomenon is the "dress illusion" (figure 9.7D). Some individuals perceive the stripes to be black and blue, whereas others perceive them to be white and gold. This is believed to reflect interindividual variability in the brain's assumptions regarding the power spectrum of the ambient light.

9.5 In Object Recognition

Marginalization over top-level nuisance variables can be found in many forms of perception. Here, we consider object recognition. Even though this example will not allow us to work out the mathematical model in detail, we present it because of its great importance to natural perception. Suppose you want to identify the object in a photograph (figure 9.8A). You care only about the object's identity, not the angle from which it was photographed. Yet, the camera angle does help determine the image and therefore the visual information received by your retina. In making the identification, your brain has to somehow *discount* camera angle, and infer only the value of the state-of-the-world variable of interest to you, object identity. In other words, your brain needs to realize that you could be viewing the object from any angle, and take into account how each object (e.g., a bicycle, a car, etc.) would look from each angle. If you are able to identify the bicycle from any angle, your identification ability is *viewpoint-invariant*.

The generative model of this task is given in figure 9.8C. Besides a node for object identity, it has a node for each top-level nuisance variable; here, only viewing angle is shown. The observation is in this case the image. We are assuming zero sensory noise; if sensory noise were present, there would be an additional node in the generative model, representing the noisy internal representation of the image. We denote object class by C, viewing angle by θ, and the image by s. We assume class and viewing angle are independent; this means

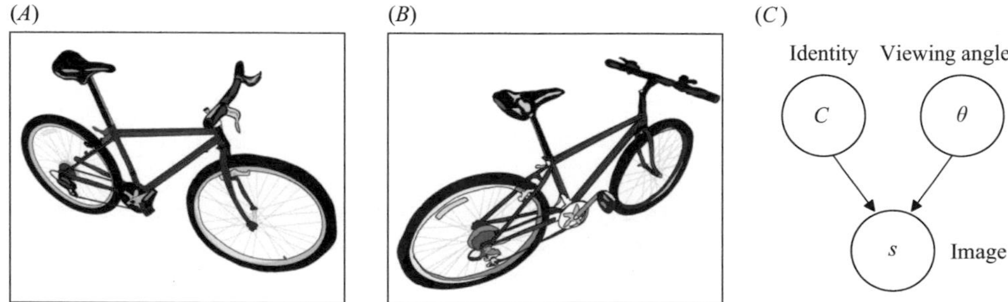

Figure 9.8
Object recognition and nuisance variables. (*A, B*) The same object when viewed from a different angle produces an image that is pixel by pixel very different. Viewing angle is a nuisance variable. Example from Kersten and Yuille [92]. (*C*) Generative model of the object recognition task.

that it is not the case that certain object classes are photographed more often from particular angles than others. The probability distributions in the generative model are the distribution over class, $p(C)$, the distribution over viewing angle, $p(\theta)$, and the distribution of the stimulus conditioned on both class and viewing angle, $p(s|C,\theta)$. The posterior distribution over C is $p(C|s)$. This is obtained by first applying Bayes' rule,

$$p(C|s) \propto p(s|C)p(C), \tag{9.21}$$

and then writing out the class likelihood as a marginalization over θ:

$$p(s|C) = \int p(s|C,\theta)p(\theta|C)d\theta \tag{9.22}$$

$$= \int p(s|C,\theta)p(\theta)d\theta. \tag{9.23}$$

In the last equality, we have used the information that C and θ are independent, so that $p(\theta|C) = p(\theta)$. To appreciate the meaning of equation (9.23), let us consider a particular class C, namely bicycle. The equation states that the probability of the visual image, given the object is a bicycle, is the probability the photographer chose to shoot at an angle 0° relative to the object AND that a bicycle shot at that angle would produce the visual image we are seeing, OR that the photographer chose to shoot at an angle 1° AND that a bicycle shot at that angle would produce the visual image we are seeing, and so on, for all angles. By computing equation (9.23) for many different classes of objects, C (bicycle, car, person, etc.), the observer can in principle generate a class likelihood function and therefore a posterior probability distribution whose mode is the most probable object identity.

Exercise 9.3 Ask yourself how you would build a technical system if you want it to be viewpoint-invariant. Why do people working on artificial neural networks not explicitly marginalize? Do a search on the term "data augmentation."

9.6 Summary

In this chapter, we introduced top-level nuisance variables. We learned:

- In depth perception, the size of an object often appears as a top-level nuisance variable.
- In color perception, the color of the light illuminating an object is a top-level nuisance variable.
- In object perception, the angle of view appears as a top-level nuisance variable.
- In general, most real-world problems are affected by countless top-level nuisance variables.
- For all nuisance variables, top-level nuisance variables require the Bayesian observer to consider all values that it might take. The observer does this through marginalization, as an integration over all possible values weighted with the relevant probabilities.
- For top-level nuisance variables in particular, marginalization involves transforming the prior over the nuisance variable into a likelihood function over the variable of interest.
- Marginalization produces integral expressions that are not analytically solvable except in the simplest cases.

9.7 Suggested Readings

- David H. Brainard and William T. Freeman. "Bayesian Color Constancy." *Journal of the Optical Society of America* 14, no. 7 (1997): 1393–1411.
- Daniel Kersten, Pascal Mamassian, and Alan Yuille. "Object Perception as Bayesian Inference." *Annual Review of Psychology* 55 (2004): 271–304.
- Daniel Kersten and Alan Yuille. "Bayesian Models of Object Perception." *Current Opinion in Neurobiology* 13 (2003): 1–9.
- David C. Knill. "Mixture Models and the Probabilistic Structure of Depth Cues." *Vision Research* 43, no. 7 (2003): 831–854.
- Rosa Lafer-Sousa, Katherine L. Hermann, and Bevil R. Conway. "Striking Individual Differences in Color Perception Uncovered by 'the Dress' Photograph." *Current Biology* 25, no. 13 (2015): R545–R546.
- James V. Stone. *Vision and Brain: How We Perceive the World*. Cambridge, MA: MIT Press, 2012.
- Pascal Wallisch. "Illumination Assumptions Account for Individual Differences in the Perceptual Interpretation of a Profoundly Ambiguous Stimulus in the Color Domain. 'The Dress.'" *Journal of Vision* 17, no. 4 (2017): 5.

9.8 Problems

Problem 9.1 In each of the following forms of behavior, mention one nuisance variable:

(a) (Vision) Estimating how heavy an object is that you are about to pick up.

(b) (Vision) Estimating the time it will take for an approaching car to reach you.

(c) (Olfaction) Determining whether food has gone bad.

(d) (Cognition) Estimating how someone will respond to your criticism.

(e) (Cognition) Estimating your capabilities based on your success in a particular task.

Problem 9.2 As you read the text of this book, you ultimately consume words. But the input to your brain is the visual scene you see. What do you marginalize over as you convert the picture of the page of the book to words and meanings?

Problem 9.3 Someone tells you, "I slept a total of only ten hours during the past two nights." This immediately makes you wonder how much they slept during each of those nights.

(a) The variables are the number of hours slept last night and the number of hours slept the night before. Based on the person's statement, what is the two-dimensional likelihood over both variables?

(b) What would your two-dimensional prior look like? Explain.

(c) As a result, what would the two-dimensional posterior look like?

(d) If you got a single guess of how many hours the person slept last night, and you were trying to minimize the squared error, what would you guess? Explain.

Problem 9.4 In section 9.2, we used a lognormal distribution over the width w (i.e., $p(w) = \text{Lognormal}(w; \mu_w, \sigma_w^2)$). Using 1 meter as the implicit unit in all cases, let us choose $\mu_w = 2$ (cars may be 2 meters wide), $\sigma_w = 0.3$ (most cars are between 1.7 and 2.3 meters wide), and eye diameter $l = 0.025$ (the eye is roughly 2.5 cm in diameter).

(a) Consider three retinal observations of width, $x = 1.6$, $x = 2$, and $x = 2.4$. For each retinal width, plot the posterior probability density over distance (single plot, three curves, color-coded).

(b) Discuss how these posteriors compare to each other, and why so.

Problem 9.5 In section 9.2, we discussed an observer who inferred car distance with car width being a nuisance variable. Now consider the opposite: an observer who infers car width w, with car distance D being a nuisance variable. Assume a flat distribution over log w, and a normal distribution over $\log D$ with mean $\mu_{\log D}$ and standard deviation $\sigma_{\log D}$.

(a) Step 2: Derive an expression for the estimate of w in terms of x.

(b) Step 3: Derive an expression for the estimate of w in terms of true w and true distance D.

(c) Interpret this expression in a way analogous to our interpretation in section 9.2.

Problem 9.6 In section 9.2, we assumed a flat prior over $\log D$. We will now consider a more realistic extension in which this prior is a normal distribution,

$$p(\log D) = \mathcal{N}(\log D; \mu_D, \sigma_D^2). \tag{9.24}$$

(a) Show that the posterior over $\log D$, from equation (9.11), now becomes

$$p(\log D|\log x) = \mathcal{N}(\log D; \mu_{\text{post}}, \sigma_{\text{post}}^2), \tag{9.25}$$

where

$$\mu_{\text{post}} = \frac{J_D\mu_D + J_w(\mu_w + \log l - \log x)}{J_D + J_w}; \tag{9.26}$$

$$\sigma_{\text{post}}^2 = \frac{1}{J_D + J_w}, \tag{9.27}$$

with $J_D \equiv \frac{1}{\sigma_D^2}$ and $J_w \equiv \frac{1}{\sigma_w^2}$.

(b) Show that the estimate of D, from equation (9.15), now becomes

$$\hat{D} = \text{Median}[D]^{\frac{J_D}{J_D+J_w}} \left(D\frac{\text{Median}[w]}{w}\right)^{\frac{J_w}{J_D+J_w}}. \tag{9.28}$$

(c) Interpret this expression.

Problem 9.7 In this problem, we work out a technical aspect in section 9.3. By combining equations (9.18) and (9.19), find the posterior over D. The answer should be equation (9.11).

Problem 9.8 In the following image, you probably see protruding (convex) half-spheres; however, if you flip the image upside down, you see hollow (concave) half-spheres. This can be explained using a prior favoring light coming from above.

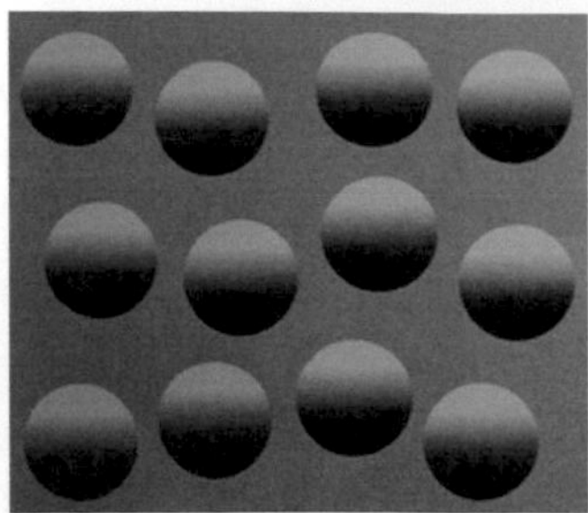

(a) Draw the generative model. For the observation, write “image” (I).

(b) Write down the equation for the log posterior ratio (LPR) for convex relative to concave given I, and apply Bayes’ rule. Assume equal priors.

(c) Evaluate the log likelihood ratio (LLR) by marginalizing over light direction. Assume for simplicity that light can only come from above or below. This should reduce the marginalization of each likelihood to two terms, for a total of four terms in the LLR.

(d) Each of the four terms contains a likelihood of the form $p(I|\ldots)$. Which two of these four are nearly zero, and why?

(e) Simplify the LLR accordingly.

(f) How does the simplified expression explain the percept described above?

Problem 9.9 We build on section 9.4. We consider the problem of estimating surface reflectance ρ when light intensity I is unknown. We start from the equation for the log of the retinal intensity, $\log x = \log\rho + \log I$. We assume that the observer has a prior over light intensity, which can be described by a lognormal distribution with parameters $\mu_{\log I}$ and $\sigma_{\log I}$.

(a) The distribution of $\log x$ given $\log\rho$ is normal with mean $\log\rho + \mu_{\log I}$ and standard deviation $\sigma_{\log I}$. Explain why.

(b) Show that the likelihood of ρ based on a measured retinal intensity x is proportional to $e^{-\frac{\log\rho-\log x+\mu_I)^2}{2\sigma_I^2}}$.

The surface reflectance ρ is a proportion and therefore a number between 0 and 1. Therefore, we assume a prior over ρ that is uniform between 0 and 1, and 0 elsewhere.

(c) Suppose $x = 10$, $\mu_{\log I} = 3$, and $\sigma_{\log I} = 1$. Choose a grid for ρ from 0 to 1 in steps of 0.001. Plot the posterior mass function over ρ, making sure that the distribution is normalized.

(d) Numerically compute the posterior mean of ρ.

(e) Vary $\mu_{\log I}$ from 1 to 5 in steps of 0.1. For each value of $\mu_{\log I}$, repeat part (d). Plot the posterior mean of ρ as a function of $\mu_{\log I}$.

(f) Interpret the plot and relate it to the simultaneous-contrast illusion in figure 9.7.

10

Same-Different Judgment

How do we tell whether two stimuli are the same or different?

In this chapter, we consider how observers infer the relationships between objects. We focus on a fundamental and useful aspect of relatedness: same-different judgment. As an example of the importance of such judgments, consider that, in order to accurately segment a visual scene, an observer can rely on knowledge that elements of an individual object will more often have the same color or orientation, whereas different objects typically differ in these features. While judging sameness sounds simple, it often requires considerable calculation. When two stimuli are affected by noise, they may seem the same despite being different or different despite being the same. Furthermore, the observer does not know the actual stimulus that occurred. Hence, marginalization is often necessary.

Plan of the Chapter

We first discuss same-different judgment using stimuli that can take only two values (analogous to chapter 7). We then consider stimuli that are drawn from continuous distributions (analogous to chapter 8). Thus, this chapter will synthesize material from multiple previous chapters, but with a new twist: central is the relation between two stimuli, rather than the identity or category of a single stimulus.

10.1 Example Tasks

Examples of same-different judgments include:

- You are a prehistoric hunter-gatherer. You encounter a berry bush. You want to determine whether its berries are the same as those that you normally eat, which you know to be nontoxic.
- A generalization of cue combination (chapter 5): two cues do not necessarily have the same source, as we assumed in chapter 5. They could instead have two different sources. You want to know whether they are from the same source or not. The inference in this problem, within perception and sensorimotor research, is also called *causal inference*.

- Contour integration: Do two line segments belong to the same continuous contour, or are they independent? Many other forms of perceptual organization can be framed as a similar inference: Do elements belong together in a coherent whole, or not?
- Change detection: Are two images, separated in time, identical or not? A related working memory task is the delayed match-to-sample task.
- At a more cognitive level, judging whether two quantities are the same underlies judgments of fairness and also forms the basis of mathematics.

10.2 Binary Stimuli

The inference of this chapter involves determining whether two stimuli, which we will denote by s_1 and s_2, are the same or different. In this section, we consider stimuli that can only take two values, as in chapter 7. We choose the values to be $-\mu$ and μ. The task is made hard by the presence of measurement noise. We will now study how a Bayesian observer makes the sameness judgment.

10.2.1 Step 1: Generative Model

The generative model is shown in two equivalent ways in figure 10.1A–B. The top-level variable is a binary variable C, which is 1 for "same" and 2 for "different"; the two depictions differ by whether or not the two possibilities for C are explicitly distinguished. For the distribution of C, we write

$$p(C=1)=p_{\text{same}}; \tag{10.1}$$

$$p(C=2)=1-p_{\text{same}}. \tag{10.2}$$

Since s_1 and s_2 each can be only equal to $-\mu$ and μ, there are four possible combinations of s_1 and s_2. The class-conditioned stimulus distributions can simply be specified explicitly for these four combinations (figure 10.1C). On "same" trials:

$$p(s_1=-\mu, s_2=-\mu)=0.5; \tag{10.3}$$

$$p(s_1=-\mu, s_2=\mu)=0; \tag{10.4}$$

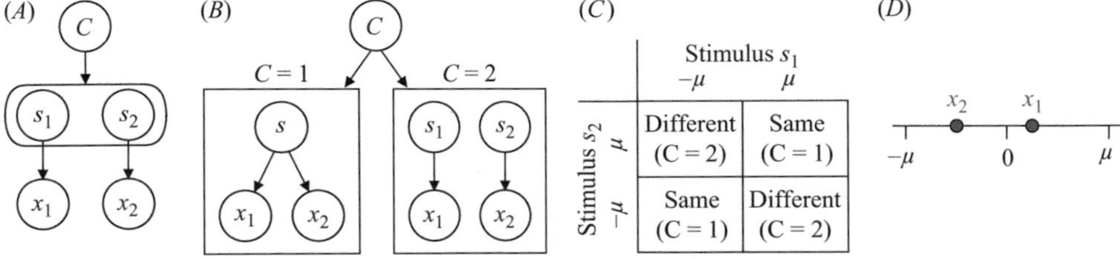

Figure 10.1
Same-different judgment. (*A*) Generative model. C is the sameness variable, s_1 and s_2 are the stimuli, and x_1 and x_2 are the noisy measurements. (*B*) Equivalent depiction, in which the $C=1$ (same) and $C=2$ (different) scenarios are made explicit. (*C*) Table of world states. The first stimulus s_1 and the second stimulus s_2 can each take on the values of μ and $-\mu$, producing four possible combinations. (*D*) Example trial. The possible stimulus values are $-\mu$ and μ, and the measurements on this trial are x_1 and x_2.

$$p(s_1 = \mu, s_2 = -\mu) = 0; \tag{10.5}$$

$$p(s_1 = \mu, s_2 = \mu) = 0.5. \tag{10.6}$$

And on "different" trials:

$$p(s_1 = -\mu, s_2 = -\mu) = 0; \tag{10.7}$$

$$p(s_1 = -\mu, s_2 = \mu) = 0.5; \tag{10.8}$$

$$p(s_1 = \mu, s_2 = -\mu) = 0.5; \tag{10.9}$$

$$p(s_1 = \mu, s_2 = \mu) = 0. \tag{10.10}$$

Finally, we assume that the measurements x_1 and x_2 are noisy versions of the stimuli s_1 and s_2, with the noise being independent between the measurements, and normally distributed with standard deviation σ:

$$p(x_1, x_2|s_1, s_2) = p(x_1|s_1)p(x_2|s_2); \tag{10.11}$$

$$p(x_1|s_1) = \mathcal{N}(x_1; s_1, \sigma^2); \tag{10.12}$$

$$p(x_2|s_2) = \mathcal{N}(x_2; s_2, \sigma^2). \tag{10.13}$$

This concludes the specification of the generative model.

10.2.2 Step 2: Inference

An example trial is shown in figure 10.1D: the observer measures x_1 and x_2, and has to determine whether they both came from the same stimulus. If they did, there are still two possibilities: that stimulus was $-\mu$, or it was μ. There are also two possibilities if the measurements were produced by different stimuli: x_1 could have come from $-\mu$ and x_2 from μ, or the other way around.

As in previous classification tasks (chapters 7 and 8), the world state variable of interest is a high-level categorical variable, C, rather than a physical stimulus. The posterior distribution over C is $p(C|x_1, x_2)$. Since C is binary, we consider for convenience the log posterior ratio (LPR), denoted by d, which is the sum of the log likelihood ratio and the log prior ratio:

$$d = \log \frac{p(C=1|x_1, x_2)}{p(C=2|x_1, x_2)} \tag{10.14}$$

$$= \log \frac{p(x_1, x_2|C=1)}{p(x_1, x_2|C=2)} + \log \frac{p(C=1)}{p(C=2)}. \tag{10.15}$$

We evaluate the likelihoods of "same" and "different," $p(x_1, x_2|C=1)$ and $p(x_1, x_2|C=2)$, by marginalizing over s_1 and s_2. Since together, these variables can take only four combinations of values, the marginalization takes the form of a sum:

$$d = \log \frac{\sum_{s_1,s_2} p(x_1, x_2|s_1, s_2)p(s_1, s_2|C=1)}{\sum_{s_1,s_2} p(x_1, x_2|s_1, s_2)p(s_1, s_2|C=2)} + \log \frac{p(C=1)}{p(C=2)}. \tag{10.16}$$

Now, we can substitute what we know from the generative model, equations (10.1)–(10.11). We do not yet substitute the normal distributions. This gives

$$d = \log \frac{p(x_1|s_1=-\mu)p(x_2|s_2=-\mu)+p(x_1|s_1=\mu)p(x_2|s_2=\mu)}{p(x_1|s_1=-\mu)p(x_2|s_2=\mu)+p(x_1|s_1=-\mu)p(x_2|s_2=\mu)} + \log \frac{p_{\text{same}}}{1-p_{\text{same}}}. \tag{10.17}$$

Although this expression is long, it is intuitive: the large numerator and the large denominator correspond directly to the four possibilities described at the beginning of this section. Their ratio is the sensory evidence that the measurements came from the same stimulus, relative to that they came from different stimuli. Finally, substituting the normal distributions from equation 10.13, we find

$$d = \log \frac{e^{-\frac{\mu}{\sigma^2}(x_1+x_2)}+e^{\frac{\mu}{\sigma^2}(x_1+x_2)}}{e^{-\frac{\mu}{\sigma^2}(x_1-x_2)}+e^{\frac{\mu}{\sigma^2}(x_1-x_2)}} + \log \frac{p_{\text{same}}}{1-p_{\text{same}}}. \tag{10.18}$$

The MAP decision rule, which maximizes accuracy, is to report "same" ($\hat{C}=1$) when $d>0$. In general, this rule cannot be further simplified. However, when $p_{\text{same}}=0.5$, the inequality $d>0$ takes an extraordinarily simple form,

$$\text{sign}(x_1) = \text{sign}(x_2). \tag{10.19}$$

In other words, the observer reports that the two measurements come from the same stimulus if they have the same sign. More generally, the decision rule is hard to interpret but can still be visualized, which we will do in problem 10.4.

Exercise 10.1 Prove equations (10.18) and (10.19).

10.2.3 Step 3: Estimate Distribution

We would like to compute the probability for the Bayesian MAP observer to report "same" ($\hat{C}=1$) for given true stimuli: $p(\hat{C}=1|s_1,s_2)$. After computing the observer's response probabilities as predicted by the model, the experimenter can compare them with empirical data. This estimate distribution can be written as a marginalization over the measurements:

$$p(\hat{C}=1|s_1,s_2) = \iint p(\hat{C}=1|x_1,x_2)p(x_1,x_2|s_1,s_2)dx_1dx_2, \tag{10.20}$$

or, since the mapping from x_1 and x_2 to $\hat{C}$ is deterministic, as

$$p(\hat{C}=1|s_1,s_2) = \iint_{d(x_1,x_2)>0} p(x_1,x_2|s_1,s_2)dx_1dx_2 \tag{10.21}$$

$$= p(d(x_1,x_2)>0|s_1,s_2), \tag{10.22}$$

the probability that the decision variable d, which is a function of x_1 and x_2, is positive, when x_1 and x_2 follow their respective distributions conditioned on s_1 and s_2. Geometrically, this probability is the volume under the two-dimensional measurement distribution in the region defined by the condition $d(x_1,x_2)>0$. Since we know the measurement distribution from equation (10.13), we can write equation (10.22) as

$$p(\hat{C}=1|s_1,s_2) = \iint_{d(x_1,x_2)>0} \mathcal{N}(x_1;s_1,\sigma^2)\mathcal{N}(x_2;s_2,\sigma^2)dx_1dx_2. \tag{10.23}$$

In the case of $p_{\text{same}} = 0.5$, when $d(x_1, x_2) > 0$ reduces to equation (10.19), we can make some progress: when the stimuli are s_1 and s_2, what is the probability that the generated measurements have the same sign? We will do this in a problem.

In general, however, because of the specific form of the decision variable in equation (10.18), we cannot make any further analytical progress. This is a new situation, since, in all Bayesian models discussed so far in this book, the response distribution could be calculated analytically or expressed in terms of a standard nonelementary function (the cumulative standard normal). In the present situation—and in fact in most Bayesian models encountered in research practice—we instead have to resort to numerical methods. There are two general classes of numerical methods for computing multidimensional integrals such as equation (10.23):

Numerical integration. The simplest numerical integration technique is Riemann integration. We first discretize the space (here consisting of x_1 and x_2) into a fine, regular grid. We then calculate the integrand, the two-dimensional measurement distribution $p(x_1, x_2|s_1, s_2)$, on that grid, and numerically normalize it. Next, we calculate d for all values on the grid. The condition $d > 0$ returns zeros and ones and serves as a so-called mask, a term from image processing. Finally, approximate the integral as a sum multiplied by that step size. In the sum, we only retain the terms that satisfy the condition $d > 0$ by multiplying the integrand by the mask. In this method, care should be taken to use a sufficiently fine grid. An improved version of Riemannian integration is the trapezoidal method.

Monte Carlo simulation. The second technique is simulation. For example, in equation (10.23), we can randomly draw a large number (of the order of one million) of pairs of measurements from their respective stimulus-conditioned distributions, namely the normal distributions inside the integral. Each pair represents a simulated trial. To each pair of measurements, we apply the decision rule to determine the simulated observer's response. The proportion of the simulated trials for which the response is "same" is an approximation of the underlying probability of a "same" response given the stimuli. This technique, of approximating a probability distribution by its samples, is a specific case of a method called Monte Carlo simulation. In a sense, a Monte Carlo simulation creates an "empirical" distribution using a computer subject.

Exercise 10.2 Can you see how Monte Carlo simulation is really a way of calculating an integral?

Box 10.1
Evaluating Multidimensional Integrals

Multidimensional integrals are common in step 3 of Bayesian modeling. We mentioned three ways of tackling such integrals: analytically, through numerical integration, and through Monte Carlo simulation. But when to use which method? Whenever an analytical approach is possible, use it, as calculating a closed-form mathematical expression is arbitrarily precise and very fast. When it is not possible, your decision depends on the computational time available. In a given amount of time, you can do only a finite number of evaluations of the decision rule, so you have to choose these wisely. Numerical integration is inefficient, as it spends many evaluations in low-probability regions of measurement space, but Monte Carlo simulation is stochastic, so your result might vary from run to run. To illustrate this, if you have two measurements and define a $1{,}000 \times 1{,}000$ grid for numerical integration, you need one million evaluations. In

Monte Carlo simulation, one million evaluations would also give you an accurate estimate of the desired probability. However, if you had three measurements, numerical integration would require one billion evaluations, take 1,000 times as long, and might be infeasible in practice, because the response probabilities need to be computed for many stimulus pairs and parameter combinations. If you had a budget of only one million evaluations, you could of course make the grid coarser (100 × 100 × 100), but Monte Carlo simulation would then generally be more reliable. As a rule of thumb, numerical integration should not be considered for more than three dimensions.

10.3 Continuous Stimuli

In standard cue combination (chapter 5), the observer has two measurements and knows that they were generated by the same stimulus. In many daily situations, however, it is unclear whether two measurements came from the same stimuli or from different stimuli. In this case, the observer has to infer the probability that there was a single stimulus from the measurements. This probability will subsequently play a role in estimating the values of the stimuli. This inference is analogous to the same-different judgment in section 10.2, but uses continuous variables. In the context of cue combination, the inference problem has also been called *causal inference*, whereas the special case in chapter 5 is called *forced fusion*.

In a laboratory experiment, we might simultaneously present an auditory tone and a visual flash (Figure 10.2A). Intuitively, when the location of the tone and the location of the flash are close to one another, observers may conclude they come from the same source; when the two stimuli are farther apart, the conclusion may be that they come from different sources.

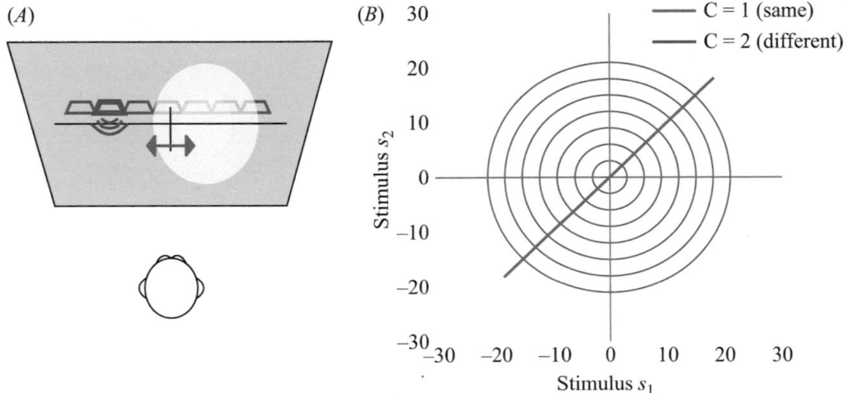

Figure 10.2
A causal inference experiment. (*A*) Sounds are presented from speakers mounted in a horizontal row behind a screen on which visual flashes are projected. The visual stimulus is a spot of light with its center on the same horizontal line as the speakers. Both auditory and visual stimuli are very brief. Visual reliability is manipulated through the size of the spot. Trials are either unisensory (auditory or visual) or multisensory. On multisensory trials, a visual and an auditory stimulus are presented simultaneously. In different blocks, the observer either localizes the auditory stimulus using a cursor on the horizontal meridian (vertical black line), or reports whether the visual and auditory stimuli shared the same location (using a key press). (*B*) Class-conditioned stimulus distributions. "Same" trials are represented by a one-dimensional Gaussian distribution on the diagonal. "Different" trials are represented by a two-dimensional Gaussian distribution.

Using this as an example, we will see how a Bayesian observer reaches the same conclusion but in a more precise way. The causal inference model was developed in 2007 by two of the authors of this book and collaborators [98] as well as simultaneously by an independent group [158].

10.3.1 Step 1: Generative Model

The generative model is the same as in figure 10.1A–B. We again start with the "sameness" variable, C. If $C=1$, then there is only a single stimulus s. If $C=2$, then there are two stimuli, s_1 and s_2, which we assume to be drawn independently:

$$p(s_1, s_2|C=2) = p(s_1|C=2)p(s_2|C=2). \tag{10.24}$$

We assume that all three stimulus variables, s, s_1, and s_2, follow the same normal distribution with mean 0 and standard deviation σ_s:

$$p(s|C=1) = \mathcal{N}(s; 0, \sigma_s^2); \tag{10.25}$$

$$p(s_1|C=2) = \mathcal{N}(s_1; 0, \sigma_s^2); \tag{10.26}$$

$$p(s_2|C=2) = \mathcal{N}(s_2; 0, \sigma_s^2). \tag{10.27}$$

A visualization of the class conditioned stimulus distributions (CCSDs) $p(s|C=1)$ and $p(s_1, s_2|C=2)$ (figure 10.2B) makes clear that this inference problem is conceptually similar to nested classification in chapter 8: one narrow category ($C=1$) that is "embedded" inside a broad category ($C=2$), except now in two dimensions.

We denote the two measurements by x_1 and x_2. As usual, we model them as conditionally independent and normally distributed, but we allow for them to have different levels of noise, σ_1 and σ_2, as in equation (10.13):

$$p(x_1, x_2|s_1, s_2) = p(x_1|s_1)p(x_2|s_2); \tag{10.28}$$

$$p(x_1|s_1) = \mathcal{N}(x_1; s_1, \sigma_1^2); \tag{10.29}$$

$$p(x_2|s_2) = \mathcal{N}(x_2; s_2, \sigma_2^2). \tag{10.30}$$

This concludes the specification of the generative model.

10.3.2 Step 2: Inference

The observer infers whether the two measurements come from the same stimulus ($C=1$) or from different stimuli ($C=2$). Thus, the posterior of interest is $p(C|x_1, x_2)$, the probability over C given the measurements x_1 and x_2. The LPR d is given by equation (10.15). As in chapter 8 and in section 10.2, we evaluate the likelihoods over C, $p(x_1, x_2|C)$, by marginalizing over the stimulus or stimuli. The likelihood of "same" is

$$\mathcal{L}(C=1; x_1, x_2) = p(x_1, x_2|C=1) \tag{10.31}$$

$$= \int p(x_1|s)p(x_2|s)p(s|C=1)ds. \tag{10.32}$$

The likelihood of "different" is

$$\mathcal{L}(C=2;x_1,x_2)=p(x_1,x_2|C=2) \tag{10.33}$$

$$=\left(\int p(x_1|s_1)p(s_1|C=1)ds_1\right)\left(\int p(x_2|s_2)p(s_2|C=2)ds_2\right). \tag{10.34}$$

Plugging in the distributions from the generative model, the LPR is equal to

$$d=\log\frac{p_{\text{same}}}{1-p_{\text{same}}}+\frac{1}{2}\log\left(1+\frac{J_1J_2}{J_s(J_1+J_2+J_s)}\right) \tag{10.35}$$

$$-\frac{1}{2}\frac{J_1J_2}{J_1+J_2+J_s}\left(\frac{J_1}{J_1+J_s}x_1^2+\frac{J_2}{J_2+J_s}x_2^2-2x_1x_2\right), \tag{10.36}$$

where we used precision notation: $J_1\equiv\frac{1}{\sigma_1^2}$, $J_2\equiv\frac{1}{\sigma_2^2}$, and $J_s=\frac{1}{\sigma_s^2}$.

Exercise 10.3 Prove this.

The final component of step 2 is the optimal decision rule:

$$\text{“report ‘same’ if } d>0.\text{”} \tag{10.37}$$

When faced with a complicated expression such as equation (10.36), it is often useful to try to plot and interpret it. This can help both to detect mistakes and to gain intuition. In figure 10.3A, we plot a heat map of the LPR d against the observations, x_1 and x_2, for a particular parameter combination. The black contours indicate $d=0$. The diagonal corresponds to trials on which the measurements happen to be identical to each other. The hypothesis $C=1$ becomes more likely relative to $C=2$ the closer to the diagonal a set of measurements lies. This is intuitive: when two measurements are similar, they are likely to have come from the same stimulus. Essentially, it would be too coincidental for the two to come from different sources. This is the same logic that we used in chapter 2 for the simultaneously

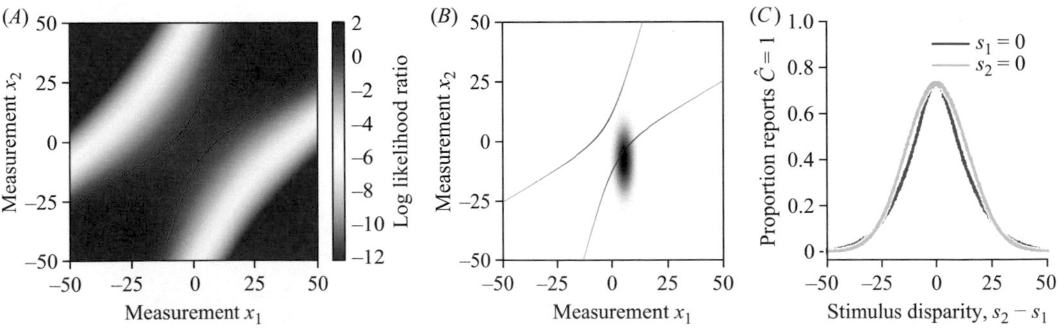

Figure 10.3
(*A*) The strength of the evidence in favor of a common cause, as expressed by the log likelihood ratio, as a function of the measurements x_1 and x_2. The $d=0$ contour lines are shown in black. Two aspects of interest are the band around the diagonal and the structure within this band. Parameters were $p_{\text{same}}=0.5$, $\sigma_1=3$, $\sigma_2=10$, $\sigma_s=10$. (*B*) The same $d=0$ contour lines as in (*A*) but with overlaid the two-dimensional measurement distribution when $s_1=5$ and $s_2=-8$. The volume under this distribution in between the two decision boundaries is equal to the probability of reporting "same" given the specific stimuli. (*C*) The proportion of "same" reports as a function of stimulus disparity (the difference between s_2 and s_1), when $s_1=0$ (so disparity $=s_2$) or when $s_2=0$ (so disparity $=s_1$).

moving dots. Formally, the current problem is more complex, but the essence is the same. Interestingly, the prior does not play any role in this argument.

Moreover, the further from 0 such a pair of measurements lies, the more likely the stimuli were the same. This is because we chose a prior that peaks at the origin. Since stimuli are drawn from this prior, even when the causes are different, the two stimuli and therefore the two measurements tend to lie close to each other near 0. When the measurements lie close to each other but far from 0, this is harder to explain away as a consequence of the prior, and it is, therefore, more likely that the stimuli were the same.

10.3.3 Step 3: Response Probabilities

The goal is to compute the probability that the observer will report "same" for a given combination of stimulus values (s_1, s_2), or in other words, $p(\hat{C}=1|s_1, s_2)$. Analogous to section 10.2.3, this probability is equal to the area under the two-dimensional measurement distribution given those stimulus values that lies in between the two boundaries (figure 10.3B). As in section 10.2.3, the response probabilities cannot be calculated analytically. The modeler instead, again, has to resort to numerical integration on a grid or to Monte Carlo simulation. We apply either method to stimulus pairs s_1, s_2 where s_1 is equal to 0 and s_2 is varied, with $p_{\text{same}}=0.5$. The resulting probability of reporting that both cues came from the same position ($\hat{C}=1$) is plotted as a function of stimulus disparity in figure 10.3C. We see that the larger the spatial disparity between the two stimuli, the less frequently the observer reports that there is a common cause. This makes sense.

Exercise 10.4 Stimulus disparity is s_2-s_1, so there are many combinations of s_1 and s_2 that produce the same disparity. In figure 10.3C, we remove this ambiguity by either setting s_1 or s_2 to 0. These choices produce slightly different curves. Why?

10.3.4 Step 2 Revisited: Inferring the Stimuli

So far we have discussed inference of the number of causes. One might also be interested in the posterior distribution over s_1 and s_2, the stimulus values. This posterior can be written as

$$p(s_1, s_2|x_1, x_2) \propto p(x_1|s_1)p(x_2|s_2)p(s_1, s_2). \tag{10.38}$$

So far, nothing special; we could have done this in chapter 5. However, in the current generative model, the prior $p(s_1, s_2)$ is not directly available. All we know is the distribution of s_1 and s_2 conditioned on the number of causes, C. To find the "prior" $p(s_1, s_2)$, we marginalize over C:

$$p(s_1, s_2) = \sum_{C=1}^{2} p(s_1, s_2|C)p(C). \tag{10.39}$$

Substituting in equation (10.38), we find

$$p(s_1, s_2|x_1, x_2) \propto p(x_1|s_1)p(x_2|s_2) \sum_{C=1}^{2} p(s_1, s_2|C)p(C) \tag{10.40}$$

$$= \sum_{C=1}^{2} p(C)p(s_1, s_2|C)p(x_1|s_1)p(x_2|s_2). \tag{10.41}$$

Thus, this protoposterior is a weighted average of the likelihood function under the hypothesis $C=1$, $p(x_1|s_1)p(x_2|s_2)p(s_1,s_2|C=1)$, and the likelihood function under the hypothesis $C=2$, $p(x_1|s_1)p(x_2|s_2)p(s_1,s_2|C=2)$. These likelihoods are weighted by the prior probabilities of $C=1$ and $C=2$, respectively. This type of weighted average appears whenever marginalization over a discrete variable (here C) is needed.

Exercise 10.5 Show that an alternative way to write the posterior over s_1 and s_2 is as a weighted average of the *posterior distributions* conditioned on C, with weights given by $p(C|x_1,x_2)$. This is pictured in figure 10.4. This is also known as "Bayesian model averaging," with the understanding that each value of C is interpreted as a "model" the observer has about the world.

This is the first time in this book that we have encountered a posterior distribution in an actual model that does not have a single local maximum (unimodal). This posterior has two peaks (is bimodal). For bimodal posteriors, MAP estimation and posterior mean estimation are not equivalent; the latter, arguably, makes more sense, since it minimizes the expected squared error.

Causal inference is an important generalization of cue combination. Kording et al. [98] showed that the causal inference accurately describes human data in an auditory-visual localization task. When observers were asked to report whether both stimuli have the same cause, their reports followed the prediction illustrated in figure 10.3C. Observers also reported the location of the auditory stimulus; their reports closely matched the Bayesian predictions that we illustrated in figure 10.4.

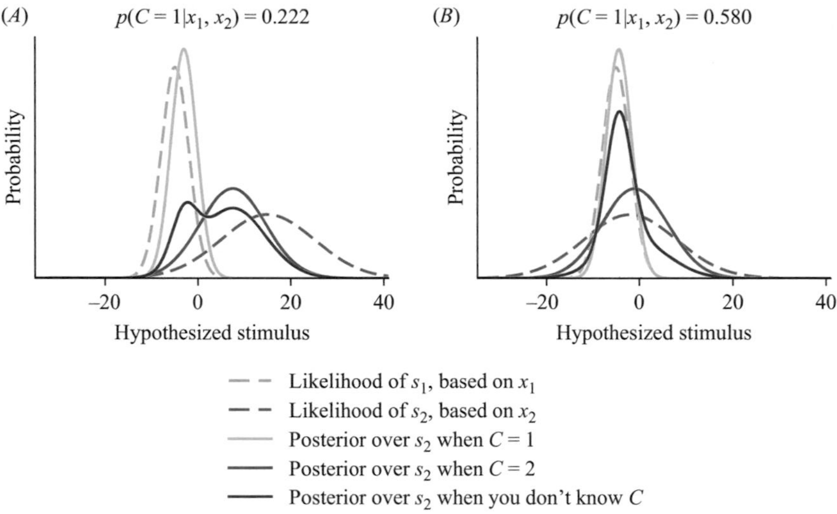

Figure 10.4
Posteriors and conditional posteriors in causal inference. (*A*) In causal inference, the posterior distribution over a stimulus can be bimodal (two-peaked). We are considering a trial on which $x_1=-5$ and $x_2=15$. Shown are the likelihoods over s_1 and s_2 the posterior over s_2 if C were known to be 1, the posterior over s_2 if C were known to be 2, and the posterior over s_2 when there exists uncertainty about C. The latter posterior is a weighted average of the conditioned posteriors, with the weights being the posterior probabilities of $C=1$ and $C=2$, respectively. Here the posterior probability of $C=1$ is only 22.2 percent, so the overall posterior is dominated by the posterior conditioned on $C=2$. (*B*) Same but with $x_1=-5$ and $x_2=-2$. Now, the posterior probability of $C=1$ is higher, so the overall posterior is more of an equally weighted average of the two conditional posteriors.

10.4 Multiple-Item Sameness Judgment

So far, we have discussed judging whether two stimuli are the same or not. This task can be generalized to any number of stimuli, say N. This generalization is important in cognitive science. William James, one of the founding fathers of psychology, called the sense of sameness "the very keel and backbone of our thinking" [81]. Judging sameness plays a role in recognizing textures, which tend to consist of elements with the same orientation. Judging sameness is also said to underlie higher cognitive concepts, such as equality and equivalence in mathematics. Many animal species, from honeybees to pigeons to dolphins, are able to detect sameness at a rather abstract level, suggesting that the concept has had substantial evolutionary importance.

We set the problem up as follows. When the stimuli are the same ($C = 1$), their common value s is drawn from a distribution $p(s)$. When the stimuli are different ($C = 2$), their values s_i, where the index i now takes values from 1 to N, are drawn independently from the same distribution $p(s)$. An example of a sameness judgment experiment is shown in figure 10.5A. Subjects judged whether or not N stimuli all shared the same orientation. The stimuli could have different elongations, chosen randomly. Elongation controlled the quality of the orientation information: the more elongated the ellipse, the lower the orientation noise would be.

Example data are shown in figure 10.5B: the greater the standard deviation of the sample of stimuli shown on a given trial, the more often people responded "different." Moreover, there was an effect of the elongation condition. The generative model of the task is shown in figure 10.5C. The variables are as follows: C is a binary variable that denotes sameness (1 for same, 0 for different), $\mathbf{s}$ denotes the vector of N orientations presented, and $\mathbf{x}$ denotes the corresponding vector of N measurements. The stimulus distribution $p(s)$ is Gaussian with

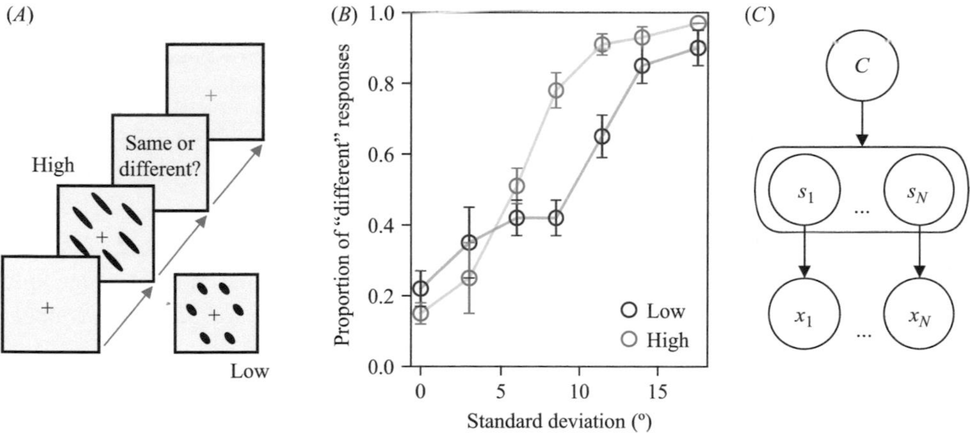

Figure 10.5
Sameness judgment (reproduced from [186]). (*A*) Experimental procedure. Subjects fixated at the cross and a display containing six ellipses was shown for 100 ms and they reported whether or not the orientations of the ellipses were all the same. Stimulus reliability was controlled by ellipse elongation. In the LOW condition, all ellipses had low elongation. In the HIGH condition, all had high elongation. Subjects reported whether or not all stimuli were the same. (*B*) Proportion of "different" responses as a function of the standard deviation of the presented set, for the two conditions. (*C*) Generative model of this task.

mean 0 and standard deviation σ_s. Each x_i is drawn from a Gaussian distribution with mean s_i and standard deviation σ.

The Bayesian observer bases their decision (same or different) on the posterior probabilities of $C=1$ and $C=2$ given the measurements $\mathbf{x} \equiv (x_1, \ldots, x_N)^T$. Since C is a binary random variable, we express that posterior as an LPR:

$$d = \log \frac{p(\mathbf{x}|C=1)}{p(\mathbf{x}|C=2)} + \log \frac{p(C=1)}{p(C=2)}. \tag{10.42}$$

Evaluating the likelihoods in this expression, $p(s\mathbf{x}|C)$; requires marginalization over the stimulus orientations, $\mathbf{s} \equiv (s_1, \ldots, s_N)$:

$$p(\mathbf{x}|C) = \int p(\mathbf{x}|\mathbf{s})p(\mathbf{s}|C)d\mathbf{s}. \tag{10.43}$$

As usual, we assume that the standard deviation, σ, of the noise associated with a stimulus is known to the observer for each stimulus and each trial. Therefore, we do not need to marginalize over σ, but can treat it as a known parameter.

When $C=1$, all elements of the vector $\mathbf{s}$ have the same scalar value s. Then the integral reduces to an integral over this scalar value. Moreover, we assumed that the measurements are conditionally independent, which means that

$$p(\mathbf{x}|\mathbf{s}) = p(x_1|s)p(x_2|s)\cdots p(x_N|s) \tag{10.44}$$

$$= \prod_{i=1}^{N} \frac{1}{\sqrt{2\pi\sigma^2}} e^{-\frac{(x_i-s)^2}{2\sigma^2}}, \tag{10.45}$$

where $\prod$ is notation for a product. Then the likelihood of "same" is

$$\mathscr{L}(C=1; \mathbf{x}) = p(\mathbf{x}|C=1) \tag{10.46}$$

$$= \int \left(\prod_{i=1}^{N} \frac{1}{\sqrt{2\pi\sigma^2}} e^{-\frac{(x_i-s)^2}{2\sigma^2}} \right) \frac{1}{\sqrt{2\pi\sigma_s^2}} e^{-\frac{s^2}{2\sigma_s^2}} ds. \tag{10.47}$$

Although this integral seems daunting, it can be evaluated using a standard equation for the product of normal distributions.

We can similarly evaluate the likelihood of the hypothesis that the stimuli are different, that is, $C=0$. In that case, all measurements are completely independent from each other, since they do not share a common s. Thus, the N-dimensional integral in equation (10.43) reduces to a product of N one-dimensional integrals, one for each measurement (see box 10.2):

$$\mathscr{L}(C=0; \mathbf{x}) = p(\mathbf{x}|C=0) \tag{10.48}$$

$$= \prod_i \int p(x_i|s_i)p(s_i|C=0)ds_i \tag{10.49}$$

$$= \prod_i \left(\int \frac{1}{\sqrt{2\pi\sigma^2}} e^{-\frac{(x_i-s)^2}{2\sigma^2}} \frac{1}{\sqrt{2\pi\sigma_s^2}} e^{-\frac{s^2}{2\sigma_s^2}} ds_i \right). \tag{10.50}$$

Box 10.2
Factorizing a Multidimensional Integral

Suppose you have a function of two variables, x and y. The function has the special property that it can be written as the product of a function $f(x)$ that depends only on x, and a function $g(y)$ that depends only on y. Then its integral over x and y can be simplified as

$$\iint f(x)g(y)dxdy = \int f(x)\left(\int g(y)dy\right)dx \tag{10.51}$$

$$= \left(\int f(x)dx\right)\left(\int g(y)dy\right). \tag{10.52}$$

We used the fact that the expression $\left(\int g(y)dy\right)$ is simply a constant number (in particular, not a function of x) and can be taken out of the integral over x. This argument works only if x and y are different variables.

Using the expressions for $p(\mathbf{x}|C=1)$ and $p(\mathbf{x}|C=0)$, we can evaluate the LPR in equation (10.42) and obtain a decision rule. The decision rule will be a quadratic function of the measurements. We do not complete the derivation here but refer to Problem 10.7. Van den Berg et al. [186] showed that people judge sameness in a way that is close to the predictions of this model. This example shows how inference of a relatively abstract quality like "sameness" can be modeled in a Bayesian way using the exact same procedure we used for inferring a physical stimulus.

10.5 Perceptual Organization

Extracting information about structure in the world from sensory input is an important part, if not the ultimate goal, of perception. The world is highly structured: the shape of an object consists of a string of small line elements, objects are ordered in depth, and a musical piece consists of delicately arranged sequences of tones. As the famous Kanizsa illusion shows (figure 10.6), the brain perceives visual structure even if there is only indirect sensory evidence for its existence. Indeed, at some level, all of our perception of structure is

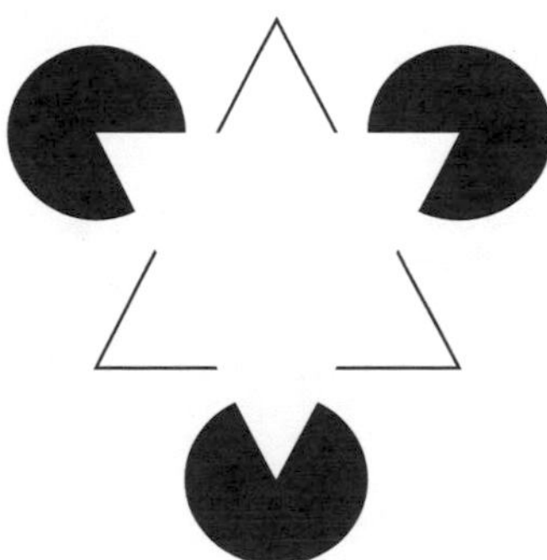

Figure 10.6
Kanizsa triangle. People tend to perceive a white triangle lying on top of three black discs.

based on indirect information. When we walk on a busy street, our brain seemingly effortlessly separates the multiple sound sources from a single continuous stream. Arguably, object recognition, no matter the sensory modality, consists of the detection of structures at multiple levels.

Since the birth of psychology, its practitioners have been intrigued by the ways in which the brain perceives structure among sets of constituent elements. For the most part, the explanations given for structure perception phenomena have been qualitative and descriptive. The leading explanatory framework has been provided by the *Gestalt laws* or Gestalt principles, which describe under which circumstances elements are perceived as belonging to a whole (as in the Kanizsa illusion). There is a long list of such laws:

- The law of good continuation: elements that suggest a continued visual line will tend to be grouped together.
- The law of closure: objects such as shapes, letters, pictures, and so on, are perceived as being whole even when they are not complete (figure 10.7A).
- The law of similarity: elements within an assortment of objects are perceptually grouped together if they are similar to each other (figure 10.7B).
- The law of common fate: objects are perceived as related if they objects point in the same direction. We encountered this in section 2.6.
- The law of proximity: objects that are close to each other are perceived as forming a group (figure 10.7C–D).
- The law of good Gestalt: objects tend to be perceptually grouped together if they form a pattern that is regular, simple and orderly.

Almost since their conception, Gestalt laws have been criticized for being vague, for example, in the principle of good Gestalt, where "regular," "simple," and "orderly" are not defined. Moreover, the Gestalt laws become murky when they conflict with each other. Bayesian models have the potential to improve on these laws, as Bayesian models are normatively motivated and allow for precise quantitative descriptions. The basic idea is that the observer considers two hypotheses—for example, the elements belong together or they do not—and evaluates their posterior probabilities. In section 2.6, we treated the Gestalt law of common fate in that way. In the following, we will examine the Gestalt law of good continuation with a Bayesian lens.

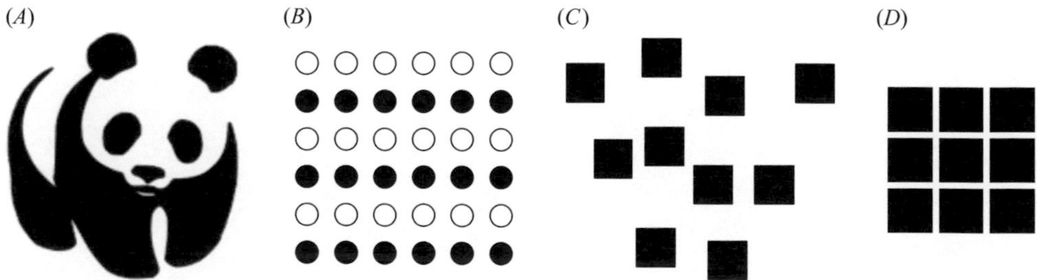

Figure 10.7
(*A*) Law of closure. (*B*) Law of similarity. (*C-D*) Law of proximity: the left set is seen as nine disjoint squares, and the right set as a single square.

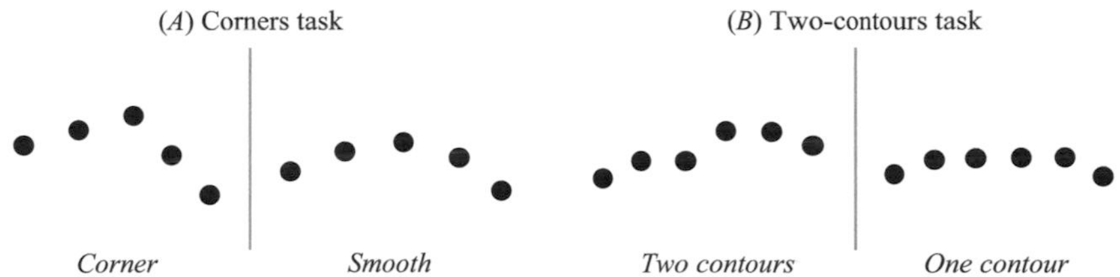

Figure 10.8
Contour integration tasks. Figure reproduced from [49] (*A*) Subjects judged whether the dots did or did not form a corner. (*B*) Subjects judged whether one or two contours were present.

10.5.1 Contour Integration

When visually segmenting a scene, an observer has to identify which line elements belong to the same boundary or contour. This task is called contour integration. In one task, Feldman had subjects judge whether five dots formed a corner or a smooth curve (figure 10.8A) [49]. In another task, subjects judged whether the dots formed one or two contours (figure 10.8B). For both tasks, Feldman developed a Bayesian model. The key ingredient of this model is the idea that on a smooth contour, the angle between two successive line segments tends to be small. The Bayesian model accounts well for the trends in human data.

Zhou, Acerbi, and Ma [214] studied the perception of alignment between horizontal line segments separated by a gap. Their main interest was the effect of uncertainty, as manipulated through retinal eccentricity. A Bayesian observer would take into account uncertainty in a specific way when setting a criterion for declaring a measured misalignment sufficient to report a misalignment in the world. Human decisions also take uncertainty into account, in a way that is close to the Bayesian predictions.

In that study, the distribution of misalignments was set by the experimenter. A different approach makes use of natural statistics to specify the generative model. Geisler and Perry [57] tested the hypothesis that natural statistics combined with a Bayesian model can predict human contour integration judgments. They first formalized the problem by introducing the relevant variables (figure 10.9). The state of the world of interest is whether two line elements belong to the same contour. This is a binary variable: $C=0$ (no) or $C=1$ (yes). The stimuli are two edge elements, and we assume that only their position relative to each other matters. The parameters used to describe this relative position are shown in figure 10.9A: distance between midpoints (d), the angle between the reference element and the line connecting the midpoints (ϕ), the angle between the reference element and the orientation of the other element (θ), and finally the contrast polarity (ρ): if one were to connect the two elements using a contour, would which side is darker change between the elements?

The generative model of this task is described by the probability distributions $p(d, \phi, \theta, \rho | C=1)$ and $p(d, \phi, \theta, \rho | C=0)$. The authors estimated these probabilities by analyzing natural scenes. A photograph of an outdoors, natural scene, such as leaves lying on the forest floor, was first analyzed automatically by an algorithm that extracted the edges. The image was then presented to a human observer with one pixel marked. The observer indicated which other pixels in the image belonged to the same contour. Humans were highly consistent in making these judgments. In this way, the generative model was estimated. Unlike the previous examples in this book, in this case the generative model was purely specified

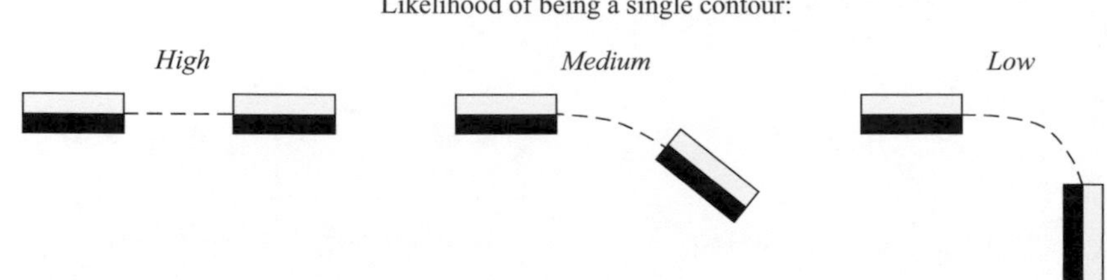

Figure 10.9
For pairs of edge elements, the likelihood that they belong to a single contour depends on their spatial separation and relative angle, among other factors. From left to right, this likelihood decreases as a single contour would produce the given configuration with increasingly lower probability.

numerically, that is, through histograms indicating the frequency of occurrence of every combination of parameters. Another difference with most generative models discussed so far is that sensory noise was assumed to be negligible. All uncertainty in the task derives from ambiguity. Finally, a difference is that human observers were used to construct the generative model; the generative model was not constructed by the experimenters.

In a subsequent experiment, human observers judged whether two edge elements passing under an occluder belonged to the same or to different contours. "Same" and "different" each occurred 50 percent of the time. A Bayesian observer would make this judgment by computing the posterior ratio. When the prior is flat, reflecting the frequencies of "same" and "different," the posterior ratio is equal to the likelihood ratio.

$$\frac{p(C=1|d,\phi,\theta,\rho)}{p(C=0|d,\phi,\theta,\rho)}=\frac{p(d,\phi,\theta,\rho|C=1)}{p(d,\phi,\theta,\rho|C=0)}. \tag{10.53}$$

The modelers were able to make predictions for human judgments based on the generative model. An illustrative example of these predictions is shown in figure 10.9B. As one might expect, this shows that contours tend to be smooth. Human observers performed close to the Bayesian observer, with similar patterns of errors.

This study is closely related to the aforementioned Gestalt law of "good continuation." We have now seen that by examining the statistics of natural scenes, this rather vague principle can be quantified: elements are grouped together if they have a higher probability of belonging to the same contour than not. This illustrates how Bayesian models can improve on qualitative observations in psychology.

10.5.2 Intersecting Lines

Consider the picture in figure 10.10A. Most people will see this as two intersecting lines, instead of two angles touching, though either interpretation is possible (see figure 10.10B; there are in fact more possible world states). The law of continuity would state in this case that individuals tend to perceive the two objects as two single uninterrupted entities because elements tend to be grouped together when they are aligned.

Here, we don't need to measure natural statistics to be able to make the general argument about why hypothesis 1 is more common. The generative model is shown in figure 10.10C. The top variable refers to the world state, corresponding to two intersecting lines ($C=1$) or to two touching angles ($C=2$). We have to parameterize the stimuli. We start with $C=1$.

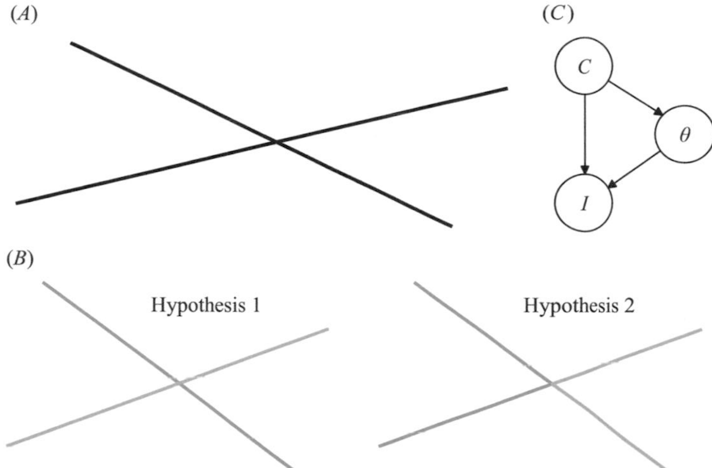

Figure 10.10
(*A*) Image. (*B*) Two possible interpretations of this image ($C=1$ and $C=2$). (*C*) Generative model. C is the world state, θ are the parameters, I is the image.

A line is parameterized by two numbers, as one can see by writing an equation of a line: $y=ax+b$. Thus, four parameters specify the two lines in hypothesis 1. Now consider $C=2$. An angle is parameterized by four numbers: two coordinates for the origin of the angle, one angle for the first leg, and one angle for the second leg. Thus, eight parameters are needed for hypothesis 2. Finally, in both interpretations, the image is uniquely determined by the parameters.

The Bayesian observer performs inference by computing the posterior ratio of the world states, based on the given image I:

$$\frac{p(C=1|I)}{p(C=0|I)}=\frac{p(I|C=1)}{p(I|C=0)}\frac{p(C=1)}{p(C=0)}. \tag{10.54}$$

We cannot say much about the prior, but we can evaluate the likelihood ratio. We denote the parameters in each hypothesis by a vector θ. Thus, θ is four-dimensional when $C=1$ and eight-dimensional when $C=2$. Since θ acts as a nuisance parameter, each of the likelihoods is computed by marginalizing over θ. To simplify the argument, let us assume that all parameters take on discrete values. The marginalization is then the sum

$$p(I|C)=\sum_{\theta}p(I|\theta,C)p(\theta|C). \tag{10.55}$$

We know that the image is uniquely determined by the parameters and the hypothesis. Therefore, $p(I|C,\theta)$ equals 0 for all parameter combinations θ except the one that produces the given image I. We denote this parameter combination by θ_I. For this combination, $p(I|\theta,C)$ equals 1. The integral then simply becomes

$$p(I|C)=p_{\theta|C}(\theta_I|C). \tag{10.56}$$

All that remains now is to evaluate the probability of θ_I under each hypothesis. The following illustrative calculation is adapted from [118]. Let us assume that all parameters are independent and each parameter takes on 100 possible values. Then the probability of θ_I (or any other parameter combination) under hypothesis 1 is $\left(\frac{1}{100}\right)^4$, whereas the probability of θ_I (or any other parameter combination) under hypothesis 2 is $\left(\frac{1}{100}\right)^8$. That means that the likelihood ratio is $\left(\frac{1}{100}\right)^{-4} = 10^8$. In other words, hypothesis 1 is 100 million times as likely as hypothesis 2.

This explains why humans observe the image as two intersecting lines rather than as two touching angles. Intuitively, the hypothesis $C=1$ requires that the opposite angles in the image share a common vertex and be equal, whereas the hypothesis $C=2$ permits this configuration but also a vast number of other configurations. The fact that the image conforms to the restricted features predicted by $C=1$ therefore favors that hypothesis.

Of course, the precise numerical value of the likelihood ratio will depend on our assumptions regarding the priors over the parameters within each hypothesis. However, any Bayesian observer who begins with broad prior distributions will favor hypothesis $C=1$ when shown the image in figure 10.10A. The essence of the argument is if two hypotheses can account for the observations equally well, a Bayesian observer will favor the hypothesis that has the lowest number of parameters. In this sense, Occam's razor is an emergent property of Bayesian inference: simpler models are better.

Often, more complex hypotheses (ones with more parameters) can account better for the data. A specific parameter combination within a complex hypothesis may (unlike the example considered here) fit the data more precisely than any parameter combination allowed by a simpler hypothesis. Thus, there is a trade-off between complexity and power. This trade-off is also captured in equation (10.55), since $p(I|\theta, C)$ is an indication of the power of the hypothesis.

Incidentally, the Bayesian observer who selects between two perceptual hypotheses is mathematically identical to a Bayesian experimenter who analyzes data in order to select between two competing models. Thus, Bayesian model comparison follows the same equations that are here discussed to describe the human brain, including the trade-off between complexity and power. This is discussed in more detail in appendix section C.5.3.

10.6 Summary

In this chapter, we introduced Bayesian models for deciding if a set of stimuli are the same or different (or if they belong together). We learned:

- Sameness judgment is about relations between stimuli, and as such represents an important type of perceptual inference.
- Mathematically, sameness judgment involves structured generative models that differ in complexity, that is, where the models have different numbers of parameters.
- Bayes' rule can be used equally to calculate a posterior over structures (models) in addition to posteriors over "internal" stimulus variables. Both of these calculations require marginalization.
- Gestalt laws have been the leading explanatory framework for visual structure perception, involving decisions such as whether or not line elements belong to a contour. Gestalt laws are, however, vague and beg for a deeper explanation.

- Visual structure perception can at least qualitatively be accounted for by Bayesian principles: elements are grouped together when the likelihood favors that interpretation.
- The trade-off between model power and model complexity, which can explain some Gestalt percepts, is also an important theme in data analysis.

10.7 Suggested Readings

- Jacob Feldman. "Bayesian Contour Integration." *Perception and Psychophysics* 63 (2001): 1171–1182.
- Wilson S. Geisler and Jeffrey S. Perry. "Contour Statistics in Natural Images: Grouping across Occlusions." *Visual Neuroscience* 26, no. 1 (2009): 109–121.
- Daniel Goldreich and Mary A. Peterson. "A Bayesian Observer Replicates Convexity Context Effects in Figure-ground Perception."*Seeing and Perceiving* 25, no. 3–4 (2012): 365–395.
- Konrad P. Kording, Ulrik Beierholm, Wei Ji Ma, Steven Quartz, Joshua B. Tenenbaum, and Ladan Shams. "Causal Inference in Multisensory Perception." *PLoS One* 2, no. 9 (2007): e943.
- David J. C. MacKay. *Information Theory, Inference and Learning Algorithms*. Cambridge: Cambridge University Press, 2003.
- Yoshiyuki Sato, Taro Toyoizumi, and Kazuyuki Aihara. "Bayesian Inference Explains Perception of Unity and Ventriloquism Aftereffect: Identification of Common Sources of Audiovisual Stimuli." *Neural Computation* 19, no. 12 (2007): 3335–3355.
- Ronald van den Berg, Michael Vogel, Krešimir Josić, and Wei Ji Ma "Optimal Inference of Sameness." *Proceedings of the National Academy of Sciences* 109, no. 8 (2012): 3178–3183.
- Max Wertheimer, "Gestalt theory." *A Source Book of Gestalt Psychology*, edited by Willis D. Ellis, 1–11. London: Kegan Paul, Trench, Trubner, 1938.
- Yanli Zhou, Luigi Acerbi, and Wei Ji Ma. "The Role of Sensory Uncertainty in Simple Contourition." *PLoS Computational Biology* 16, no. 11 (2020): e1006308.

10.8 Problems

Problem 10.1 Think of a real-world problem, not covered in this chapter, that would require same-different judgements. Formulate it in Bayesian terms.

Problem 10.2 Imagine you are a psychophysicist. Come up with an experiment, not covered in this chapter, to study whether people are Bayesian in their same-different judgments.

Problem 10.3 Refer back to section 10.2.2 on same-different judgments of binary stimuli. Consider the case where the stimuli are the same in half of the cases: $p_{\text{same}} = 0.5$.

(a) Prove that the condition $d > 0$, with d in equation (10.18) reduces to equation (10.19), $\text{sign}(x_1) = \text{sign}(x_2)$. (Hint: it might be helpful to use the definition and properties of the hyperbolic cosine function.)

(b) In this case, derive an expression for $p(\hat{C}=1|s_1,s_2)$ in terms of cumulative standard normal distribution functions, Φ_{standard} (see box 7.1).

(c) Based on your answer to (b), derive an expression for proportion correct. Simplify the expression until it has only a single Φ_{standard} in it.

Problem 10.4 Refer back to section 10.2.2 on same-different judgments of binary stimuli. Unless $p_{\text{same}}=0.5$, the decision rule $d>0$ does not simplify analytically. However, we can still visualize the decision rule in (x_1,x_2) space: in which regions of this space does the Bayesian observer respond "same," and in which regions "different"? Our goal is to get an intuition for how σ and p_{same} interact to affect optimal behavior.

(a) Choose $\mu=1$. Consider three values of σ (0.5, 1, and 2), and three values of p_{same} (0.4, 0.5, and 0.6). For each combination of σ and p_{same}, create a plot that shows the boundary between the "respond same" and "respond different" regions in (x_1,x_2) space, where either measurement can take values between -2 and 2. Display the nine plots in a 3×3 grid for easy comparison.

(b) Describe and interpret the effects of σ and p_{same} on the decision boundary.

Problem 10.5 Refer back to section 10.2.3 on the response distribution in the binary same-different task. Choose $p_{\text{same}}=0.4$, $\mu=1$, and $\sigma=1$. We will use two numerical techniques for calculating the probability that the observer will report "same" in a given stimulus condition, $p(\hat{C}=1|s_1,s_2)$. Note that we need to calculate four numbers, since there are four possible combinations of s_1 and s_2.

(a) Use Riemann integration. For both x_1 and x_2, use a grid from -5 to 5 in steps of 0.01. Calculate the four numbers we are looking for.

(b) Use Monte Carlo simulation. Compare. The results should be very similar.

Problem 10.6 We build on the causal inference model in section 10.3.

(a) Code up the causal inference model and reproduce figure 10.3A. Use the parameters given in the caption to figure 10.3.

(b) Reproduce figure 10.3C using Riemann integration.

(c) Explain why fixing $s_1=0$ and varying s_2 produces a different result than fixing $s_2=0$ and varying s_1.

(d) Reproduce figure 10.3C using Monte Carlo simulation.

Problem 10.7 We work out some details of the model for same-different judgment of multiple stimuli in section 10.4. Suppose there are N stimuli. When $C=1$, the stimuli are all identical, and their common value is drawn from a normal distribution with mean 0 and standard deviation σ_s. When $C=2$, all stimuli are drawn independently from the same normal distribution. Assume $p(C=1)=0.5$ and independent measurement noise with standard deviation σ.

(a) Derive the Bayesian MAP decision rule. The inequality representing the decision rule should have a quadratic function of $x_1,\ldots,x_N$ on the left-hand side and a constant expression on the right-hand side. (Hints: (1) Use precision instead of variance notation.

(2) Use appendix section B.7.4. The final decision rule will have on the left-hand side a function of the measurements that can be interpreted relatively easily.)

(b) Derive the probability of reporting “same” on a “different” trial, when the stimuli are $\mathbf{s} = (s_1, \ldots, s_N)$.

Problem 10.8 A musical tone has a pitch, which defines how high a tone sounds. An experimenter conducts an auditory oddity detection task, as follows. She draws two values of pitch, denoted s (we may think of it as the log of the frequency that is close to perception), independently from a Gaussian distribution $p(s)$ with standard deviation σ_s (the mean is irrelevant). She then presents to the subject a sequence of three tones, two with the first drawn value of pitch, and one with the second value. The three tones are presented in random order and the subject reports which of the three is the odd one out. Assume that the measured pitch of each tone is independently corrupted by zero-mean Gaussian noise with standard deviation σ.

(a) Draw a graphical representation of the generative model and write down expressions for the probability distributions over the variables.

(b) Derive how the Bayesian MAP observer should estimate the temporal location of the oddball (1, 2, or 3) from the measured pitches.

(c) Explain why the rule you obtained makes intuitive sense.

(d) Assume $\sigma_s = 2$. For each value of σ from 0.05 to 5 in steps of 0.05, simulate 100,000 sets of measurements and use those to estimate the probability correct of the optimal observer. Plot this probability as a function of σ.

(e) What is the value of the asymptote as $\sigma \to \infty$?

(f) Repeat step (d) for two ad hoc models. In the first ad hoc model, the observer determines which measurement lies farthest away from the average of the three measurements, and reports the location of that measurement as the oddball location. In the second ad hoc model, the observer compares the distances between pairs of measurements, finds which of these three distances is smallest, and chooses as the oddball location the location of the measurement not included in that pair. Show that these ad hoc models lead to lower performance by plotting probability correct of the three models as a function of σ in the same plot.

11

Search

How do we find a target from among many similar objects?

In this chapter, we continue our exploration of inference with multiple stimuli by studying search. Search is the task of looking for a *target* among *distractors*. If the target is known to be present, the search task is a *target localization* task: Where is the target? If the target is not known to be present, the search task is a *target-detection* task: Is the target present? Search can take place in any sensory modality. For example, you might want to know where in a crowd your friend is standing (visual), determine whether someone is calling your name among many speaking voices (auditory), figure out from which direction an offensive odor is emanating (olfactory), or find a particular coin from among several in your pocket (tactile). Interestingly, the popular N-alternative forced-choice task, while not traditionally presented as such, is a search task: in such a task, the observer localizes a target that is sure to be present among N stimuli, presented either simultaneously or sequentially. Search is being probed by a classical behavioral set of experiments and in this chapter we see how it can be modeled.

Plan of the Chapter

We begin by discussing the ecological relevance of visual search to an animal in the wild and by defining three types of search: target localization, target detection, and target categorization. We consider a specific example of target localization in which a predator attempts to localize a fish, known to be present somewhere in the scene, that hovers above a riverbed of similar appearance. We later consider the same example as a target-detection task, in which the predator is unsure whether a fish is present. We consider both target localization and detection in the presence of measurement noise. Finally, we briefly discuss visual search that involves eye movements.

11.1 Many Forms of Visual Search

To an animal in the wild, performing efficient visual search can be a matter of life or death. Frequently, animals must detect whether a predator is present in a visual scene. The predator might be hidden or camouflaged, making it difficult to distinguish from the surrounding visual elements (figure 11.1A). Predators must similarly attempt to detect and localize

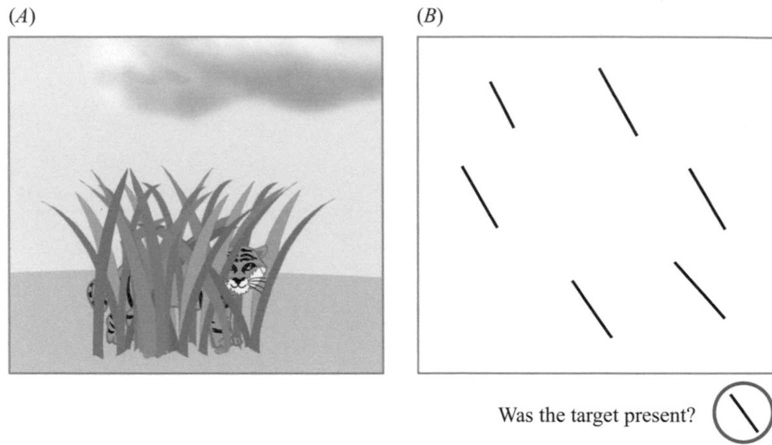

Figure 11.1
Visual search. (*A*) In the animal world. (*B*) In the laboratory.

well-camouflaged prey. Modern-day humans might want to find a particular piece of paper from among many similar notes on a cluttered desk (figure 11.1B) or locate a particular product from among many similar ones on the shelves of a grocery store.

In the laboratory, investigators use a variety of simplified tasks to study visual search. Doing psychophysics with natural scenes is difficult for several reasons. First, natural scenes are so rich in content that describing them mathematically would require a high-dimensional space. Second, it is not clear what exactly an object is: is the entire tree the object, or is the object an individual branch, or perhaps even a leaf? Third, the noise in the different dimensions of a natural scene typically has a complex and largely unknown correlation structure that goes far beyond the Gaussian noise distribution that we have considered so far. Therefore, in laboratory visual search tasks with a modeling approach, it is common to use extremely simplified search scenes that contain a relatively small number of highly distinct objects that differ only along a single stimulus dimension (figure 11.1B). Obviously, a big gap exists between such simple, artificial scenes and natural tasks, but it is hoped that the computational principles we can unveil using laboratory tasks matter in natural tasks—in essence, that the laboratory task allows us to study some of the minimal building blocks of the computations that the brain performs in the real world.

Laboratory visual search tasks come in at least three main flavors:

- *Target localization.* It is given that one or more targets are present. The observer has to decide which of the presented stimuli are the targets (discrete) or where the targets are located (continuous).
- *Target detection.* Targets may or may not be present, and the observer has to decide whether any targets are present.
- *Target categorization.* The observer is asked about the category of the target. We will only handle this third flavor in, below problem 11.12.

Computationally, these three flavors of search tasks are deeply similar, as we will see in the following pages. In each of these paradigms, there are several factors to consider:

- How many targets can there be? In this chapter, we will for simplicity mostly consider the case of a single target (but see problem 11.7).
- Are the distractors independent or somehow related to each other? We will throughout the chapter consider mainly the case of independent distractors (but see Problems).
- Is the task hard because of ambiguity, noise, or both? In many visual search tasks, ambiguity is sufficient to make the task hard. It also simplifies our treatment to assume no measurement noise, so in this chapter, we will start with that setting.

In each paradigm, the location of the target is a priori unknown to the observer. For this reason, the observer has to consider every possible location. In localization tasks, the target location is the world state variable of interest. In detection tasks, the target location is a nuisance parameter and the Bayesian observer marginalizes over this variable.

11.2 Target Localization: Camouflage

A beautiful naturalistic example of search is seen in animals attempting to perceive camouflaged predators or prey. In chapter 1, we discussed camouflage as an evolved strategy for producing broad likelihood functions in the observer. Here we consider a toy problem inspired by the image of the flounder (figure 1.9). Suppose that a dotted fish species inhabits a river and that a fish sometime hovers just above the pebble-covered riverbed. River fish often point upstream, as this orientation provides easier stability against the current, and faces the fish in the direction of potential food that is being swept downstream. We assume that the dotted fish takes this orientation (figure 11.2A). We divide the area under consideration into a 10×10 grid and consider a species of dotted fish that has size 1×10. If the fish is present, then the prior probability of it being located in any of the ten rows is equal. We assume that the fish skin color is identical to that of the riverbed mud, and that each dot on the skin closely resembles a pebble. Suppose that each grid square on the riverbed independently has a probability a of containing a pebble and $1-a$ of not containing a pebble. For the fish species, the probability of a dot within any grid square on the skin is b, and of no dot is $1-b$.

Exercise 11.1 Which factors do you think are important for camouflage. Which ones do animals in nature regularly use? Do you know of any that are missing?

We will begin our study of search with a localization task. Consider a predator that knows a fish is present in this area; perhaps the predator saw the outline of the fish as it moved in this general direction a moment earlier. The predator's task is to determine, given a visual observation such as in figure 11.2B, in which row the fish is located. Each row is a "stimulus" s_i. The predator knows that exactly one of N presented stimuli $s_1, \ldots, s_N$ is the target, and has to decide which one. We denote the index of the target by L; thus, s_L is the target stimulus.

Step 1: Generative model (figure 11.2C) The generative model contains target location L, which is an integer between 1 and 10, the number of possible y positions for the fish, and the stimuli $(s_1, \ldots, s_N)$, which we will sometimes summarize by $\mathbf{s}$ (boldface notation for a vector). The uniform prior is $p(L) = \frac{1}{N}$. We next need to specify the distribution of the stimulus vector $\mathbf{s}$ given a target location L. In our scenario, the appearances of the target (fish) and each of the distractors (rows of pebbles) are drawn independently from their respective

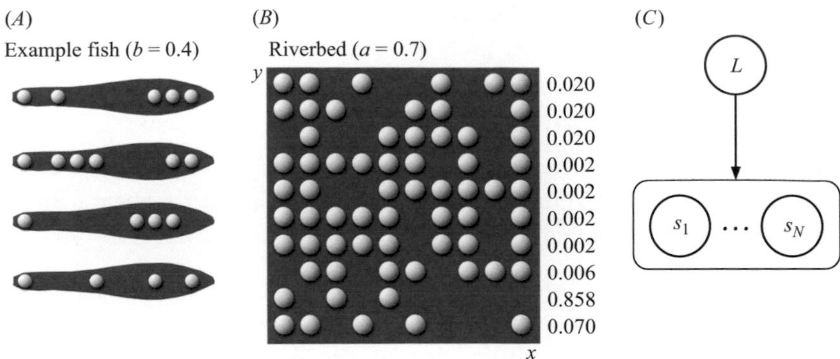

Figure 11.2
The challenge of localizing a camouflaged object. (*A*) Four members of the dotted fish species with dot probability $b = 0.4$. The fish has size 1×10. (*B*) A 10×10 grid of riverbed, with pebble probability $a = 0.7$, in which a fish is present in one row (i.e., one y-position). Next to each row, we list the posterior probability that the fish is present in that row. (*C*) Generative model. Here, the number of rows, N, is 10.

distributions. Thus,

$$p(\mathbf{s}|L) = p(s_1|L)p(s_2|L) \cdots p(s_N|L) \equiv \prod_{i=1}^{N} p(s_i|L). \tag{11.1}$$

We denote the distribution of the target stimulus by $p_{\text{target}}(s)$ and the distribution of the distractor stimulus by $p_{\text{distractor}}(s)$. Then

$$p(s_i|L) = \begin{cases} p_{\text{target}}(s_i) & \text{if } i = L; \\ p_{\text{distractor}}(s_i) & \text{if } i \neq L. \end{cases} \tag{11.2}$$

We can work out these probabilities using the proportion of dots on a fish and the proportion of pebbles in the riverbed. This calculation parallels the one of the probability of a sequence of dry and rain days in section 6.2. For example, if the stimulus is a fish, then the probability of a particular dot pattern is obtained by multiplying ten factors, one for each horizontal position, where a dot contributes a factor b and an empty position contributes a factor $1 - b$. As a result, if we denote the number of dots in row i as n_i, we have

$$p_{\text{target}}(s_i) = b^{n_i}(1 - b)^{10 - n_i}; \tag{11.3}$$

$$p_{\text{distractor}}(s_i) = a^{n_i}(1 - a)^{10 - n_i}. \tag{11.4}$$

This completes the specification of the generative model.

Step 2: Inference Even in this relatively simple case of visual search, the observer needs to combine information across the N items: after all, the target is present in *exactly one* location. That means that evidence in favor of a row containing pebbles should count as evidence in favor of one of the other rows containing the fish.

We now formalize the inference. The observation is a set of stimuli $\mathbf{s}$. The world state variable of interest is target location L, so we want to calculate the posterior over L, $p(L|\mathbf{s})$. By Bayes' rule,

$$p(L|\mathbf{s}) \propto p(\mathbf{s}|L)p(L). \tag{11.5}$$

Given the uniform prior and equation (11.1), this becomes

$$p(L|\mathbf{s}) \propto \prod_{i=1}^{N} p(s_i|L). \tag{11.6}$$

Substituting equation (11.2),

$$p(L|\mathbf{s}) \propto p_{\text{target}}(s_L) \prod_{i \neq L} p_{\text{distractor}}(s_i). \tag{11.7}$$

Since we are in step 2, L is the *hypothesized* target location and s_L is the observed stimulus value at that location. The product is over all distractors, which means all stimuli besides the target. Therefore, we can also write

$$p(L|\mathbf{s}) \propto \frac{p_{\text{target}}(s_L)}{p_{\text{distractor}}(s_L)} \prod_{i=1}^{N} p_{\text{distractor}}(s_i). \tag{11.8}$$

Now, the product is over all stimuli and therefore does not depend on L. Since our relationship is only a proportionality, we can write

$$p(L|\mathbf{s}) \propto \frac{p_{\text{target}}(s_L)}{p_{\text{distractor}}(s_L)}. \tag{11.9}$$

Interestingly, the fraction is the likelihood ratio of target presence if there had been only a single stimulus, or in other words, the "local" likelihood ratio of classifying s_L as a target versus a distractor.

Finally, to obtain a proper posterior, the right-hand side of equation (11.9) needs to be normalized by dividing by the sum of that expression over all L. Equation (11.9) is the fundamental relationship for localization of a single target among independent distractors.

So far, we have used the following aspects of the generative model: (a) that distractors are drawn independently from a distractor distribution; (b) that there is only a single target, drawn independently from a target distribution; (c) that if the target is present, it has the same probability of being at each location (although this is easily generalized, by replacing $\frac{1}{N}$ by $p(L)$ everywhere). We did not yet use the specifics of p_{target} and $p_{\text{distractor}}$ in our camouflage scenario. We now do so by substituting the expressions from equations (11.3) and (11.4) into equation (11.9) for the posterior over target location:

$$p(L|\mathbf{x}) \propto \left(\frac{b}{a}\right)^{n_i} \left(\frac{1-b}{1-a}\right)^{10-n_i}. \tag{11.10}$$

Exercise 11.2 Show that this expression is correct.

We calculated these posterior probabilities in the example of figure 11.2B.

If the predator maximizes accuracy, they would perform maximum-a-posteriori estimation of L. This means picking the value of L for which the right-hand side of equation (11.9) is highest. We can write this as

$$\hat{L} = \underset{L}{\operatorname{argmax}} \frac{p_{\text{target}}(s_L)}{p_{\text{distractor}}(s_L)}. \tag{11.11}$$

An interesting aspect of this rule is that each candidate target stimulus, s_L, should be considered individually. The observer computes for each individual stimulus the likelihood ratio as if they were *classifying* (categorizing) that stimulus as a target or a distractor. Thus, the stimulus with the highest categorization likelihood ratio is also the one with the highest posterior probability of being the target in this search task. This is not a general law; it crucially depends on the independence assumption we made here.

Since there is no measurement noise, step 3 is unnecessary: for the same set of stimuli **s**, the Bayesian observer will always make the same response. (This could be a curse rather than a blessing, since actual humans would not necessarily always make the same response. To account for such variability, decision noise can be added to the model.) However, it is not trivial to calculate performance metrics (such as proportion correct) for given a and b, as this would require marginalizing over the **s** associated with these parameters. We will do this in problem 11.11.

11.3 Target Localization with Measurement Noise

In the previous example, stimuli were discrete. We will now consider target localization with continuous stimuli, in which we also introduce our familiar concept of measurement noise. The basic structure of the target localization tasks we consider is still the same as in the fish example: the observer searches for an always-present single target, whose value is drawn from a target distribution, among $N-1$ distractors, whose values are drawn independently from a distractor distribution. Now, however, the stimuli are all one-dimensional continuous variables. Moreover, the observer has measurement noise. The uncertainty now does not come only from the overlap between the target and distractor distributions but also from the measurement noise.

Incidentally, target localization of this kind is computationally identical to an N-alternative forced-choice paradigm, also called N-interval forced choice, in which a target is present among N stimuli presented sequentially. For example, in a two-interval forced-choice paradigm, the subject might be presented with two images in sequence, one of which contains one pure noise and the other noise plus a faint target object. The subject reports whether the first or the second image contained the target object. (Note that some researchers use the term "N-alternative forced choice" in a broader way, namely for any task with N choice options.) Thus, in the rest of this section, one can replace space by time to obtain a model for N-interval forced-choice tasks.

We now go through our usual three steps.

Step 1: Generative model (figure 11.3A) We again denote the location of the unique target by L and the vector of stimuli by $\mathbf{s}=(s_1,\ldots,s_N)$. We now allow for a general prior distribution over target location, $p(L)$. We assume independence of the stimuli conditional on L, as in equation (11.1). Given L, we also again have

$$p(s_i|L)=\begin{cases} p_{\text{target}}(s_i) & \text{if } i=L; \\ p_{\text{distractor}}(s_i) & \text{if } i\neq L. \end{cases} \tag{11.12}$$

In contrast to the fish example, $p_{\text{target}}(s)$ and $p_{\text{distractor}}(s)$ are now continuous distributions; examples are shown in figure 11.3B.

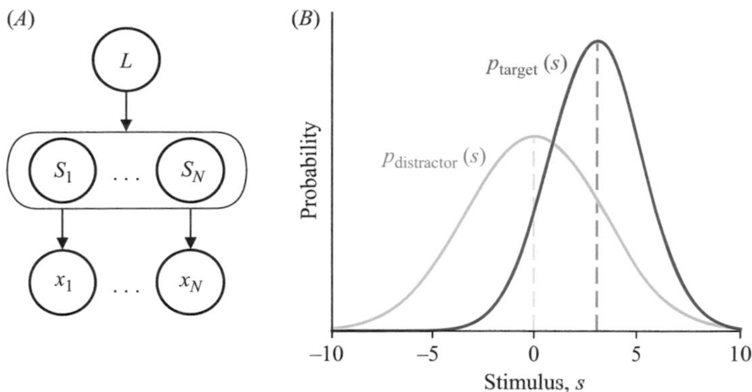

Figure 11.3
Visual search with a single target, independent distractors, and measurement noise. (*A*) Generative model when there are N stimuli. L is the index of the target; it is an integer between 1 and N. (*B*) Example target and distractor distributions. Here, the target distribution is normal with mean 3 and standard deviation 2, and the distractor distribution is normal with mean 0 and standard deviation 3. The two distributions overlap, so distractors can be confused with the target.

For the measurements, we will make our usual assumption that they are independent when conditioned on the stimuli. Formally, we can write for the probability of the measurements $\mathbf{x}$ given stimuli $\mathbf{s}$:

$$p(\mathbf{x}|\mathbf{s}) = \prod_{i=1}^{N} p(x_i|s_i). \tag{11.13}$$

Each individual measurement follows a Gaussian distribution with mean equal to the respective stimulus:

$$p(x_i|s_i) = \mathcal{N}(x_i; s_i, \sigma^2). \tag{11.14}$$

We will shortly that the two forms of conditional independence go well together.

Step 2: Inference In this task, Bayes' rule takes the form

$$p(L|\mathbf{x}) \propto p(L)p(\mathbf{x}|L). \tag{11.15}$$

The likelihood over target location can be written as a marginalization over the stimuli, so that

$$p(L|\mathbf{x}) \propto p(L) \int p(\mathbf{x}|\mathbf{s})p(\mathbf{s}|L)d\mathbf{s}. \tag{11.16}$$

Thanks to the conditional independence of the measurements, we can further evaluate the likelihood as an integral over each stimulus variable $s_1, \ldots, s_N$:

$$p(L|\mathbf{x}) \propto p(L) \int \cdots \int \left(\prod_{i=1}^{N} p(x_i|s_i) \right) \left(\prod_{i=1}^{N} p(s_i|L) \right) ds_1 \cdots ds_N. \tag{11.17}$$

The two products combine, and each s_i appears in only one factor in the resulting product. As a result, the N-dimensional integral reduces to a product of one-dimensional integrals (see box 10.2):

$$p(L|\mathbf{x}) \propto p(L) \prod_{i=1}^{N} \int p(x_i|s_i) p(s_i|L) ds_i. \tag{11.18}$$

Thus, comparing to equation (11.6), we have replaced the individual-item likelihoods $p(s_i|L)$ with marginals. The rest of the derivation proceeds analogously to the previous section, and we find for the likelihood

$$p(L|\mathbf{x}) \propto p(L) \frac{\int p(x_i|s_i) p_{\text{target}}(s_i) ds_i}{\int p(x_i|s_i) p_{\text{distractor}}(s_i) ds_i}. \tag{11.19}$$

As in the fish example, the second factor is reminiscent of the likelihood ratio in a binary classification task, in particular equation (8.25): it is the "local" likelihood ratio of stimulus s_L being the target versus a distractor.

Special Case: Single Target Value, Single Distractor Value

We now consider the special case where the target can take only one particular value, s_T, and the distractor can take only one particular other value, s_D. Formally, this is written as a set of delta functions (see appendix section B.8),

$$p_{\text{target}}(s) = \delta(s - s_T); \tag{11.20}$$

$$p_{\text{distractor}}(s) = \delta(s - s_D). \tag{11.21}$$

Because we now have measurement noise, localizing the target can be a hard problem: the target is confusable with the distractors because of this noise. We assume that the measurement x_i follows a Gaussian distribution with mean s_i and variance σ_i^2. Then in equation (11.19), the delta functions and the integrals "cancel" each other out and we are left with

$$p(L|\mathbf{x}) \propto p(L) e^{\frac{s_T - s_D}{\sigma_L^2}\left(x_L - \frac{s_T + s_D}{2}\right)}. \tag{11.22}$$

This target location likelihood has a close connection to the discrimination task of chapter 7, specifically to equation (7.10): it is equal to the likelihood *ratio* of the stimulus s_L in isolation being the target versus a distractor.

11.4 Target Detection: Camouflage

We mentioned in section 11.1 that localization is only one of various forms of search. Another important form of search is target detection, the problem of determining whether a target object is (or multiple target objects are) present or absent in a scene. In this chapter, we assume that there is no more than a single target in the scene. We now continue with the camouflage example from section 11.2 but treat it as a detection problem. Suppose that the predator has newly arrived on the scene and we can no longer assume that a fish is present. The predator's task, given a visual observation of the riverbed, is to determine whether a fish is present ($C = 1$) or not ($C = 0$).

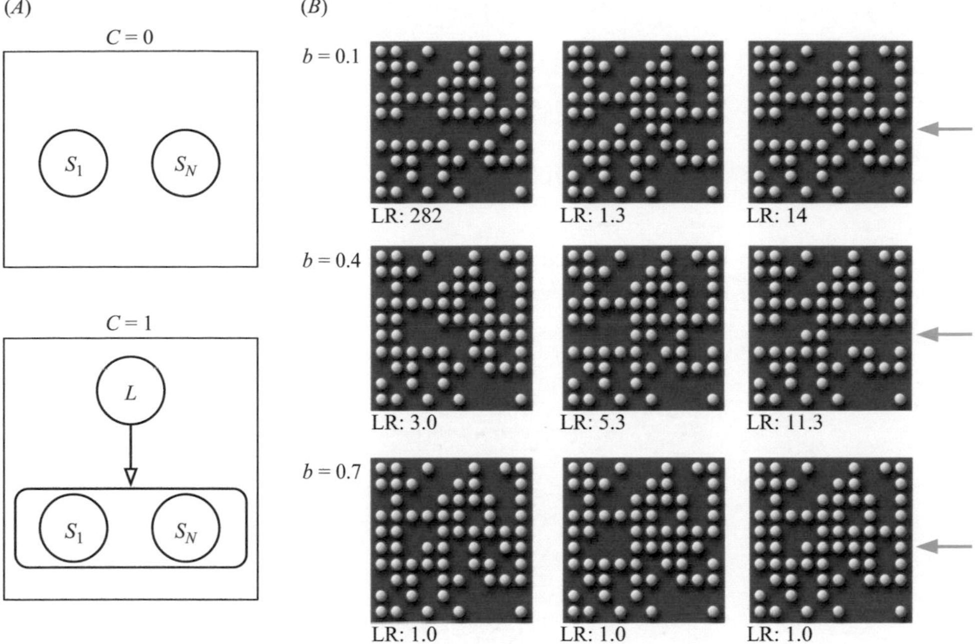

Figure 11.4
The challenge of detecting a camouflaged object. (*A*) Generative model diagram. For the current example, $N = 10$. (*B*) A riverbed for which $a = 0.7$. Three fish were randomly selected from species characterized by either $b = 0.1$, 0.3, or 0.7 (nine fish in total). The arrows indicate the true location of the fish. The likelihood ratio for fish presence is shown below each image.

Step 1: Generative model The generative model (figure 11.4A) is similar to the one in section 11.2 but now contains an extra variable, C, indicating the absence or presence of a target. The binary variable C comes with a prior $p(C)$. It also causes the generative model to have two components, just as in chapter 10. If the target is absent ($C = 0$), all stimuli s_i are distractors, that is, the riverbed with a pebble probability of a for each square. If the target is present ($C = 1$), the generative model is the same as in figure 11.2C, with L denoting target location and $\mathbf{s} = (s_1, \ldots, s_N)$ the set of dot patterns. When $C = 1$, only a single target is present, and each location is equally probable, $p(L) = \frac{1}{N}$.

Step 2: Inference The observer (predator) is now interested in inferring whether a fish is present, that is, the binary variable, C, from an observation of the riverbed, $\mathbf{s}$. Figure 11.4 shows the observations for nine randomly selected fish, with $b = 0.1$, 0.4, or 0.7 (three fish each), hovering above the same riverbed characterized by $a = 0.7$. Following our usual procedure when we encounter a binary variable, we express the posterior over C using the log posterior ratio:

$$d = \log \frac{p(C = 1|\mathbf{s})}{p(C = 0|\mathbf{s})} \tag{11.23}$$

$$= \log \frac{\mathcal{L}(C = 1; \mathbf{x})}{\mathcal{L}(C = 0; \mathbf{x})} + \log \frac{p(C = 1)}{p(C = 0)}. \tag{11.24}$$

We examine the likelihoods separately, starting with $C=0$. Using the independence of the observations, we immediately find for the likelihood of $C=0$:

$$\mathscr{L}(C=0;\mathbf{x}) \equiv p(\mathbf{s}|C=0) \tag{11.25}$$

$$= \prod_{i=1}^{N} p(s_i|C=0). \tag{11.26}$$

The likelihood of $C=1$ is a bit more complicated because the target location, L, serves as a nuisance variable and therefore we need to marginalize over it:

$$\mathscr{L}(C=1;\mathbf{x}) = p(\mathbf{s}|C=1) \tag{11.27}$$

$$= \sum_{L=1}^{N} p(L)p(\mathbf{s}|L, C=1) \tag{11.28}$$

$$= \frac{1}{N}\sum_{L=1}^{N} p(\mathbf{s}|L, C=1) \tag{11.29}$$

$$= \frac{1}{N}\sum_{L=1}^{N}\left(\prod_{i=1}^{N} p(s_i|L, C=1)\right). \tag{11.30}$$

where we have used conditional independence of the stimuli for the last equality. We recognize the product inside the sum, $\prod_{i=1}^{N} p(s_i|L, C=1)$, from equation (11.6). We can borrow from the derivation following that equation to further evaluate equation (11.30). This yields

$$\mathscr{L}(C=1;\mathbf{x}) = \frac{1}{N}\left(\prod_{i=1}^{N} p_{\text{distractor}}(s_i)\right)\sum_{L=1}^{N}\frac{p_{\text{target}}(s_L)}{p_{\text{distractor}}(s_L)}. \tag{11.31}$$

Combining equations (11.26) and (11.31), we can now calculate the likelihood ratio (LR) of the target being present versus absent,

$$\text{LR} \equiv \frac{\mathscr{L}(C=1;\mathbf{x})}{\mathscr{L}(C=0;\mathbf{x})} = \frac{1}{N}\sum_{L=1}^{N}\frac{p_{\text{target}}(s_L)}{p_{\text{distractor}}(s_L)}. \tag{11.32}$$

In parallel to equation (11.9), the fraction inside the sum is the "local" likelihood ratio of classifying s_L as a target versus a distractor. We could denote the corresponding likelihood ratio by LR_L,

$$\text{LR}_L \equiv \frac{p_{\text{target}}(s_L)}{p_{\text{distractor}}(s_L)}, \tag{11.33}$$

so that equation (11.32) becomes

$$\text{LR} = \frac{1}{N}\sum_{L=1}^{N}\text{LR}_L. \tag{11.34}$$

In other words, under the assumptions that we made, the "global" likelihood ratio, LR, is equal to an average of the local likelihood ratios. If we had used a general prior over location, $p(L)$, this average would have been a weighted average, with the weights given by $p(L)$.

The log of the global likelihood ratio can be written as

$$\log \mathrm{LR} = \log\left(\frac{1}{N}\sum_{L=1}^{N} e^{\mathrm{LLR}_L}\right), \tag{11.35}$$

where $\mathrm{LLR}_L = \log \mathrm{LR}_L$. Equation (11.35) shows that in this form of target detection, the "global" log likelihood ratio of target presence is a nonlinear function of the "local" log likelihood ratios. One can think of this as a spatial integration rule for evidence. Keep in mind that in this derivation, we have made essential use of the specific statistical properties of the task modeled here: that the distractors are independent, and that if there is a target, there is only one target. If these properties do not apply, then equation (11.35) might not either.

Finally, we can use the specifics of p_{target} and $p_{\text{distractor}}$ in our camouflage scenario. Doing so, equation (11.32) for the LR of the fish being present versus absent becomes

$$\mathrm{LR} = \frac{1}{10}\sum_{L=1}^{10}\left(\frac{b}{a}\right)^{n_L}\left(\frac{1-b}{1-a}\right)^{10-n_L}, \tag{11.36}$$

where n_L is the number of dots within the Lth row. We explore the implications of this expression in problem 11.11.

11.5 Target Detection with Measurement Noise

We now consider target detection with measurement noise. This section can be seen as a combination of section 11.3 on target localization with measurement noise, and section 11.4 on target detection without measurement noise.

Step 1: Generative model (figure 11.5) The generative model is the same as in section 11.4 except (a) we use a general prior, $p(L)$, instead of a uniform one; (b) there is now a layer of noisy measurements generated from the stimuli. We assume that the measurements are independent when conditioned on the stimuli.

Step 2: Inference Inference proceeds as in section 11.4 except that now the measurements are given, not the stimuli; the latter are unknown to the observer. Thus, the log posterior ratio is

$$d = \log\frac{p(C=1|\mathbf{x})}{p(C=0|\mathbf{x})} \tag{11.37}$$

$$= \log\frac{p(\mathbf{x}|C=1)}{p(\mathbf{x}|C=0)} + \log\frac{p(C=1)}{p(C=0)}. \tag{11.38}$$

The likelihoods are obtained by marginalizing over the stimuli, $\mathbf{s}$. The likelihood of $C=0$ becomes

$$\mathcal{L}(C=0;\mathbf{x}) \equiv p(\mathbf{x}|C=0) \tag{11.39}$$

$$= \int p(\mathbf{x}|\mathbf{s})p(\mathbf{s}|C=0)d\mathbf{s} \tag{11.40}$$

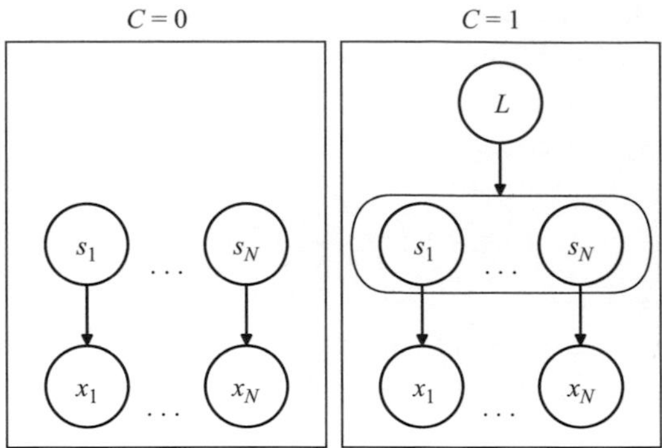

Figure 11.5
Graphical depiction of the generative model for detection of a single target. $C=0$: target absent. $C=1$: target present.

$$=\prod_{i=1}^{N}\int p(x_i|s_i)p(s_i|C=0)ds_i \tag{11.41}$$

$$=\prod_{i=1}^{N}\int p(x_i|s_i)p_{\text{distractor}}(s_i)ds_i, \tag{11.42}$$

and similarly for $C=1$. Other than that, we follow section 11.4 and find for the likelihood ratio of target presence,

$$\text{LR}\equiv\frac{\mathcal{L}(C=1;\mathbf{x})}{\mathcal{L}(C=0;\mathbf{x})} \tag{11.43}$$

$$=\sum_{L=1}^{N}p(L)\frac{\int p(x_L|s_L)p_{\text{target}}(s_L)ds_L}{\int p(x_L|s_L)p_{\text{distractor}}(s_L)ds_L}. \tag{11.44}$$

This means that equation (11.34) for relating the "global" likelihood ratio, LR, to the local ones, LR_L, still holds.

Relationship to embedded-class task from section 8.6 We now consider the case that $N=1$, the target value follows a Gaussian distribution with mean 0 (or any arbitrary value) and variance σ_1^2, and the distractor value follows a Gaussian distribution with the same mean and variance σ_2^2. Then, the target-detection task reduces to the embedded-class task from section 8.6 (see in particular figure 8.5). An important special case is that the target has a single value ($\sigma_T=0$). For general N, with otherwise the same assumptions, the embedded-class task is still the building block of the target-detection task.

Special case: single target value, single distractor value As in section 11.3, we consider the case of a single target value, s_T, and a single distractor value, s_D. The local log likelihood ratio of target presence is then equal to

$$\text{LLR}_L=\frac{s_T-s_D}{\sigma_L^2}\left(x_L-\frac{s_T+s_D}{2}\right). \tag{11.45}$$

11.6 Applications

There exists a large cognitive psychology literature on visual search, with many descriptive and arguably ad hoc models. A much smaller community has built probabilistic models of visual search. To create the conditions under which the models would apply, experimenters would typically ask subjects to fixate on the center of the screen, present items at a fixed distance from that center, equip items with only a single task-relevant feature, and present the stimulus display for a very short period of time (e.g., 100 ms) [141]. Such experiments sacrifice some real-world relevance for experimental control and modelability. Bayesian models of simple visual search date back at least to [143]. However, for many decades, starting with [136], Bayesian models were overshadowed by signal detection theory models, in particular the maximum-of-outputs or max model. The max model is reasonable when distractors are homogeneous (identical to each other), a case not considered in this chapter; it is then also very similar to the Bayesian model. However, the max model fails when distractors are heterogeneous. Then, Bayesian models of search still describe human data well as long as it is additionally assumed that the amount (standard deviation) of the measurement noise increases as the number of stimuli increases [30, 123, 154]. This increase in noise can also be thought of as a "resource limitation."

In natural visual search, eye movements are crucial. In laboratory experiments, the role of eye movements is sometimes deliberately minimized by requiring the participant to fixate. This has both experimental and modeling advantages [141]. Nevertheless, Bayesian models can also be applied to visual search tasks in which eye movements are allowed, such as free viewing a scene while localizing a target. In such tasks, the goal is often to predict the next fixation location. Posterior distributions can be used to evaluate the expected benefits of different candidate fixation locations [132, 211]. These scenes are still not natural but have carefully controlled statistics. There is a lot of work on modeling visual search in natural scenes. This work is more useful for applications than anything we discussed in this chapter, but it is beyond the scope of this book because it is not Bayesian.

11.7 Summary

In this chapter, we introduced Bayesian inference in target detection and target localization problems. We learned:

- Search is an ecologically important task, for example taking place when an animal hunts for camouflaged prey.
- Search leads to a statistical dependence between all potential targets. This produces a more complex inference problem, which we can still treat using Bayesian methods.
- In visual search with heterogeneous distractors, inference requires marginalizing over distractor values (and in the case of detection, also over target location).

11.8 Suggested Readings

- Miguel P. Eckstein. "Visual Search: A Retrospective." *Journal of Vision* 11, no. 5 (2011): 14.

- Wei Ji Ma, Vidhya Navalpakkam, Jeffrey M. Beck, Ronald van den Berg, and Alexandre Pouget. "Behavior and Neural Basis of Near-Optimal Visual Search." *Nature Neuroscience* 14, no. 6 (2011): 783–790.
- Wei Ji Ma, Shan Shen, Gintare Dziugaite, and Ronald van den Berg. "Requiem for the Max Rule?" *Vision Research* 116 (2015): 179–193.
- Helga Mazyar, Ronald Van den Berg, and Wei Ji Ma. "Does Precision Decrease with Set Size?" *Journal of Vision* 12, no. 6 (2012): 10.
- Jiri Najemnik and Wilson S. Geisler. "Optimal Eye Movement Strategies in Visual Search." *Nature* 434, no. 7031 (2005): 387–391.
- John Palmer, Preeti Verghese, and Misha Pavel. "The Psychophysics of Visual Search." *Vision Research* 40, no. 10–12 (2000): 1227–1268.
- Ruth Rosenholtz. "Visual Search for Orientation among Heterogeneous Distractors: Experimental Results and Implications for Signal-Detection Theory Models of Search." *Journal of Experimental Psychology: Human Perception and Performance* 27, no. 4 (2001): 985–999.
- Shan Shen and Wei Ji Ma. "A Detailed Comparison of Optimality and Simplicity in Perceptual Decision Making." *Psychological Review* 123, no. 4 (2016): 452–480.
- Scott Cheng-Hsin Yang, Mate Lengyel, and Daniel M. Wolpert. "Active Sensing in the Categorization of Visual Patterns." *Elife* 5 (2016): e12215.

11.9 Problems

Problem 11.1 Come up with an example of a search problem in everyday life that has not been used as an example in this chapter.

Problem 11.2 We mentioned that in experiments testing the models of this chapter the subject is typically asked to fixate on the center of the screen, items are presented on a fixed distance from that center, items have a single relevant feature, and the display is presented for a very short period of time (e.g., 100 ms). Explain how deviations from each of these design elements could create the need for more complex models.

Problem 11.3 Consider the case of section 11.3 but without measurement noise. Assume that the target distribution is normal with mean μ_T and variance σ_T^2, and that the distractor distribution is normal with mean μ_D and variance σ_D^2, as in figure 11.3B.

(a) Show that the posterior over target location L is

$$p(L|s) \propto p(L) e^{-\frac{1}{2}(J_T - J_D)\left(s_L - \frac{J_T\mu_T - J_D\mu_D}{J_T - J_D}\right)^2}. \tag{11.46}$$

(b) Assume $N=3$, $p(L)=(0.3, 0.3, 0.4)$, and the parameters of figure 11.3B. Suppose the observations are $\mathbf{s}=(0.9, 6.1, -0.2)$. Calculate the posterior over L.

(c) Assume that $p(L)$ is uniform. Assume moreover that $\sigma_T < \sigma_D$, as is reasonable in a search task (targets are usually more narrowly defined than distractors). Show that MAP estimation then amounts to choosing the location L for which s_L is closest to $\frac{J_T\mu_T - J_D\mu_D}{J_T - J_D}$.

(d) Assume that $p(L)$ is uniform and use the parameters of figure 11.3B. Vary N from 1 to 8. For each value of N, calculate proportion correct. Plot proportion correct as a function of N.

(e) Explain intuitively why in this model, proportion correct decreases as a function of N.

Problem 11.4 Refer back to section 11.3 on target localization with measurement noise. Derive equation (11.19) step by step, starting from equation (11.15).

Problem 11.5 In this chapter, we considered tasks in which the distractors are independent. In reality, they might not be. For example, if oriented line segments are part of a textured background, they will tend to point in the same direction. This matters for the decision rule and observer performance. Here, we consider an extreme form of dependence, namely that all distractors are identical to each other. However, their common stimulus value, which we denote by s_D, still varies from trial to trial, following a distribution $p_{\text{distractor}}(s_D)$. Further assume a distribution over target location $p(L)$, a target distribution $p_{\text{target}}(s)$, and measurement distributions $p(x_i|s_i)$. Derive an expression for the posterior over target location.

Problem 11.6 Assume $N=2$, $p(L=1)=p_1$, independent distractors, a target that always has value 0, a distractor that has a normal distribution with mean 0 and variance σ_D^2, and measurement noise with variance σ^2. The observer has to localize the target.

(a) Show that the Bayesian observer reports location 1 when

$$x_1^2 - x_2^2 < \frac{2\log\frac{p_1}{1-p_1}}{\frac{1}{\sigma^2} - \frac{1}{\sigma_d^2+\sigma^2}}. \tag{11.47}$$

(b) Assume $p_1 = 0.6$, $\sigma_D = 10$, $\sigma = 2$. By using a grid over x_1 and x_2, numerically calculate and plot the following psychometric curves, which show the proportion of "location 1" reports as a function of the stimulus value of the distractor.

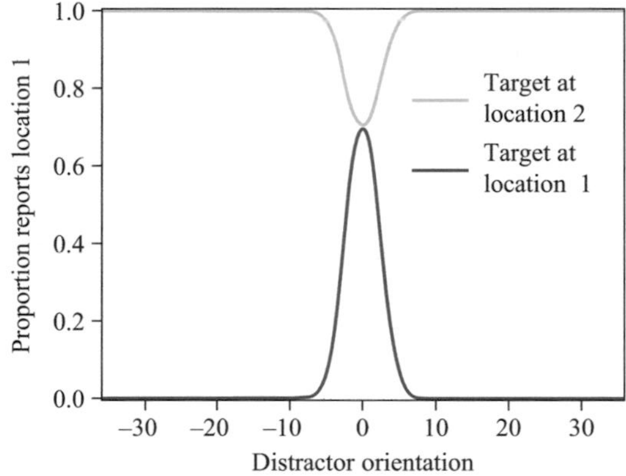

(c) Repeat using sampling of x_1 and x_2 instead of a grid.

Problem 11.7 How would equation (11.32) change if in the $C=1$ condition, instead of there always being exactly one target, each stimulus individually is a target with probability ϵ?

Problem 11.8 In this problem, we examine the distributions of the log posterior ratio (LPR). An observer has to detect whether a target orientation is present or absent among N oriented line segments. The target has orientation 0° and each distractor has orientation 10°. On each trial, the experimenter chooses whether the target will be present or absent with equal probability. When the target is present, each stimulus is equally likely to be the target. Assume that the measurement at each location is corrupted by Gaussian noise with standard deviation 5°.

(a) Suppose $N=2$. Simulate the measurement on 5,000 target-present trials and 5,000 target-absent trials. For convenience and without loss of generality, you can assume that when the target is present, it is at the first location.

(b) On each simulated trial, compute the LPR.

(c) Plot the resulting histograms of the LPR, one for target present and one for target absent, in the same plot. Is either histogram symmetric?

(d) Assume the observer performs MAP estimation. Calculate proportion correct.

(e) Repeat your simulation for set sizes from one to twenty. Plot proportion correct as a function of set size.

Problem 11.9 In this problem, we study search with a circular world state variable. An observer detects whether a target, defined by orientation, is present among N line segments. The prior probability that the target is present is 0.5. The target always has orientation s_T. Each distractor orientation is drawn independently from a uniform distribution on $[0, \pi)$. The observation at the ith location, x_i, is drawn from a Von Mises distribution with circular mean s_i (the true orientation) and concentration parameter κ,

$$p(x_i|s) = \frac{1}{\pi I_0(\kappa)} e^{\kappa \cos 2(x_i - s_i)}, \tag{11.48}$$

where I_0 is the modified Bessel function of the first kind of order 0. For background on the Von Mises distribution, see appendix section B.7.6.

(a) Show that the MAP observer responds "target present" if

$$\sum_{i=1}^{N} e^{\kappa \cos 2(x_i - s_T)} > N I_0(\kappa). \tag{11.49}$$

(b) Simulate 10^5 trials with $N=2$, $s_T=0$, and $\kappa=10$. On each trial, draw observations from the generative model. We want to point out that for drawing from Von Mises distributions there are likely to be implementations already in your favorite programming language. Then compute the LPR of target presence.

(c) Plot the empirical distributions of this LPR when the target is present and when it is absent (using a smooth curve might be better for presentation than using histograms).

(d) Plot the ROC.

(e) Repeat parts (c) and (d) for two different set sizes, $N=4$ and $N=8$. Plot LPR distributions and ROCs in a way that you can easily compare across N. Interpret the effects of N.

Problem 11.10 Derive the LLR at set size N in each of the following visual search scenarios. Assume independent Gaussian noise.

(a) The target is always vertical. Distractors are homogeneous but their value is drawn on each trial from a uniform distribution over orientation. The observer reports whether the target is present.

(b) The target is always vertical. Distractors are drawn from a Gaussian distribution around vertical with variance σ_D^2. The observer reports whether the target is present.

(c) The target is always vertical. Each distractor is independently chosen to be tilted an amount Δ to the left or to the right of vertical. The observer reports whether the target is present.

(d) Distractors are always vertical. The target is drawn on each trial from a Gaussian distribution around vertical with variance σ_T^2. The observer reports whether the target is present.

(e) The target is drawn from a symmetric distribution around 0. Distractors are all vertical. The observer reports whether the target is tilted to the right or left of vertical.

(f) Distractors are homogeneous but their value is drawn on each trial from a uniform distribution over orientation. The target, if present, has a value such that the target-distractor difference is Δ on each trial. The observer reports whether the target is present.

(g) The target is always vertical and always present. Distractors are drawn from a uniform distribution. The observer reports which location contained the target.

Problem 11.11 This problem builds on the detection version of the camouflage scenario in section 11.4, and in particular on equation (11.36) for the LR of the fish being present versus absent.

(a) Calculate the numerical value of LR for each panel and verify your answers against the numbers in figure 11.4.

(b) Vary b from 0.1 to 0.7 in steps of 0.1. Instead of assuming the specific observations (combinations of fish plus riverbed) in figure 11.4, we now simulate, for every value of b, 10,000 observations randomly generated with that value of b (and still $a=0.7$). Represent each observation by a vector of ten n_L values. Calculate the mean and standard deviation of log LR as a function of b, and plot these values with error bars whose sizes represent standard deviations.

(c) Modify the derivation leading up to equation (11.36) to treat the case of a species of smaller fish of 3 units long. Hint: you will need to marginalize over both the horizontal and vertical coordinates.

(d) Repeat (b) for this new expression and plot in the same plot with a different color.

(e) Based on the plot in (d), argue that the visibility of the smaller fish is less affected by suboptimal camouflage.

Problem 11.12 This problem involves a complex search task and model fitting. In the task by Shen and Ma [167], subjects judge whether a target stimulus is tilted leftward (counterclockwise) or rightward (clockwise) with respect to vertical. The only way the target differs from the distractors is that the distractors are identical to each other on a given trial, whereas the target value is drawn independently. We will fit the optimal Bayesian model described in the paper. This model has two parameters, sensory noise level σ and the lapse rate λ. For background on model fitting, see appendix C.

(a) For $\sigma = 1$ and $\lambda = 0$, calculate, separately for each trial, the probability that the observer will report "target tilted right" through Monte Carlo simulation. For each trial, draw 1,000 measurement vectors $\mathbf{x}$.

(b) Define a fine grid for σ and λ. Repeat part (a) for every parameter combination on this grid. Save the results in a three-dimensional matrix, with one dimension corresponding to σ, one to λ, and one to trials.

Download visual_search.csv from https://osf.io/84kpb/. These are data from one subject in the Shen and Ma experiment. The matrix in the file has 2,000 rows (trials) and three columns. The first column is the target orientation, the second the distractor orientation, the third the subject's response (-1 for leftward, 1 for rightward).

(c) Use the result of (b) to calculate the log likelihood of the subject responses.

(d) Find the maximum-likelihood estimates of σ and λ on the grid.

(e) Using these parameter estimates, reproduce the Opt model fits in figure 3C–D of the paper by Shen and Ma [167]. The model curves could differ slightly from the figure, depending on the grid you choose and due to sampling noise.

12

Inference in a Changing World

How do we estimate the current state of a changing world?

So far in this book, we have considered only world states that do not change. The real world, however, is often changing, and the observer typically wants to estimate the world state at the current time or at a time in the future based on a history of observations.

Plan of the Chapter

Here, we work out the Bayesian model for two common tasks with a changing world state: one in which the world state is continuously changing in a lawful manner, the other one in which changes occur at discrete moments. The generative model of the former is an example of a *hidden Markov model* (HMM).

12.1 Tracking a Continuously Changing World State

Observers often have to make inferences while the world is changing in a continuous (grad ual) manner. For example, when you are playing a ball sport, you want to know where the moving ball is now (or arguably a little bit into the future). In many sports, you also want to track team mates or opponents. When you are trying to find the light switch in a dark room, your own hand moves and you want to know where it is now (relative to the light switch). When you are trying to understand a spoken sentence, the meaning evolves while you gather auditory information. When an animal forages, the rate of return of a food source will decrease as the food becomes scarcer. In these examples, the world changes in a continuous and lawful manner. Here, we will examine how such changes affect the inference done by a Bayesian observer.

12.1.1 Step 1: Generative Model

To make the math simpler, we will discretize time: time moves in unit steps, for example, seconds. We will denote time by t; thus, t is an integer, and we will make it start at $t=1$. The generative model is shown in figure 12.1. The top row shows the world state, which now changes across time (columns). We denote by s_t the world state at time t. The bottom row contains the measurements at the different time points. We denote by x_t the measurement at time t. This generative model is different from section 5.5, where the observer was making a sequence of measurements, but the world state did not change.

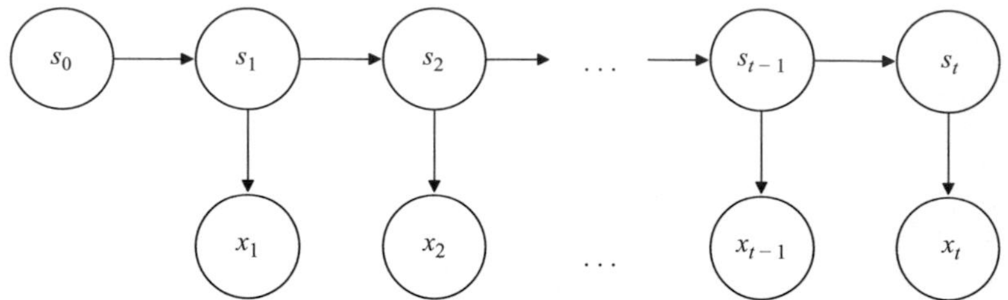

Figure 12.1
Generative model for a world state that changes continuously in time. The subscript refers to time.

In this chapter, we use the term *world state* instead of *stimulus*. The latter term suggests a physically presented world state that is subsequently corrupted by internal noise. In our experience, change point detection—both in the real world and in the laboratory—is most often done on data that have been corrupted by external noise (noise in the world). Thus, the underlying world state is never physically presented. This nuance in terminology has no consequences for the formalism.

For the measurements, we make our usual assumption that they are independent given the world states, and normally distributed:

$$p(\mathbf{x}|\mathbf{s}) = \prod_{i=1}^{t} p(x_i|s_i); \tag{12.1}$$

$$p(x_i|s_i) = \mathcal{N}(x_i; s_i, \sigma^2). \tag{12.2}$$

The measurement variance, σ^2, does not depend on t.

In the top row of figure 12.1, we drew arrows only from the world state at one time step to the world state at the next time step. There are no arrows that bridge states more than one time step apart. Formally, we can therefore write

$$p(\mathbf{s}) = p(s_0)p(s_1|s_0)p(s_2|s_1)\cdots p(s_t|s_{t-1}) \tag{12.3}$$

$$= p(s_0)\prod_{i=1}^{t} p(s_i|s_{i-1}). \tag{12.4}$$

The conditional probability $p(s_i|s_{i-1})$ is called the *state transition model*. It describes the dynamics of the world, namely how much we expect the state at time i to have a value s_i given that the state at time $i-1$ had value s_{i-1}. The assumption that the distribution of s_i depends only on s_{i-1}, and not on s_{i-2}, s_{i-3}, etc., is called the *Markov property*, and the corresponding generative model for $s_1, s_2, \ldots s_T$ (top row of figure 12.1) is called a Markov model or, more precisely, a first-order Markov process. A Markov process $s_1, s_2, \ldots, s_T$ is defined by the property

$$p(s_t|s_1, \ldots, s_{t-1}) = p(s_t|s_{t-1}). \tag{12.5}$$

In a sense, a Markov process has *no memory*: the expectation for what happens next depends only on the current state, not on how the current state was arrived at. Physical systems that obey Newton's laws of motion are Markov processes, since those laws can be stated as *differential equations*. For example, to predict the trajectory of a ball, we need to know only the current position and velocity, not its past positions or velocities.

We now have to make specific distributional assumptions. Throughout this chapter, we will consider only the simplest possible state transition model: that the state increases by a fixed value Δ from time step to time step: $s_i = s_{i-1} + \Delta$. However, we do allow for noise. We assume that this noise is Gaussian and has standard deviation σ_s, so that

$$p(s_i|s_{i-1}) = \mathcal{N}(s_i; s_{i-1} + \Delta, \sigma_s^2). \tag{12.6}$$

This could, for example, describe linear motion of your hand, at constant velocity but with some variability. Finally, we will assume a form for the prior at time 0, $p(s_0)$ in equation (12.4). We will assume a Gaussian with mean μ_0 and standard deviation σ_0:

$$p(s_0) = \mathcal{N}(s_0; \mu_0, \sigma_0^2). \tag{12.7}$$

This concludes step 1. Since the states $s_1, s_2, \ldots$ are not directly observable, thus "hidden," the generative model we defined here is called a hidden Markov model (HMM).

12.1.2 Step 2: Inference

The observer is interested in inferring the current state, s_t, based on the time series of measurements $x_1, x_2, \ldots, x_t$. The Markov property makes this calculation simpler. In particular, we will be able to write down a recursive relation for the posterior distribution: the posterior over s_t given $x_1, x_2, \ldots, x_t$ can be expressed in terms of the posterior over s_{t-1} given $x_1, x_2, \ldots, x_{t-1}$. We now build up the logic of the inference process step by step, using the auxiliary generative models in figure 12.2.

Suppose first that the only world state is the current one, s_t, and a corresponding measurement x_t (figure 12.2A). The world state s_t obeys the distribution $p(s_t)$, which the observer uses as a prior. Then we are in the same situation as in chapter 3. The posterior over s_t is

$$p(s_t|x_t) \propto p(x_t|s_t)p(s_t). \tag{12.8}$$

If there were only one state, we would just need to use Bayes' rule to calculate the posterior distribution. Now we add the previous world state, s_{t-1}, to the generative model (figure 12.2B). Its distribution is $p(s_{t-1})$. Then the prior over s_t is not readily available to the observer, but has to be obtained through marginalization over s_{t-1}:

$$p(s_t) = \int p(s_t|s_{t_1})p(s_{t-1})ds_{t-1}. \tag{12.9}$$

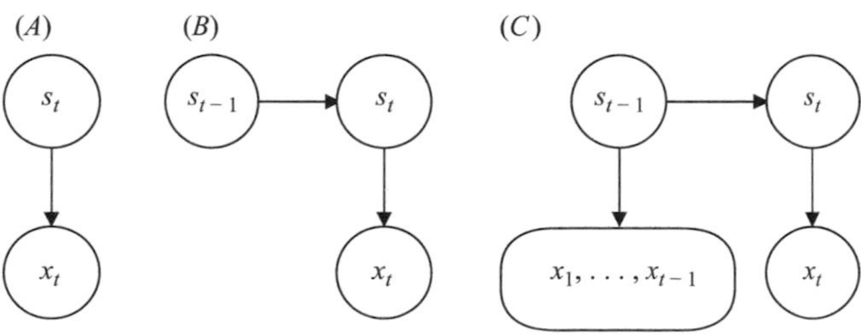

Figure 12.2
Simplified generative models in the problem of tracking a continuous change.

This expresses that the observer's belief about the current state, s_t, is obtained by "extrapolation" or "prediction" from the previous state, s_{t-1}. This extrapolation or prediction is captured by the conditional probability distribution $p(s_t|s_{t-1})$. If from one time step to the next, noise is added to the state, then marginalization has the effect of widening the distribution. Specifically, if $p(s_{t-1})$ is a normal distribution with mean $\mu_{\text{post},t-1}$ and variance $\sigma^2_{\text{post},t-1}$, and we assume equation (12.6) for the state transitions, then $p(s_t)$ will also follow a normal distribution with mean $\mu_{\text{post},t-1}+\Delta$ and variance $\sigma^2_{\text{post},t-1}+\sigma_s^2$: the distribution is shifted over by Δ and is wider.

Substituting equation (12.9) into equation (12.8), we find

$$p(s_t|x_t) \propto p(x_t|s_t)\int p(s_t|s_{t-1})p(s_{t-1})ds_{t-1}. \tag{12.10}$$

In the special case of a static world, in which $s_t=s_{t-1}$, we have $p(s_t|s_{t-1})=\delta(s_t-s_{t-1})$, and equation (12.10) reduces to equation (12.8).

In equation (12.10), we assumed that the observer uses a prior distribution $p(s_{t-1})$. In reality, though, instead of a *prior* distribution at the previous time, we have a *posterior* distribution, $p(s_{t-1}|\text{ past})$, where "past" is the sequence of all previous measurements, $x_1,x_2,\ldots,x_{t-1}$. We can explicitly add those into the generative model (figure 12.2C). Equation (12.10) changes by conditioning on the past measurements on both sides. This gives

$$p(s_t|x_t) \propto p(x_t|s_t)\int p(s_t|s_{t_1})p(s_{t-1}|x_1,\ldots,x_{t-1})ds_{t-1}. \tag{12.11}$$

This is a *recursive* relationship: the posterior over the state at time t given the measurements through time t is expressed as a function of the posterior over the state at time $t-1$ given the measurements through time $t-1$.

In the special case of a static world, in which $s_t=s_{t-1}$, we have $p(s_t|s_{t-1})=\delta(s_t-s_{t-1})$, Then equation (12.11) becomes

$$p(s_t|x_t) \propto p(x_t|s_t)p(s_t|x_1,\ldots,x_{t-1}), \tag{12.12}$$

which is equivalent to evidence accumulation as discussed in section 5.5, specifically equation (5.31).

If we again postulate that the posterior at time $t-1$ is normal with mean $\mu_{\text{post},t-1}$ and variance $\sigma^2_{\text{post },t-1}$, then the "prior" at time t is normal with mean

$$\mu_{\text{prior},t}=\mu_{\text{post},t-1}+\Delta \tag{12.13}$$

and variance

$$\sigma^2_{\text{prior},t}=\sigma^2_{\text{post},t-1}+\sigma_s^2. \tag{12.14}$$

Moreover, the posterior at time t, $p(s_t|x_1,\ldots,x_t)$, is also normal, with mean

$$\mu_{\text{post},t}=\frac{\frac{x_t}{\sigma^2}+\frac{\mu_{\text{prior},t}}{\sigma^2_{\text{prior},t}}}{\frac{1}{\sigma^2}+\frac{1}{\sigma^2_{\text{prior},t}}} \tag{12.15}$$

and variance

$$\sigma^2_{\text{post},t}=\frac{1}{\frac{1}{\sigma^2}+\frac{1}{\sigma^2_{\text{prior},t}}}. \tag{12.16}$$

Exercise 12.1 Derive these equations. You will need equations (12.2) and (12.6).

These equations show an interesting combination of combining a measurement with a prior (chapter 3) and that prior widening, as discussed above. These "forces" are counteracting: the former will make the posterior narrower, the latter wider. We see this in figure 12.3. Eventually, the variance of the posterior (and thus uncertainty) will asymptote at a fixed value (see problem 12.5).

Equations (12.15) and (12.16) define the posterior at time t in terms of the posterior of time $t-1$. However, we have to specially consider what happens at $t=1$, because at $t-1$ we do not have a posterior, only a prior, equation (12.7). However, that prior serves as the posterior: we can simply write

$$\mu_{\text{post},0} = \mu_0; \tag{12.17}$$

$$\sigma^2_{\text{post},0} = \sigma^2_0. \tag{12.18}$$

These equations define what can be called the "initial condition" of the problem.

To conclude step 2, we simply recall that the estimate of a continuous world state that minimizes expected squared error is the posterior mean, which in our case is given by equation (12.15).

While we have studied an observer who infers the present state, s_t, the formalism can be extended to prediction of a future state. This is simply done by removing the current measurement, x_t, from the generative model and accordingly from equations (12.10) and (12.11). Then, those equations describe the probabilistic prediction for the future state s_t from past measurements $x_1, \ldots, x_{t-1}$.

12.2 Change Point Detection

So far, we have examined continuous change of a stimulus. Sometimes, however, there is a discontinuous change or "jump" in the world state. For example, the ownership of a restaurant might change hands, causing the quality of the food to suddenly improve. A neurologist

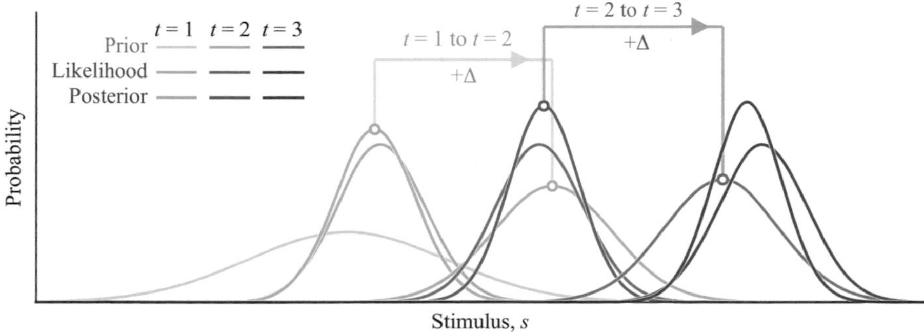

Figure 12.3
Evolution of the posterior over a world state that changes at a constant rate. Shown are the prior (yellow), likelihood (red), and posterior (blue) at three subsequent time steps, $t = 1, 2, 3$. At each time point, the posterior is calculated just like in chapter 3. The posterior becomes the prior at the next time step through a shift by Δ and a widening, per equations (12.13) and (12.14). We used the following parameter values: $\mu_0 = -5$, $\sigma_0 = 5$, $\Delta = 4$, $\sigma_s = 1$, $\sigma = 1$. The model observer has the measurements $x_1 = -0.24$, $x_2 = 3.34$, and $x_3 = 8.36$.

may want to detect a seizure on an EEG in a patient who does not exhibit any outwardly visible symptoms of a seizure. Or a friend might have experienced a life event that drastically changes the nature of their interactions with you. These situations require a different treatment than gradual changes. We will examine two cases: (a) the observer knows that a single change occurred over the entire observable time period, but has to determine when (this section); (b) a change could independently have occurred at any time point during the observable time period (next section).

12.2.1 Single Change Point

Intuitively, change point detection can be hard because the noise in the measurements can create large "apparent changes" that are not due to a change in the underlying state. When, then, to declare that a true change point occurred? The Bayesian observer solves this problem in an optimal manner.

Step 1: Generative model The generative model is shown in figure 12.4. We assume that every time point between $t = 1$ and $t = T$ has the same probability of being the change point:

$$p(t_{\text{change}}) = \frac{1}{T}. \tag{12.19}$$

Next, we define what it means to have a change point at t_{change}:

$$s_t = \begin{cases} s_{\text{pre}} & \text{for } t < t_{\text{change}}; \\ s_{\text{post}} & \text{for } t \geq t_{\text{change}}, \end{cases} \tag{12.20}$$

where s_{pre} and s_{post} are fixed and assumed known to the observer. Finally, we make the usual assumption that measurements are conditionally independent and normally distributed (with a time-independent variance):

$$p(\mathbf{x}|\mathbf{s}) = \prod_{t=1}^{T} p(x_t|s_t) \tag{12.21}$$

$$= \prod_{t=1}^{T} \mathcal{N}(x_t; s_t, \sigma^2). \tag{12.22}$$

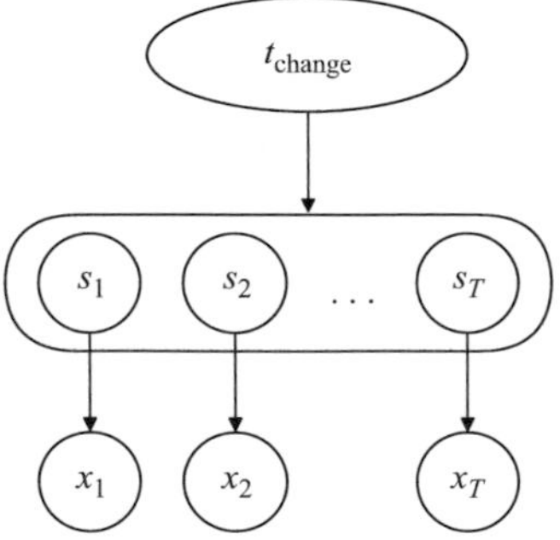

Figure 12.4
Generative model for change point detection with a single change point. The elongated shape in the middle denotes the nonlocal effect introduced by the change point.

Step 2: Inference The world state of interest is the change point time t_{change}. Therefore, the Bayesian observer computes the posterior over t_{change} given a sequence of measurements, $\mathbf{x}$. We first apply Bayes' rule with a uniform prior:

$$p(t_{\text{change}}|\mathbf{x}) \propto p(\mathbf{x}|t_{\text{change}})p(t_{\text{change}}) \tag{12.23}$$

$$\propto p(\mathbf{x}|t_{\text{change}}). \tag{12.24}$$

Then, we marginalize over $\mathbf{s}$:

$$p(t_{\text{change}}|\mathbf{x}) \propto \sum_{\mathbf{s}} p(\mathbf{x}|\mathbf{s})p(\mathbf{s}|t_{\text{change}}). \tag{12.25}$$

Here, the sum is over all sequences $\mathbf{s}$. There are a total of 2^T such sequences but, given t_{change}, only one is allowed, as specified by equation (12.20) . Then, the sum reduces to a single term, which becomes

$$p(t_{\text{change}}|\mathbf{x}) \propto \left(\prod_{t=1}^{t_{\text{change}}-1} p(x_t|s_t = s_{\text{pre}})\right)\left(\prod_{t_{\text{change}}}^{T} p(x_t|s_t = s_{\text{post}})\right). \tag{12.26}$$

This can be written as

$$p(t_{\text{change}}|\mathbf{x}) \propto \prod_{t_{\text{change}}}^{T} \frac{p(x_t|s_t = s_{\text{post}})}{p(x_t|s_t = s_{\text{pre}}}. \tag{12.27}$$

Exercise 12.2 Show this mathematically.

So far, we have not used the form of the measurement distribution. Using that, the posterior over t_{change} becomes

$$p(t_{\text{change}}|\mathbf{x}) \propto \prod_{t_{\text{change}}}^{T} e^{\frac{\Delta s}{2\sigma^2}\sum_{t_{\text{change}}}^{T}(x_t - \bar{s})}, \tag{12.28}$$

where $\Delta s = s_{\text{post}} - s_{\text{pre}}$ and $\bar{s} = \frac{s_{\text{pre}} + s_{\text{post}}}{2}$, by analogy with the discrimination task in chapter 7. Equation (12.28) has an intuitive explanation: the evidence for a change point at t_{change} increases the more measurements following t_{change} are on the same side of the mean as s_{post}. However, the evidence decreases if one goes so far back in time that one includes measurements coming from the original s_{pre}. To obtain the actual posterior probabilities, compute the value of the right-hand side of equation (12.28) for all values of t_{change}, and divide them by their total (normalization).

The inference concludes with a readout stage. Since t_{change} is discrete, MAP estimation—picking the mode of the posterior, in order to maximize accuracy—is the most obvious strategy. MAP estimation amounts to maximizing $\sum_{t_{\text{change}}}^{T}(x_t - \bar{s})$.

Exercise 12.3 Why is this the case?

Step 3: Response distribution The response distribution has to be simulated and we leave this to problem 12.9.

12.2.2 Random Change Points

We studied how to localize (in time) a single change point in a time series. In real-world inference, it will be rare for an observer to know that there is exactly one change point. There could be zero or there could be more than one. While this opens up a vast space of possibilities, we would be remiss in not discussing at least one case. Therefore, we now consider the scenario in which change points randomly occur and the observer infers their times.

Step 1: Generative model The generative model is shown in figure 12.5. We assume that each C_t takes values 0 (no change) and 1 (change), that all C_ts are independent, and that $p(C_t=1)=\epsilon$. Thus, there can be anywhere between 0 and T change points. We next assume that each s_t takes values -1 and 1. If $C_t=0$, s_t is equal to s_{t-1} while if $C_t=1$, s_t has the opposite sign from s_{t-1}. We further assume the initial state $s_0=1$. Then, to each vector $\mathbf{C}$ corresponds a specific vector of stimuli, which we will denote by $\mathbf{s_C}$. We make the same assumptions about the measurements as before (see equation (12.22)).

Step 2: Inference Since the observer no longer has any constraints on when the change points occur, they have to consider every possible time series $\mathbf{C}=(C_1,\ldots,C_T)$. Since each C_T is binary and all are independent, the hypothesis space consists of 2^T binary vectors. The posterior is

$$p(\mathbf{C}|\mathbf{x})\propto p(\mathbf{x}|\mathbf{C})p(\mathbf{C}). \tag{12.29}$$

Here, the prior over $\mathbf{C}$ cannot be assumed uniform, since the construction of the generative model makes some $\mathbf{C}$ more probable than others. To be precise, we have

$$p(\mathbf{C})=\epsilon^{||\mathbf{C}||}(1-\epsilon)^{T-||\mathbf{C}||}, \tag{12.30}$$

where $||\mathbf{C}||$ is the total number of 1s in $\mathbf{C}$. The notation $||\cdot||$ is called the *norm* of $\mathbf{C}$. In this case, we are using the so-called L_1 norm, which is simply the sum of the elements.

Exercise 12.4 Why?

The likelihood function over $\mathbf{C}$ is

$$\mathcal{L}(\mathbf{C};\mathbf{x})=p(\mathbf{x}|\mathbf{C}) \tag{12.31}$$

$$=\sum_{\mathbf{s}}p(\mathbf{x}|\mathbf{s})p(\mathbf{s}|\mathbf{C}) \tag{12.32}$$

$$=p(\mathbf{x}|\mathbf{s}=\mathbf{s_C}), \tag{12.33}$$

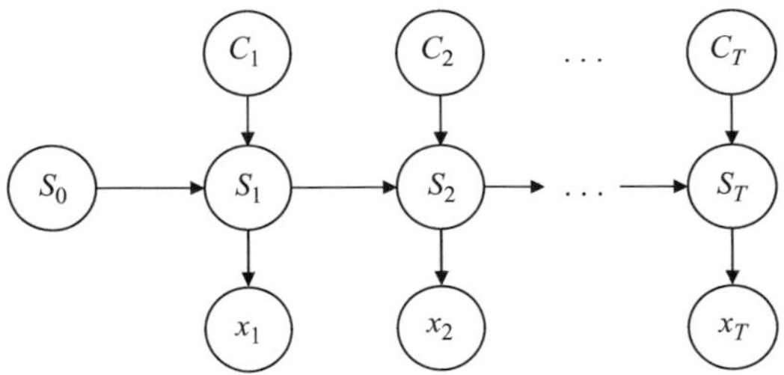

Figure 12.5
Generative model for random change point detection.

where we recall that $\mathbf{s}_C$ is the unique world state vector $\mathbf{s}$ that corresponds to the change-point vector $\mathbf{C}$. We leave closer examination of this model to problems 12.11, 12.12, and 12.13.

12.2.3 More Realistic Change Point Detection

We made several simplifying assumptions in this section. We assumed that change points are either unique (one per trial) or random (independent across time points). Neither is realistic. In many real-world change detection problems, change points occur "every now and then," which means that there is some prior over the interval between two change points. Moreover, change point detection can usually not be done with the benefit of hindsight, that is, based on a complete time series of observations; instead, a decision has to be made instantly about whether a change occurred. This is called *online* change point detection [4, 138, 206]. Finally, the change is often not between two specific world state values but between two categories. None of these aspects makes inference conceptually different from the examples discussed in this section; however, each can introduce substantial technical complications.

In view of such complications, and in particular considering how bad they get as T is large, change point detection is a domain in which exact Bayesian models quickly become implausible as models of human behavior. Nevertheless, the Bayesian model can serve as a reference model or as a starting point for building more plausible models. We will discuss this further in chapter 15.

12.3 Summary

In this chapter, we introduced methods to enable inference in a world that is constantly changing. We learned the following:

- Inference often takes place in a changing world. To handle this, a Bayesian decision maker needs an statistical model of how the world is changing.
- If the dynamics of the world are characterized by a Markov process, then the resulting update equations are tractable. Inference boils down to alternating between updating our belief given the temporal changes of the world and updating it given the observations.
- It is often not possible to explicitly "solve for" the mean and variance of the posterior at each time. The recursive equations, combined with the initial condition, are usually the best we can do.
- We derived the recipe for optimal inference for a linear state transition model and Gaussian measurement noise. This recipe is also called a Kalman filter.
- A change point is a moment when the world state changes suddenly. This happens in many real-world situations.
- We derived recipes for inferring change points from time series of noisy observations, in two cases: when there is only one change point in the entire time series, or when change points occur at random.

12.4 Suggested Readings

- Ryan Prescott Adams and David J. C. MacKay. "Bayesian Online Changepoint Detection." *arXiv preprint arXiv:0710.3742* (2007).
- J. Yu Angela. "Adaptive Behavior: Humans Act as Bayesian Learners." *Current Biology* 17, no. 22 (2007): R977–R980.
- Kathryn Bonnen, Johannes Burge, Jacob Yates, Jonathan Pillow, and Lawrence K. Cormack. "Continuous Psychophysics: Target-Tracking to Measure Visual Sensitivity." *Journal of Vision* 15, no. 3 (2015): 14.
- Daniel Goldreich and Jonathan Tong. "Prediction, Postdiction, and Perceptual Length Contraction: A Bayesian Low-Speed Prior Captures the Cutaneous Rabbit and Related Illusions." *Frontiers in Psychology* 4 (2013): 221.
- Konrad P. Kording, Joshua B. Tenenbaum, and Reza Shadmehr. "The Dynamics of Memory as a Consequence of Optimal Adaptation to a Changing Body." *Nature Neuroscience* 10, no. 6 (2007): 779–786.
- Elyse H. Norton, Luigi Acerbi, Wei Ji Ma, and Michael S. Landy. "Human Online Adaptation to Changes in Prior Probability." *PLoS Computational Biology* 15, no. 7 (2019): e1006681.
- Kunlin Wei and Konrad P. Körding. "Uncertainty of Feedback and State Estimation Determines the Speed of Motor Adaptation." *Frontiers in Computational Neuroscience* 4 (2010): 11.
- Robert C.Wilson, Matthew R. Nassar, and Joshua I. Gold. "Bayesian Online Learning of the Hazard Rate in Change-Point Problems." *Neural Computation* 22, no. 9 (2010): 2452–2476.
- Daniel M. Wolpert. "Computational Approaches to Motor Control." *Trends in Cognitive Sciences* 1, no. 6 (1997): 209–216.

12.5 Problems

Problem 12.1 The muscles in our bodies change over time. Sometimes they get stronger, for example, after we work out, sometimes they get weaker. How would you formalize changing muscles, and inference about their strength?

Problem 12.2 Change points, where an aspect of the world undergoes a sudden change, happen in many domains. Think of three real-world examples not mentioned in this chapter.

Problem 12.3 Refer to the HMM in section 12.1. Formally derive the posterior $p(s_t|x_1, \ldots, x_t)$ using the structure of the generative model. "Follow the arrows." You do not need to substitute any specific distributions.

Problem 12.4 We build on the HMM in section 12.1. Suppose the observer wants to make a prediction for the future state s_{t+1}, given the measurements $x_1, \ldots, x_t$ (i.e., the measurement at time $t+1$ has not been made yet). Derive the mean and variance of the posterior. You can assume the mean and variance of $p(s_t|x_1, \ldots, x_{t-1})$ to be known.

Problem 12.5 A time series $y_1, \ldots, y_t$ is said to *asymptote* if it monotonically increases or decreases, but as t grows very large, approaches a finite value.

(a) Explain intuitively why the standard deviation asymptotes.

(b) Prove that in our treatment of the HMM in section 12.1, the variance of the posterior asymptotes at

$$\sigma^2_{\text{posterior}} = \frac{\sigma_s^2}{2}\left(\sqrt{1+\frac{4\sigma^2}{\sigma_s^2}}-1\right). \tag{12.34}$$

Problem 12.6 In this problem, we implement the HMM in section 12.1. Take $\mu_0 = 0$, $\sigma_0 = 1$, $\sigma_s = 1$, $\sigma = 2$, $\Delta = 1$, and t running from 1 to 30.

(a) Draw a sequence of world states $s_0, s_1, \ldots, s_t$ according to the distributions in the top row of the generative model.

(b) Draw a sequence of corresponding measurements $x_1, \ldots, x_t$ (no measurement at time 0).

We will now simulate an observer who uses the sequence of measurements you just drew to infer s_t at every time step.

(c) Create a movie consisting of thirty frames in which the tth frame shows:

  a. the posterior distribution over s_t,

  b. a vertical dashed black line at the true s_t (from part (a)),

  c. a vertical dashed blue line at the measurement x_t (from part (b)).

  Make sure that the scales of the axes remain fixed. Also label the axes.

(d) Plot in a single plot, separately from the movie:

  a. the true world state as a function of time (black line),

  b. the measurement as a function of time (blue line),

  c. the posterior mean as a function of time (red line).

(e) Plot in a separate plot the posterior standard deviation as a function of time.

Problem 12.7 Look up an external source (such as Wikipedia) on the dynamical systems model underlying the Kalman filter. This is a description of a generative model that includes the generative model of this chapter as a special case. Here, we explore how.

(a) Identify the variables in the dynamical systems model with variables or constants in our chapter.

(b) Making use of the correspondences from part (a), simplify the prediction and update equations until you arrive at our two recursive equations, equation (12.15) and equation (12.16). Keep in mind that our state estimate is the posterior mean.

Problem 12.8 An observer observes the time series (−0.46, 0.83, −3.26, −0.14, −0.68, −2.31, 0.57, 1.34, 4.58, 3.77). The observer knows that one change occurred and performs inference under the model in section 12.2.1, with $s_{\text{pre}} = -1$, $s_{\text{post}} = 1$, and $\sigma = 1$. Calculate the observer's posterior distribution over change point time.

Problem 12.9 In the model of section 12.2.1, assume $s_{\text{pre}} = -\mu$, $s_{\text{post}} = \mu$, and $\sigma = 1$. We expect that as μ grows larger, values of s before and after the change point become easier to distinguish. In this problem, we simulate this process.

(a) Set $\mu = 1$. Randomly draw a change point. The change point determines the sequence of world states. Now draw a corresponding sequence of measurements. Then apply the MAP decision rule to this sequence, and record whether the MAP observer was correct or not.

(b) In part (a), you did a single Monte Carlo simulation. Now, do 10,000 Monte Carlo simulations for the same μ. Calculate proportion correct across all simulations.

(c) Repeat parts (a–b) for all values of μ between 0 and 2 in steps of 0.1. Plot proportion correct as a function of μ.

Problem 12.10 In the model for detection of a single change point in section 12.2.1, we assumed that s_{pre} and s_{post} were known. Derive the decision rule in the case when one of these takes the value s_-, the other the value s_+, but the observer does not know which is which. In other words, the change could be from s_- to s_+ or vice versa.

Problem 12.11 In this problem, we implement the Bayesian model for detecting randomly occurring change points from section 12.2.2. You observe the time series $\mathbf{x} = (-0.25, -1.66, -0.34, -0.41, -0.55, -1.88, -2.63, -0.79, 1.54, 0.85, 2.12, 1.22, -0.85, -0.61, -1.14)$.

(a) Using the model with $s_0 = 1$, $\epsilon = 0.3$, and $\sigma = 1$, calculate the posterior distribution over the change point vector $\mathbf{C}$.

(b) How many $\mathbf{C}$ vectors have a posterior probability greater than 1 percent?

(c) What is the MAP estimate of $\mathbf{C}$?

(d) Plot in a single plot $\mathbf{x}$ as a function of time as dots connected by lines, and the MAP estimate of $\mathbf{C}$ as a set of dashed vertical lines at the change point times). Does your inference look plausible?

Problem 12.12 Simulate 1000 synthetic trials from the random change points model in section 12.2.2, with $T = 10$, $s_0 = 1$, $\epsilon = 0.3$, and $\sigma = 1$.

(a) With what frequency does the MAP estimate of $\mathbf{C}$ have the same number of change points as the true $\mathbf{C}$?

(b) Repeat part (a) for values of σ on a grid from 0.1 to 2 in steps of 0.1. Plot the frequency as a function of σ.

Problem 12.13 In the model of section 12.2.2, we assumed that s_t could take values -1 and 1 and changed sign when $C_t = 1$. Derive the decision rule in the case when s is drawn from a normal distribution with standard deviation σ_s and a mean μ_t that can take values -1 and 1 and that changes sign when $C_t = 1$.

13

Combining Inference with Utility

How can we make optimal decisions when potential rewards and costs are involved?

So far, we have modeled perceptual decision-making as a process in which the observer computes a posterior distribution over the world state variable of interest, then produces an estimate that is as close as possible to the truth. In real life, however, we often do not simply estimate world states; instead, we perform actions that have consequences, which in turn lead to rewards or costs.

Plan of the Chapter

We begin the chapter by showing how the optimal decision depends not only on our probability distribution over world states but also on our preferences over outcomes. We first consider how to make an optimal binary decision (e.g., should I take my umbrella?). We next consider how to choose among several actions (e.g., where should I search for my lost keys?) or even a continuum of actions. Finally, we use Bayesian decision theory to revisit the deceptively simple process of deciding which value to readout from a posterior distribution.

13.1 Example Tasks

We start the chapter with range of examples where utility matters on top of probabilities:

- In deciding when to cross a road, you are not only computing a posterior distribution over the speeds of oncoming cars but you are also combining this information with the utility that you derive from saving time and the negative utility associated with causing an accident or ending up in the hospital.
- In deciding whether to greet a person approaching you, you are not only computing a posterior distribution over the familiarity or identity of the person but you are also weighing the awkwardness of ignoring someone you know and the other awkwardness (in some locales) of greeting a stranger.
- When you decide whether to buy travel insurance, you are computing not only the posterior probability that something will go wrong on your trip but also the cost of the insurance and the cost of having to pay out of pocket when not insured.

- In deciding whether to drink old milk, you are computing not only the posterior probability that the milk has gone bad, but also the time, effort, and financial costs of buying new milk, as well as the cost of falling sick after drinking milk that has gone bad.
- A physician often has to decide whether to order a bothersome or expensive medical test to rule out a low-probability, but very serious diagnosis.
- Suppose you are hiking in foggy weather on a mountain trail with a cliff on your right side. From your limited visual information, you can estimate your distance from the cliff with some uncertainty. You then need to decide where along the trail to walk, choosing from among a continuum of possible headings. The center of the path may be most comfortable on your feet. Making an error by veering off toward the left is relatively harmless; perhaps the path becomes rockier. An error to the right, in contrast, could be fatal. Because the costs are asymmetric, the optimal decision will be to bias your position toward the side of the path that is further from the cliff.

These examples illustrate that the computation of a posterior probability distribution is only one part of a decision process that also takes into account costs and rewards. The theory of taking into account costs and rewards in step 2 of the Bayesian modeling process is also called *Bayesian decision theory.*

Note on terminology: The scientific fields of optimal control, economics and decision theory each make use of cost functions in one form or another. In economics and decision theory, researchers often specify a *utility function*. This may be related to economics' traditional focus on utility, which measures the trade-off between costs and benefits: the higher the utility, the better for the agent. In optimal control theory, researchers usually specify a *cost* or *loss function*. This may be related to the field's focus on minimizing the energetic cost of producing movements or the fuel costs of rockets. However, in each of the fields, there are positive and negative terms contributing to the function. Ultimately the two formulations are equivalent, as cost can be thought of as negative utility.

13.2 Deciding between Two Actions

Suppose that, as you prepare to leave home, you wonder whether it will rain. Combining a quick assessment of the cloudy sky (visual data) with knowledge of weather patterns in your area, you estimate the posterior probability of rain at 30 percent. Your posterior distribution, $p(\text{it will rain} \mid \text{visual data, background knowledge}) = 0.3$ and $p(\text{it will not rain} \mid \text{visual data, background knowledge}) = 0.7$, represents your belief about the world state of interest, but it does not dictate how you should *behave*. You need to make a decision: should you carry an umbrella or not?

It might at first seem that, since you believe that upcoming rainfall has a lower probability than 0.5, you should simply leave your umbrella at home. However, it should become clear on reflection that your decision whether to carry an umbrella will be based not only on your estimate of the chance of rain but also on the value you attach to different possible *outcomes* that could result from your choice of action. If you decide not to carry the umbrella and it rains, then you will suffer the undesirable consequence of becoming wet. On the other hand, if you decide to carry the umbrella and it does not rain, then you may feel inconvenienced by holding the unnecessary umbrella. An outcome may be undesirable, in which case we associate it with a *cost*, or desirable, in which case we associate it with a

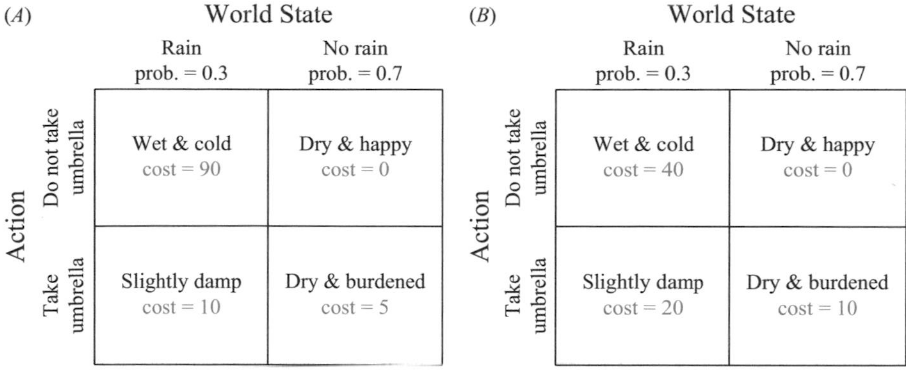

Figure 13.1
On a 100-point scale, two people rank the unpleasantness (cost) of four possible outcomes. (*A*) Walking in the rain without an umbrella results in a wet-and-cold outcome of cost 90; walking in the rain with an umbrella leaves one only slightly damp (cost 10); and so on. The optimal decision for this individual is to carry his umbrella. (*B*) Another individual assigns different costs. The optimal action for this individual is to leave her umbrella at home.

utility. Figure 13.1A illustrates the costs, specified by one individual, of the four possible outcomes in the umbrella problem.

Under the framework of Bayesian decision theory, the optimal behavior is the decision that minimizes expected cost, or, equivalently, maximizes expected utility. The expected cost is the cost associated with each possible outcome multiplied by the probability of that outcome. Referring to figure 13.1A, we see that the expected cost of carrying the umbrella is:

$$\text{EC(umbrella)} = 0.3 \cdot 10 + 0.7 \cdot 5 = 6.5. \tag{13.1}$$

That is, if we choose to carry the umbrella, we have a 30 percent chance of incurring a cost of 10, and a 70 percent chance of incurring a cost of 5. The expected cost of 6.5 can be thought of as the average cost that would result, if we chose to take the umbrella each day, over many days with weather identical to the current day. By contrast, the expected cost of not carrying the umbrella is:

$$\text{EC(no umbrella)} = 0.3 \cdot 90 + 0.7 \cdot 0 = 27. \tag{13.2}$$

If we choose not to carry the umbrella, we have a 30 percent chance of incurring a cost of 90, and a 70 percent chance of incurring a cost of 0. The expected cost of 27 can be thought of as the average cost that would result, if we chose not to take the umbrella each day, over many days with weather identical to the current day. Because $6.5 < 27$, the action that minimizes expected cost is to carry the umbrella; that is the optimal action for this individual.

Importantly, the cost or utility placed on an outcome reflects a personal preference; it is inherently subjective. Indeed, two people with the identical posterior distribution over the world state may choose opposite courses of action, because of the distinct values that they place on particular outcomes. To illustrate this point, consider a second individual who agrees that the chance of rain is 30 percent, but for whom the costs of the outcomes are different (figure 13.1B). Note that the two people agree in their *ranking* of the outcomes from highest to lowest in cost, but they assign different numerical costs to the outcomes. For this second person, the optimal action is to leave the umbrella at home.

13.3 Deciding among Several Actions

The procedure that we have described for selecting one of two actions applies equally to situations that require the selection of one of several actions. As an illustration, suppose that, on getting a ride home from a picnic in the park with a group of friends, you realize that your house keys are not in the right pocket of your pants, where you customarily keep them. Based on knowledge of your recent activities, you quickly generate a probability distribution over the location of your lost keys. They might have fallen out in your friend's car; you might have placed them in a different pocket; or you might have lost them at the park. Depending on where your keys actually are, and where you decide to search for them, nine outcomes are possible, and each of these has a particular cost to you (figure 13.2). The decision you need to make is: where should you search first?

The optimal decision will be the one associated with minimal expected cost. Computing the expected cost of each action, we have:

$$\text{EC(search in car)} = 0.1 \cdot -75 + 0.1 \cdot 15 + 0.8 \cdot 15 = 6 \tag{13.3}$$

$$\text{EC(search in pockets)} = 0.1 \cdot 11 + 0.1 \cdot -79 + 0.8 \cdot 11 = 2 \tag{13.4}$$

$$\text{EC(search in park)} = 0.1 \cdot 90 + 0.1 \cdot 90 + 0.8 \cdot 0 = 18. \tag{13.5}$$

Thus, despite the fact that you consider it most probable that the keys are in the park, it is optimal to search first in your pockets.

World State (location of keys)

Action	Your friend's car prob. = 0.1	Another pocket prob. = 0.1	The park prob. = 0.8
Search car	Inconvenience friend, find keys! cost = –75	Inconvenience friend, disappointing result cost = 15	Inconvenience friend, disappointing friend cost = 15
Search pockets	Easy to do, disappointing result cost = 11	Easy to do, find keys! cost = –79	Easy to do, disappointing result cost = 11
Search park	Big inconveneince, disappointing result cost = 90	Big inconveneince, disappointing result cost = 90	Big inconveneince, find keys! cost = 0

Costs:
- Find keys: –80
- Easy to do: 1
- Inconvenience friend: 5
- Disappointing result: 10
- Big inconvenience: 80

Figure 13.2
On a -100 to $+100$ point scale, the costs of different outcomes in the key search problem. Costs for each outcome were calculated by addition of the costs associated with each feature of the outcome (inset). If you decide to search the car, you will need to call your friend and ask them to do that for you, which imposes an inconvenience on your friend and you do not like that (cost 5); if you decide to search the grass at the park where you picnicked, you will need to take a long trip back to the park, and probably spend a lot of time searching there, a big inconvenience (cost 80). Thoroughly searching the other pocket of your pants and the pockets in your coat is easy to do (cost 1). Searching and not finding your keys would be disappointing (cost 10). Finally, finding your keys would be very rewarding (cost -80).

Box 13.1
The Drunkard and the Lamppost

The tale of the drunkard who loses his keys walking from the bar to his car, but decides to search for them under a lamppost, presents an amusing case of suboptimal decision making. From a Bayesian decision theoretic perspective, the drunkard has correctly assigned low cost (high reward value) to the outcome (search under lamppost, find keys), because if his keys are there he is likely to encounter them easily and quickly given the light. However, he has failed to take into consideration that the probability is zero that his keys are in that location, since he never was near the lamppost to begin with. Being biased toward searching where it is easiest is also called the *streetlight effect.*

We won't find our keys if we search only under the streetlight.

Box 13.2
Bayesian Search

The example we have been considering is one of Bayesian search. This is similar conceptually to the visual search examples that we discussed in chapter 11, but in those examples, we did not consider a cost function. Bayesian search has been used successfully to discover valuable lost objects; it is commonly used to search for ships lost at sea. A famous example was the loss of the U.S. Navy's nuclear submarine, USS *Scorpion*, which disappeared during a voyage in the Atlantic in May 1968, with ninety-nine crew aboard. Interviews with Navy experts were used to assign probabilities to different scenarios that might have caused the sinking of the sub, and computer simulations were then run to define a prior probability distribution for the sub's location on the ocean floor. A search grid was constructed, with a prior probability assigned to each square on the map. In search operations, each grid square can be associated with a particular cost, based in part on the difficulty of finding the sub if it were at that location; the seafloor differs in depth, and in some areas has narrow canyons that would increase the difficulty of the search. If a grid square is searched and the sub not found, the probability distribution over the map can be updated, and the optimal search location recalculated. The USS *Scorpion* was found in October 1968, in about 10,000 feet of water, approximately 400 miles southwest of the Azores islands. Bayesian search was similarly helpful in searching for Air France flight 447, which perished over the Atlantic Ocean in 2009.

13.4 Mathematical Formulation: Expected Utility

Notation for this chapter	
s	World state
a	Action
x	All observations/measurements
$U(s,a)$	Utility of action a when the world state is s

We consider an *agent* who considers actions a while the world state s is not known. In previous chapters, the action was an estimate of the world state s, but in this chapter, it does not have to be; for example, if s is whether the milk is still good then a could be "drink" or "not drink." The term "agent" is used instead of "observer" to emphasize the action.

We define a *utility function* $U(s,a)$, which is a function of both state s and action a. Reward corresponds to positive utility, cost to negative utility. U is typically real-valued. In previous chapters, when $a=\hat{s}$, then U is typically a function of the difference between s and $\hat{s}$.

Suppose now that the observer has a posterior $p(s|x)$ over s. Here, x stands for the collective set of measurements or observations made by the observer; it does not have to be a single scalar measurement, as in previous chapters. Under this posterior, the expected value of the utility of action a given x, more concisely called the *expected utility* of a and x, is

$$\text{if } s \text{ is discrete:}\quad \mathrm{EU}(a,x)=\sum_s U(s,a)p(s|x); \tag{13.6}$$

$$\text{if } s \text{ is continuous:}\quad \mathrm{EU}(a,x)=\int U(s,a)p(s|x)ds. \tag{13.7}$$

The sum and the integral are over s, so the result does not depend on s. The simplest version of Bayesian decision theory postulates that the agent maximizes expected utility: they choose the action that—averaged over all possible world states weighted by their respective posterior probability—yields the highest possible utility:

$$a_{\text{optimal}}=\underset{a}{\operatorname{argmax}}\ \mathrm{EU}(a). \tag{13.8}$$

Box 13.3
The Multiple Faces of "Decision Making."

In psychology and neuroscience, different subcommunities have different notions of what constitutes the field "decision making." Here are three:

- Decision making must involve a **nontrivial cost/reward structure**. People holding this notion typically work in behavioral economics, neuroeconomics, or in the field called "judgment and decision making," which is to a large extent concerned with cognitive biases. Decisions studied in these disciplines rarely involve perceptual uncertainty; for example, subjects choose between two gambles/lotteries presented on the screen. With this notion of decision making, the current chapter is the first one that touches on true decision making.
- Decision making must involve **accumulation of evidence**. People holding this notion typically measure subjects' reaction times in perceptual or cognitive tasks: the subject can decide when to respond. A prominent model in this field is the *drift-diffusion model*.

The work in this book is only tangentially related to this field, since we do not consider reaction-time paradigms. A point of contact is the accumulation of evidence discussed in section 5.5.

- In this book, we use an inclusive notion: decision making is any process that maps stimuli to response. This includes purely perceptual decisions, purely utility-based decisions, and anything in between.

13.4.1 Binary Classification

To start, we will work out the expected utility framework in a simple case: Consider the generative model of chapter 7 or 8, where a class C (0 or 1) has to be inferred from a measurement x. The observer reports their estimated class. Thus, $a=\hat{C}$ as in those chapters, and the utility function is a function $U(C,\hat{C})$. But here is the twist: a correct response of $C=0$ yields utility $U(0,0)=U_0$, while a correct response of $C=1$ yields utility $U(1,1)=U_1$. An incorrect response yields no utility: $U(0,1)=U(1,0)=0$. If U_0 were equal to U_1, maximizing expected utility would reduce to maximizing accuracy, for which we already know MAP estimation to be the solution. Here, however, we allow U_0 and U_1 to be different from each other. Intuitively, we would expect that if, for example, $U_1>U_0$, then we would expect the observer to be more inclined to report class 1. The expected utility of the action "report class 1" for given x is then

$$\mathrm{EU}(\hat{C}=0,x)=\sum_{C=0}^{1}U(C,\hat{C}=0)p(C|x) \tag{13.9}$$

$$=U(C=0,\hat{C}=0)p(C=0|x)+U(C=1,\hat{C}=0)p(C=1|x) \tag{13.10}$$

$$=U_0 p(C=0|x)+0\cdot p(C=1|x) \tag{13.11}$$

$$=U_0 p(C=0|x). \tag{13.12}$$

This is simply the probability of being correct multiplied by the reward following a correct response. Similarly, $\mathrm{EU}(\hat{C}=1)=U_1 p(C=1|x)$. If the agent maximizes expected utility, then they will choose $\hat{C}=1$ if the expected utility of this choice is greater than of the alternative, $\hat{C}=0$. Substituting the expressions for EU, we find that this is the case when

$$\frac{U_1 p(C=1|x)}{U_0 p(C=0|x)}>1. \tag{13.13}$$

The left-hand side is an *expected-utility ratio*, a generalization of the posterior ratio. We now write out the posterior ratio as the product of a prior ratio and an LR:

$$\frac{U_1}{U_0}\frac{p(C=1)}{p(C=0)}\frac{p(x|C=1)}{p(x|C=0)}>1. \tag{13.14}$$

We learn from this equation that utilities act like priors: we can consider the product of the utility ratio and the prior ratio, $\frac{U_1}{U_0}\frac{p(C=1)}{p(C=0)}$, as a modified prior ratio. In other words, the agent's bias toward class 1 changes in the same way when the prior ratio changes by a factor as when the utility ratio changes by that factor. A difference in utility could arise from peripheral factors such as a difference in effort it takes to press the buttons associated with $C=0$ and $C=1$ reports. In practice, this is an argument to fit the prior in a binary task as a free parameter instead of assuming that it reflects the experimental statistics.

13.4.2 Continuous Estimation

In the examples discussed above, both the world state variable and the action choice could take on only a limited number of discrete values. In contrast, many daily examples involve continuous world states and/or choices among a continuum of possible actions. We can no longer apply costs individually to every outcome in such cases; rather, we need to use a utility function over the continuum of possible world states or actions.

For instance, returning to our umbrella example, we could consider a more nuanced scenario in which the world state can take a continuum of values reflecting the rate of the rainfall, from 0 (no rain) to 10 (extremely heavy rainfall). This results in a continuum of possible outcomes, each with a specific cost, even when we consider only two actions (to take or not to take the umbrella). If we tried to create an outcome table to represent this situation, we would have two rows (as in figure 13.1) but an infinite number of columns. Clearly, a different approach is needed here.

In such cases, we need to construct a function that reflects cost over the continuous space of outcomes for each action. Perhaps the cost to us of being in the rain grows linearly with the amount of water that lands on us. The utility function could then be expressed as $U(s,a) = -(A+Ba)s$, where s represents the rainfall rate, a represents our action (0 for carrying, and 1 for not carrying the umbrella), and A and B are constants. Thus, whether we have our umbrella or not, we would be increasingly displeased by greater rainfall, but the rise in our displeasure, as a function of rainfall rate, is greater when we lack the umbrella. We would then replace the sum with an integral:

$$\begin{aligned} a_{\text{optimal}} &= \underset{a}{\operatorname{argmax}}\, \mathrm{EU}(a,x) \\ &= \underset{a}{\operatorname{argmax}} \int U(s,a)p(s|x)ds. \end{aligned} \tag{13.15}$$

To make the problem even more realistic, we could consider not just a continuum of rainfall rates, but also a continuum of possible actions reflecting not just our decision to take or leave the umbrella but also our walking speed. We might walk slowly or attempt to minimize our exposure time to a possible rainfall by sprinting to our location (or go at any speed in between). If we tried to create an outcome table to represent this situation (similar to figure 13.1), we would have an infinite number of rows and columns. Again, to deal with this situation we would need to specify a cost function over the space of outcomes.

The specification of a cost function for real-life decision problems is a difficult task, although there are multiple obvious ingredients of the cost function. We want to satisfy our immediate needs and desires as well as progress toward more distal goals. This implies that obtaining food, drink, and shelter, mating, and other factors will have utility. At the same time, we do not want to overly exert ourselves, we want to minimize our energy consumption, the risk of damage to our body, and other factors that are negative to our well-being.

Exercise 13.1 Describe a cost function that could account for much of your behavior today.

13.5 The Cost Functions of "Pure Perception"

Before this chapter, we did not consider explicit costs or rewards associated with outcomes. Instead, we worked in the "purely perceptual" domain. Nevertheless, costs and rewards were playing a role in the background. Namely, we assumed that the observer's perceptual system aims to generate a decision (or percept) that is as close as possible to the truth. This aim takes different forms for discrete and continuous tasks. For discrete (categorical) tasks, the natural goal is to maximize the proportion of correct responses—accuracy. For continuous estimation tasks, one possible goal is to minimize the expected squared estimation error. We mentioned before that these objectives give rise to MAP and posterior mean estimation, respectively. However, we did not yet prove those statements. With the background we now have on utility functions, we can correct this omission. We will conclude with a comment on the relation between perception and action.

13.5.1 Discrete Tasks

In previous chapters, in discrete tasks, the observer cared about maximizing accuracy. If the discrete world state is s, then this corresponds to a utility function that is 1 (or any other positive number) when $\hat{s}=s$ and 0 otherwise. We can write such a "correctness" utility function as

$$U(s,\hat{s})=\begin{cases} 1 & \text{if } \hat{s}=s \\ 0 & \text{otherwise.} \end{cases} \tag{13.16}$$

This is also called the *0-1 utility function* (or its negative the 0-1 cost function). When the posterior is $p(s|x)$, expected utility is

$$\text{EU}(\hat{s},x)=\sum U(s,\hat{s})p(s|x) \tag{13.17}$$

$$=p(s=\hat{s}|x). \tag{13.18}$$

To maximize this quantity, the observer should choose $\hat{s}$ to be the value of s that maximizes $p(s|x)$—in other words, MAP estimation.

13.5.2 Continuous Estimation

In continuous estimation tasks, the 0-1 utility function is not very sensible. For example, in a location estimation task, it is impossible to get the position *exactly* right, $\hat{s}=s$. Moreover, large errors are worse than small ones. We need to catch this in the utility function. One reasonable try is to postulate that the cost of reporting the estimate $\hat{s}$ when the true stimulus is s is the squared difference between the two: a quadratic cost function. The utility function then takes the form

$$U(s,\hat{s})=-(s-\hat{s})^2. \tag{13.19}$$

When the posterior is $p(s|x)$, expected utility is

$$\text{EU}(\hat{s},x)=\int U(s,\hat{s})p(s|x)ds \tag{13.20}$$

$$=-\int(s-\hat{s})^2 p(s|x)ds. \tag{13.21}$$

To find the value of the estimate that maximizes EU, we compute the partial derivative of EU with respect to $\hat{s}$ (the derivative is partial because EU also depends on x):

$$\frac{\partial \text{EU}}{\partial \hat{s}} = \int 2(s-\hat{s})p(s|x)ds \tag{13.22}$$

$$= \int 2sp(s|x)ds - \int 2\hat{s}p(s|x)ds \tag{13.23}$$

$$= 2\int sp(s|x)ds - 2\hat{s}\int p(s|x)ds \tag{13.24}$$

$$= 2\mathbb{E}[s|x] - 2\hat{s}\cdot 1, \tag{13.25}$$

where $\mathbb{E}[s|x]$ is the posterior mean. The derivative equals 0 when $\hat{s}=\mathbb{E}[s|x]$, i.e., when the estimate is equal to the posterior mean. Thus, the posterior mean is the optimal readout given the quadratic cost function. We already mentioned this as early as section 3.3.6 but so far, we have not provided a proof.

For every utility function, the procedure of maximizing expected utility will lead to a different decision rule. The quadratic cost function is often used in practice as the mean is easy to calculate, and the utility function often approximates what subjects care about (be close to the target). However, other cost functions are also plausible. Perhaps most notable is the *absolute error* cost function,

$$U(s,\hat{s}) = -|s-\hat{s}|. \tag{13.26}$$

Maximizing expected utility now leads the decision maker to report the *median* rather than the mean of the posterior (see problem 13.3).

In the context of perceptual estimation problems, the choice of utility function is particularly important when the posterior distribution is asymmetric. For symmetric unimodal distributions such as the Gaussian distribution, the mean, median and mode are identical. For asymmetric distributions, however, they are distinct (figure 13.3). Furthermore, some

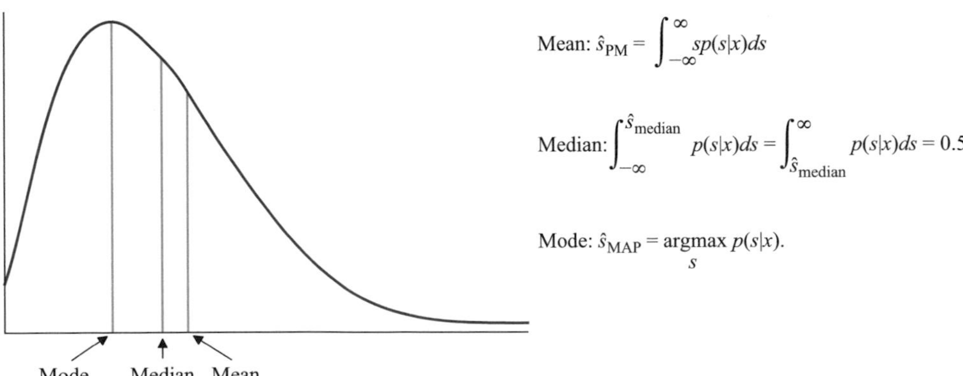

Figure 13.3
An asymmetric probability density; this might represent an observer's posterior probability distribution over sound source location on a particular trial. The three lines, from left to right, represent the mode, median, and mean of the distribution. Also shown: definitions of mean, median, and mode read-out of a posterior distribution.

distributions, while symmetric, are bimodal. In chapter 10, when discussing causal inference, we encountered bimodal (two-peaked) posterior distributions (see figure 10.4). For such distributions, the mean and mode are generally distinct, and research has suggested that people's estimates are closer to the mean of such a bimodal posterior [98].

13.5.3 Perception and Action

In section 3.9, we discussed the propositions that percepts and perceptual decisions (responses in perceptual tasks) are not that different and that they can be modeled in the same way. In terms of Bayesian decision theory, these propositions would imply that the cost functions that are activated—or assumed to be activated—when asking human subjects for a perceptual decision are quite similar to the ones that the mind uses in unqueried perception. These cost functions would presumably be the ones discussed in this section, or very similar to them. Moving away from this limited set of cost functions would make the task less perceptual and increase the conceptual distance between percepts and responses. We saw a minimal example in section 13.4.1, where different types of correct responses were rewarded differently. The disconnect can be further increased by decoupling the set of available actions from world state inference. One might infer whether it will rain, but decide whether to bring an umbrella. One might infer how much time a car will take to reach you, but decide whether to cross the road. In these examples, perception is only a stage in action selection. Nevertheless, the "natural" cost functions of perception might continue to operate alongside the cost functions used in action selection. This would lead to a view in which percepts are automatically generated even though they do not determine the action.

13.6 What It Means to Decide Optimally

In Bayesian models, optimal decision making means maximizing expected utility. Here, we ask why the correct posterior distribution is a necessary ingredient of deciding optimally. Why could the observer not use some other probability distribution over the world state computed from the observations, say $q(s|x)$? A situation where this might happen is *model mismatch*, introduced in section 3.5, where the observer makes wrong assumptions about one or more distributions in the generative model. The resulting incorrect posterior distribution could be used for decision making.

Using a $q(s|x)$ that differs from $p(s|x)$ will in general fail to maximize expected utility. To understand this, suppose the observer is free to follow *any* strategy to map observations x to an action a. We will denote this strategy by a function F; therefore, $a = F(x)$. The utility obtained on a single trial is then $U(s, F(x))$ and the expected utility for given x is

$$\text{EU}_F(x) = \int U(s, F(x))p(s|x)ds. \tag{13.27}$$

The Bayes-optimal strategy is the function F that, *for any* x, maximizes EU under the true posterior, $p(s|x)$. Now consider the quantity formed by averaging EU over all observations x:

$$\mathbb{E}[\text{EU}_F] = \int \text{EU}_F(x)p(x)dx, \tag{13.28}$$

which can also be written as a grand average over x and s:

$$\mathbb{E}[\mathrm{EU}_F] = \iint U(s, F(x))p(x, s)dxds. \tag{13.29}$$

Thus, the expected value of EU_F is the overall expected utility, the amount of utility that can be expected in the long run as world states and measurement x are sampled from the generative model. Because the Bayes-optimal strategy maximizes $\mathrm{EU}_F(x)$ for every x, it must also maximize overall expected utility. In particular, that means that constructing F based on an alternative posterior distribution $q(s|x)$ can never yield a higher overall expected utility than using the true posterior distribution, $p(s|x)$, which is obtained from the generative model. Thus, the correct computation of the Bayesian posterior is the essence of optimal decision making.

Exercise 13.2 How does equation (13.29) follow from equation (13.28)?

A related argument is the *Dutch Book argument*. It is possible to show that, under a plausible set of axioms, if someone does not use proper probability calculus—that is, Bayesian inference—in a situation in which uncertainty is involved, then a set of bets (the Dutch Book) can be proposed that the person should accept but that will end up losing them money. In other words, non-Bayesian beliefs are costly.

13.7 Complications

The expected utility framework laid out in previous sections is a simplification. Here, we discuss a few common complications.

13.7.1 Cost Functions for Uncertain Outcomes

In the examples we have considered so far, each combination of world state and action has mapped deterministically onto a single outcome, to which we assigned a cost. In reality, however, it is often the case that many outcomes could result—with different probabilities—from a particular combination of action and world state. For example, in figure 13.1, we assumed that, if we were to leave the house without our umbrella (action) and it were to rain (world state), then we would get wet (outcome). In reality, however, a range of possible outcomes might result from this combination of action and world state, depending probabilistically on the availability of places along our path under which we could take cover from the rain. In such situations, given a particular combination of action and world state, one can define a probability distribution across outcomes: $p(o|a, s)$ (figure 13.4). Now, each *outcome* (o) is associated with a utility, and the optimal decision maker would choose the action a that maximizes expected utility. Thus, the formula for the optimal action (equation (13.8)) generalizes to

$$a_{\text{optimal}} = \underset{a}{\operatorname{argmax}}\ \mathrm{EU}(a) \tag{13.30}$$

$$= \underset{a}{\operatorname{argmax}} \int \left(\int U(o)p(o|s, a)do \right) p(s|x)ds. \tag{13.31}$$

In other words, the agent averages the utility function both over possible outcomes given your hypothesized action and over possible states of the world.

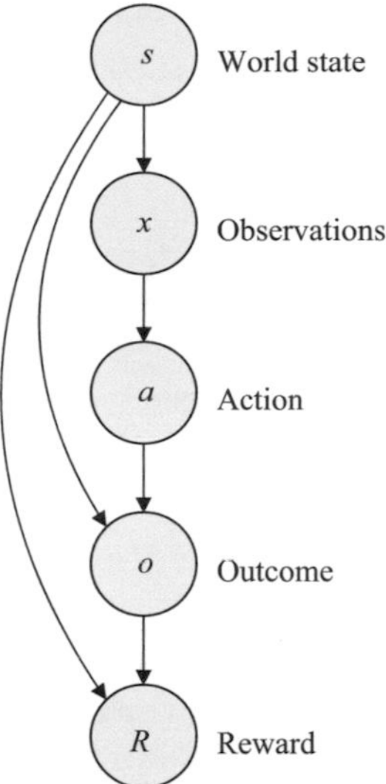

Figure 13.4
General framework for Bayesian decision making.

13.7.2 Nonlinearity between Reward and Utility

If an experimenter gives double the money as a reward to a human subject, or double the juice to a monkey subject, then it does not mean that their utility is doubled. The same increment in objective reward might yield a smaller increase in subjective utility when added to a larger base amount. This can be modeled by relating utility U to reward R through a power law:

$$U = R^{\alpha}. \tag{13.32}$$

This provides a resolution to the famous St. Petersburg paradox (see box 13.4).

Box 13.4
History of Utility Theory

Gabriel Cramer *Daniel Bernoulli* *Jeremy Bentham*

Daniel Bernoulli and before him Gabriel Cramer worked on the St. Petersburg paradox. This paradox relates to a simple game of chance: on every round, a fair coin is thrown, and the game ends as soon as the coin lands tails (T). The subject then gets 2^h dollars, where h is the number of heads (H) observed. For example, if the sequence of landings observed was T, the player would win \$1; HT, \$2; HHT, \$4; HHHT, \$8; and so on. It turns out that the expected payoff of this game is infinite. The expected utility is the sum over all possible outcomes of the payoff times the probability of the outcome. Representing each possible outcome by its number of heads, we have:

$$\mathrm{EU} = \sum_{h=0}^{\infty} p(h) 2^h = \sum_{h=0}^{\infty} \left(\frac{1}{2}\right)^{h+1} 2^h = \sum_{h=0}^{\infty} \frac{1}{2} = \infty. \tag{13.33}$$

The paradox is that, although the expected utility is infinite, people would not pay more than a few dollars to enter the game. A way to resolve this paradox is through the introduction of a utility function that reflects the decreasing marginal utility of money: 1,001,000 dollars is only slightly more valuable than 1,000,000 dollars, whereas 1,000 dollars is much more valuable than nothing.

Jeremy Bentham, another thinker who considered the idea of utility, applied it more directly to the pleasures and pains of humans. He used these ideas to derive how society should be organized—namely by maximizing the utilities of all the citizens, the philosophy of utilitarianism. Bentham saw all moral and legal norms as derivable from this simple principle using methods from logic and experimentation.

13.7.3 Distortions of Probability

In *prospect theory*, which mostly deals with economic decisions such as whether to take \$10 or a 50 percent chance of winning \$25, it is sometimes alleged that in the computation of expected utility, the probabilities of the outcomes are not linearly taken into account. However, the probabilities under consideration are then usually explicitly presented like the 50 percent in the example. Whether posterior probabilities are nonlinearly weighted is unknown.

13.7.4 Decision Noise

Subjects do not always make the same decision in the same situation. In perception, such behavioral variability can sometimes be explained as a consequence of measurement noise, as we have done throughout the book. In value-based decisions without a perceptual component, that is not an option. Therefore, often a form of *decision noise* is introduced to account for human behavior: given a set of actions a with corresponding expected utilities $\mathrm{EU}(a)$, the observer does not always choose the same a. The most common way to implement decision noise is by postulating that the *probability* of choosing a is proportional to an exponential function of $\mathrm{EU}(a)$:

$$p(a) = \frac{e^{\beta \mathrm{EU}(a)}}{\sum_a e^{\beta \mathrm{EU}(a)}}. \tag{13.34}$$

When β grows very large, the action with the highest expected utility gets a huge boost, so that its probability approaches 1: this is the original case of the EU-maximizing agent. When $\beta = 0$, the observer randomly chooses an action with equal probability, regardless of EU. For any other β, the agent does something in between randomly choosing and maximizing. For this reason, the decision rule is also called a *softmax* rule and the parameter β is also called the *inverse temperature* (by analogy with thermodynamics): the lower β, the higher the "temperature," and the noisier the system. When a can take only two values, the softmax rule simplifies to a logistic mapping (see problem 13.3).

Please note that decision noise is not inherently a Bayesian concept. It is a generic, optional addition to any utility-based model, including models in which expected utility involves a posterior distribution.

13.8 Applications

When we are examining the behavior of a human subject or animal, we often want to compare it against what is optimal. We can compare predictions from Bayesian decision theory with actual behavior. For this type of analysis to be useful, we must know beforehand what it is the subject is trying to achieve, that is, what their cost function is. With this in mind, researchers have designed experiments where the cost of the task is relatively explicit. We discuss three example studies from different domains: visual discrimination, reaching movements, and confidence intervals. Finally, we discuss approaches for estimating an unknown cost function.

13.8.1 Visual Discrimination

Whiteley and Sahani [201] tested the expected utility model of section 13.4.1 in a visual classification task (figure 13.5A). Subjects reported whether one briefly presented visual pattern was offset to the left or to the right of another pattern. They received point rewards for correct responses and point penalties for incorrect responses. Crucially, in most blocks of trials, an incorrect "left" response incurred a different penalty from an incorrect "right" response. In this task, a Bayesian observer would compute and compare, given a measurement x, the expected utilities of responding "left" and "right" according to equation (13.13). This would amount to applying a decision rule of the type

$$\text{"report 'right' if } x > k\text{,"} \tag{13.35}$$

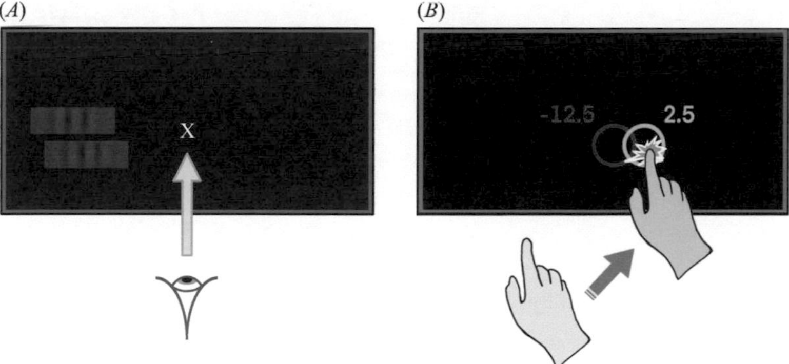

Figure 13.5
Utility interacting with perception or action. (*A*) Schematic of an experimental design to test whether reaching movements under noise maximize expected reward. Subjects fixated on the X and pressed a key to indicate whether the lower patch was offset to the left or to the right of the upper patch. Incorrect "left" and "right" decisions where penalized differently. Figure reproduced from [201]. (*B*) Schematic of an experimental design to test whether reaching movements under noise maximize expected reward. Subject made a speeded pointing movement to gain rewards associated with the inside of the green circle, while trying to avoid the penalties associated with the inside of the red circle. Figure reproduced from [183].

where k depends both on the level of sensory uncertainty and on the costs and rewards of the block. The authors found that people approached optimal behavior even though feedback in the experiment was insufficient to learn this strategy. This suggests that sensory uncertainty is computed and flexibly combined with varying costs and rewards.

13.8.2 Reaching Movements

Trommershäuser and colleagues used the following experiment to study movement under uncertainty (figure 13.5B) [180, 182]. Subjects were instructed to make a pointing movement to inside a small green circle on a screen, while avoiding the inside of a small red circle. If they hit inside the green circle but not inside the red, they earned 2.5 points. If they hit inside the red circle but not inside the green, they lost 12.5 points. If they hit in the intersection of both circles, they lost 10 points. If they hit outside both circles, they earned 0 points. If they took too long, they earned 0 points as well, so they were forced to move quickly. Importantly, the place where a subject hit the screen was not exactly where they aimed, because it was corrupted by movement noise; thus, this experiment contained outcome uncertainty (section 13.7.1). In the experiment there is a point where the subject should aim to maximize the expected number of points they earn. This optimal point will depend on the costs and rewards, as well as on the separation of the circles and their own motor noise (see problem 13.8). The researchers found that people learn to be close to the optimal choices prescribed by decision theory. Similar experiments have used force-producing tasks [99, 102]. This demonstrates people's ability to integrate utility with outcome uncertainty.

13.8.3 Incentivized Confidence Intervals

In section 3.4.2, we described different ways of obtaining a confidence rating from a posterior distribution $p(s|x)$. Here, we study a method to incentivize meaningful confidence *intervals* (figure 13.6). Variations of this method have, for example, been used in confidence judgments of future location [196], memorized location [212], and memorized color [76]. The idea is simple: after reporting a point estimate $\hat{s}$ (e.g. the PME or the MAP estimate), the observer sets a symmetric interval around that estimate (figure 13.6A). This interval acts like a "catcher": if it contains the true world state value, the observer is rewarded. Crucially, the magnitude of the reward is a decreasing function of the size of the interval. Thus, the subject could set a larger interval to be more sure to include s, but would earn a lower reward if successful. In figure 13.6B, we show one example:

$$R(L) = 100\, e^{-0.5L}, \tag{13.36}$$

where L is the half-length of the interval. Many other functional forms could be used. Following equation (13.32), the associated utility is

$$U(L) = R(L)^{\alpha}. \tag{13.37}$$

While utility decreases as L increases, the probability of being rewarded at all, denoted by $p_{\text{reward}}(L)$, increases. We can calculate this probability for a Bayesian observer: given a plot of the posterior distribution (figure 13.6A), the probability of being rewarded is equal to the area under the distribution within the interval $[\hat{s} - L, \hat{s} + L]$ (figure 13.6C). Expected utility is then

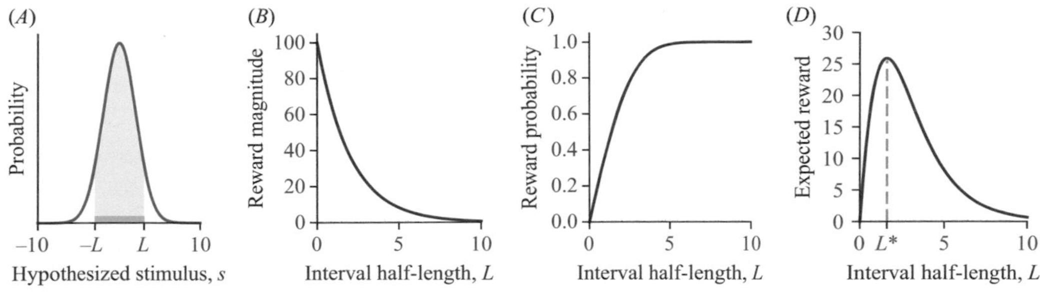

Figure 13.6
Incentivized confidence intervals. (*A*) Example Gaussian posterior distribution over a hypothesized world state s. We assume that the observer reports the posterior mean. The green line segment on the x-axis represents an interval (symmetric around the estimate) that the observer sets to "catch" the true value of s. (*B*) If the catch is successful, the observer gets rewarded by an amount that monotonically decreases with the size of the interval. (*C*) The probability of successful catch monotonically increases with the size of the interval. (*D*) Expected utility (here with $\alpha = 1$) is the product of the functions in (*B*) and (*D*), and shows an optimal interval size, indicated by L^*.

$$\mathrm{EU}(L; x) = U(L) p_{\text{reward}}(L) \tag{13.38}$$

$$= U(L) \int_{\hat{s}-L}^{\hat{s}+L} p(s|x) ds. \tag{13.39}$$

In this expression, the first factor monotonically decreases as L increases, whereas the second one monotonically increases. As a result, $\mathrm{EU}(L)$ has a maximum (figure 13.6D), representing the optimal trade-off between reward magnitude and reward probability. The optimal interval length serves as an incentivized measure of confidence: if the posterior is narrower (the observer is more confident), they should set a smaller interval length. We work this out in problem 13.9.

We would like to mention two possible modifications of this model:

- While the pure utility-maximizing observer would set the interval half-length deterministically, one can add softmax noise to this model by applying equation (13.34) with $a = L$.
- We have described setting $\hat{s}$ and L through separate processes. This implies different objectives in the two stages of the decision, for example minimizing expected squared error to set $\hat{s}$ and maximizing EU from equation (13.38) to set L. Instead, the observer might set $\hat{s}$ and L jointly so as to maximize EU. Depending on the shape of the posterior distribution, this process could lead to different predictions (see problem 13.11).

Finally, an experimental modification could be considered: instead of asking the subject to set a symmetric interval around $\hat{s}$, the experimenter could allow them to set an asymmetric interval. In terms of reward structure, but not necessary in terms of subject behavior, this would be equivalent to forgoing the point estimate altogether and *only* asking the subject to set an interval. Of course, the posterior distribution would have to be more interesting than Gaussian for this modification to be worth doing.

13.8.4 Inferring Utility Functions

Utility functions are hypothesized functions that people are supposed to optimize. In section 13.8.2 we assumed utility was equal to earned reward, and in equation (13.32)) we postulated a power law dependency. Thus, we have assumed a specific functional form with at most one parameter. Research practice is more complicated. First, any linear transformation of utilities leaves behavioral predictions invariant. Second, and more profoundly, researchers may have no prior constraints on functional forms. This problem arises in many contexts. How much better is it to have \$2 versus \$1? How much better is it to use our muscles to push strongly for a shorter versus a longer time? It is easy to figure out experimentally that people prefer more money and less effort, but that finding does not nail down the functional form.

Several approaches exist to address this challenge. One generic approach is to use large numbers of behaviors and ask which utility function can best describe the measured behaviors following a strategy called inverse decision theory [97, 135]. This will usually also include terms that bias the solutions towards simpler inferred utility functions. For movements, these inferred cost functions have highly nonlinear and nontrivial forms [176]. We can, given some assumptions, experimentally make progress towards identifying which cost functions are being optimized. This approach is similar in spirit to methods that infer the priors used by the observer [168]. This approach is quite general, but it is hard to know for which set of measured behaviors it will produce good estimates of underlying utility functions.

If we have a cost function that depends on more than one variable (e.g. time consumed and energy consumed for movement, probability of missing a target versus being fast), we can ask subjects which of two possible outcomes they prefer. From asking many such questions we can then estimate which combinations of variables have equal utility. From many such measurements and some additional assumptions, such as smoothness, it is then possible to estimate the full utility function used, for example, in the case of forces of varying strengths and duration [99]. Because we primarily focus on identifying pairs of outcomes with identical utility using preference judgments, it is relatively straightforward to choose comparisons that allow a good identification of underlying utility functions.

13.9 Summary

In this chapter, we formulated the problem of optimal decision making in the context of utility maximization. We learned:

- A utility (or cost) function is the starting point for defining optimal decision making. Utility depends on action and world state.
- Expected utility combined utility with posterior probability.
- When actions correspond to world state estimates, differences in reward between different actions influence decisions in the same way as priors.
- Computing posterior distributions correctly is necessary for maximizing overall expected utility, no matter the form of the utility function.
- In purely perceptual tasks, the posterior readout can be derived from minimizing a cost function.

- The notion of expected utility maximization can be generalized to incorporate outcome uncertainty, a nonlinearity between reward and utility, distortions of probability, and decision noise.
- Experimental tests suggest that people are good at combining utility with perceptual or motor uncertainty in making decisions.
- To induce an observer to set meaningful confidence intervals, a reward can be associated with "capturing" the stimulus in the interval, with reward magnitude decreasing with increasing interval size.

13.10 Suggested Readings

- Maija Honig, Wei Ji Ma, and Daryl Fougnie. "Humans Incorporate Trial-to-Trial Working Memory Uncertainty into Rewarded Decisions." *Proceedings of the National Academy of Sciences* 117, no. 15 (2020): 8391–8397.
- Konrad P. Kording, Izumu Fukunaga, Ian S. Howard, James N. Ingram, and Daniel M. Wolpert. "A Neuroeconomics Approach to Inferring Utility Functions in Sensorimotor Control." *PLoS Biology* 2, no. 10 (2004): e330.
- Konrad P. Kording and Daniel M. Wolpert. "The Loss Function of Sensorimotor Learning." *Proceedings of the National Academy of Sciences* 101, no. 26 (2004): 9839–9842.
- Laurence T. Maloney, Julia Trommer shäuser, and Michael S. Landy. "Questions without Words: A Comparison between Decision Making under Risk and Movement Planning under Risk." *Integrated Models of Cognitive Systems* 29 (2007): 297–313.
- Chris R. Sims. "The Cost of Misremembering: Inferring the Loss Function in Visual Working Memory." *Journal of Vision* 15, no. 3 (2015).
- Lawrence D. Stone, Colleen M. Keller, Thomas M. Kratzke, and Johan P. Strumpfer. "Search Analysis for the Underwater Wreckage of Air France Flight 447." *14th International Conference on Information Fusion*. IEEE (2011), 1–8.
- Emanuel Todorov. "Optimality Principles in Sensorimotor Control." *Nature Neuroscience* 7, no. 9 (2004): 907–915.
- Julia Trommershäuser, Sergei Gephstein, Laurence T. Maloney, Michael S. Landy, and Martin S. Banks "Optimal Compensation for Changes in Task-Relevant Movement Variability." *Journal of Neuroscience* 25, no. 31 (2005): 7169–7178.
- Julia Trommershäuser, Laurence T. Maloney, and Michael S. Landy. "Statistical Decision Theory and the Selection of Rapid, Goal-Directed Movements." *Journal of the Optical Society of America A* 20, no. 7 (2003): 1419–1433.
- Paul A. Warren, Erich W. Graf, Rebecca A. Champion, and Laurence T. Maloney. "Visual Extrapolation under Risk: Human Observers Estimate and Compensate for Exogenous Uncertainty." *Proceedings of the Royal Society B: Biological Sciences* 279, no. 1736 (2012): 2171–2179.
- Louise Whiteley and Maneesh Sahani. "Implicit Knowledge of Visual Uncertainty Guides Decisions with Asymmetric Outcomes." *Journal of Vision* 8, no. 3 (2008): 1–15.

13.11 Problems

Problem 13.1 Suppose that you are walking in the dark in an area that is occasionally inhabited by lions. You hear a suspicious noise that may indicate the presence of a lion. In deciding whether to run away or not, you apply the following cost structure to each of four possible outcomes:

(a) Given the cost structure shown, if you believe that there is only a 30 percent chance that the lion is present, should you run or stay? Give a complete calculation.

(b) How low would your belief in the lion's presence have to be before you decided to stay rather than to run? Give a complete calculation.

		World state	
		Lion	No lion
Action	Run	Physical effort no injury cost = 2	Physical effort no injury cost = 2
	Stay	No effort severe injury cost = 100	No effort no injury cost = 0

(c) The outcomes listed in the table above represent a simplistic view of the problem. To make the problem more realistic, consider the following. In reality, if the lion is present, it may catch you even if you run; if you stay, the lion might (with some probability) decide not to attack you; if you run, you have some probability of injuring yourself by falling or colliding with objects (trees, boulders). Try to modify your solution to the problem, taking into account these realistic considerations. In your calculations, you may use what you believe to be realistic probabilities for each of these contingencies.

Problem 13.2 We build on the case study of looking for your keys in section 13.3. Suppose that, on searching your pockets, you fail to find the key. Where would you search next? To find out, use Bayes' rule to update your probability distribution over location, given the new data that your keys are not in your pockets, then once again find the action that minimizes expected cost.

Problem 13.3 We build on our discussion of cost functions in continuous estimation in section 13.5.2. Suppose we use absolute error instead of squared error as the cost function: $U(s, \hat{s}) = -|s - \hat{s}|$. Prove that then the posterior *median* rather than the posterior mean maximizes expected utility.

Problem 13.4 Show mathematically that if there are only two possible actions, equation (13.34) for the probability of choosing one action reduces to a logistic function of the scaled difference in expected utility between that action and the alternative action.

Problem 13.5 We consider a variant of the task of chapter 7—discrimination between s_+ and s_-. Assume a flat prior: $p(s_+) = p(s_-) = 0.5$. Further assume that the observer makes a measurement x that follows a Gaussian distribution with mean the stimulus and standard deviation σ. We now also allow for different amounts of reward (utility) for the four combinations of stimulus and the subject's report of the stimulus:

		Subject report	
		s_+	s_-
Stimulus	s_+	U_{++}	U_{-+}
	s_-	U_{+-}	U_{--}

(a) Derive the optimal decision rule and express it in terms of a cumulative standard normal distribution, Φ_{standard} (see section 7.1).

(b) Derive expressions for the probability of a correct response when the true stimulus is s_+, and when it is s_-.

Problem 13.6 Repeat problem 13.5 for the task of section 8.2: binary classification of a continuous stimulus. We denote the classes by $C=-1$ and $C=1$. Assume the following (improper) class-conditioned stimulus distributions:

$$p(s|C=-1)=\begin{cases} k & \text{if } s<0; \\ 0 & \text{if } s>0; \end{cases} \tag{13.40}$$

$$p(s|C=1)=\begin{cases} 0 & \text{if } s<0; \\ k & \text{if } s>0, \end{cases} \tag{13.41}$$

where k is a constant.

Problem 13.7 We build on section 13.5 on "purely perceptual" cost functions. Consider a stimulus variable s that takes values on the circle, such as motion direction. The posterior is $p(s|x)$.

(a) For estimation of a circular variable, using squared error as a cost function does not make sense. Why not? Explain using a concrete example.

(b) A sensible utility function is the cosine of the estimation error, $U(s,\hat{s})=\cos(\hat{s}-s)$. Show that the estimate that then maximizes expected utility on a given trial is the circular mean of the posterior, denoted μ_{post}, which is defined by the equations

$$\cos\mu_{\text{post}}=\mathbb{E}[\cos s\,|x]; \tag{13.42}$$

$$\sin\mu_{\text{post}}=\mathbb{E}[\sin s\,|x]. \tag{13.43}$$

Problem 13.8 This problem refers to the speeded reaching task under time pressure in section 13.8.2. Assume the radius of the circles is 1 and the distance between their centers is D. Also assume that the movement noise has a two-dimensional Gaussian distribution with standard deviation σ. For simplicity, we regard this as a one-dimensional problem: we consider only aim points a that lie on the (infinite) line running through the centers of the two circles. We define the point in the middle of the two centers as the origin. Thus, the centers of the two circles are at $-D/2$ and $D/2$. Use the rewards and costs specified in section 13.8.2.

(a) Derive an expression for the expected utility of aiming at point a on this line.

(b) Choose a reasonable range of values for D and a reasonable range of values for σ. Numerically compute the optimal aim point as a function of both D and σ. Plot this optimum as a function of D and σ using a color plot.

(c) Interpret your plot.

Problem 13.9 We build on section 13.8.3 on incentivized confidence intervals. Assume a Gaussian posterior with standard deviation σ_{post}. The observer reports the mean of the posterior as an estimate of the stimulus, and then sets a symmetric interval of half-length L around this estimate. The observer is rewarded if the true stimulus falls within this interval, with utility function $U(L) = \max(0, 100 - 10L)$.

(a) Plot the utility function.

(b) Set $\sigma_{\text{post}} = 2$. Calculate and plot the probability of receiving reward as a function of L.

(c) Calculate and plot expected utility as a function of L.

(d) Calculate the optimal interval half-length, L^*.

(e) Now define a set of σ_{post} values from 0.1 to 5 in steps of 0.1. Repeat part (d) for each value in this set. Plot L^* as a function of σ_{post}.

(f) Does this relationship make sense? Explain.

(g) Is L^* a reasonable measure of confidence? Explain.

Problem 13.10 In section 13.8.3 on incentivized confidence intervals, we focused on the situation where the observer makes a deterministic point estimate, such as the posterior mean, before setting the confidence interval. We know from section 4.7 that some Bayesian models postulate a stochastic readout of the posterior. Here, we consider an observer who samples $\hat{s}$ from the posterior distribution. Otherwise, we make the same assumptions as in problem 13.9: a Gaussian posterior, a symmetric interval, and a utility function $U(L) = \max(0, 100 - 10L)$.

(a) Set $\sigma_{\text{post}} = 2$. Draw a single $\hat{s}$ from the posterior distribution. For that draw, calculate the optimal interval half-length, L^*.

(b) Repeat part (a) for 1,000 draws. Calculate the average value of L^*.

(c) Repeat part (b) for σ_{post} values from 0.1 to 5 in steps of 0.1. Plot average L^* as a function of σ_{post}.

(d) If you also did problem 13.9: Calculate for each value of σ_{post} in part (c) the ratio of the average L^* under the sampling readout to L^* under the posterior mean readout. Plot the ratio against σ_{post}.

(e) If you also did problem 13.9: Calculate for each value of σ_{post} in part (c) the ratio of the average realized utility under the sampling readout to the realized utility under the posterior mean readout. Plot the ratio against σ_{post}.

Problem 13.11 We further build on the concept of incentivized confidence intervals in section 13.8.3. At the end of that section, we mentioned an alternative model in which the observer sets $\hat{s}$ and L jointly so as to maximize EU from equation (13.38). Think of a situation in which that would lead to a very different response from a model in which $\hat{s}$ is the posterior mean and L is set to maximize EU.

Problem 13.12 In section 4.5, we discussed why the posterior mean estimator minimizes overall mean squared error, MSE[$\hat{s}$], in equation (4.12); this argument did not involve x directly. In section 13.5.2, we discussed why the posterior mean estimator minimizes the expected squared error under the posterior, per equation (13.20); this argument was conditioned on x. What is the mathematical connection between the two statements?

Problem 13.13 The woodchuck does not actually chuck wood—it rather lives off grass, herbs, and insects. Every day, again and again, it has to allocate its resources between various possible activities. Let us say you have recorded what a woodchuck does, for example, where it walks, how fast, and what it eats and when. How would you build a normative model of why the woodchuck behaves the way it does and how would you test that with data?

Problem 13.14 Research in education has shown that to become an effective learner, students have to reflect on what they know and how confident they are about their knowledge. New forms of testing and grading can be designed to incorporate such confidence judgments. One example [79]:

Students are asked a question that requires them to indicate whether they are absolutely sure, fairly sure, or just guessing at their answer. The points for their response are dependent on the correctness of their answer and the confidence they state. For example, if students indicate they are absolutely sure, they earn nine points if they are correct, but no points if they are wrong. However, if they indicate they are unsure or just guessing, they earn three points if they are correct and two points if they are wrong. Students quickly learn that carefully reflecting on their confidence improves their grade.

Assume that confidence is the student's posterior probability of being correct, which we will denote by p.

(a) Given p, what is the expected number of points of indicating that you are absolutely sure? And of indicating that you are unsure or just guessing?

(b) For which p should a points-maximizing student indicate that they are absolutely sure?

(c) Why does this grading scheme incentivize students to veridically report their confidence?

(d) The specific point values given to the six possible responses make up a grading scheme. Specify the conditions that a grading scheme has to meet in order to incentivize students to both be correct and veridically report their confidence.

14

The Neural Likelihood Function

How can we incorporate neural variability into a generative model?

To describe sensory noise, we have so far used the concept of a measurement, which has the same unit as the stimulus variable that is being measured. Ever since Chapter 3, however, we have known that this concept was an abstraction. The biological reality is that sensory information comes in the form of action potentials (spikes) fired by groups of neurons working in concert. While for the purpose of modeling behavior, in a sort of separation of scales, the abstraction suffices, we would be remiss to ignore the biological substrate altogether. The Bayesian framework continues to apply, with neural variability taking the place of measurement noise. That being said, our discussion of inference in this chapter will be limited to having neural activity as input, not on the neural implementation of the inference computation itself.

Plan of the Chapter

We provide a crash course in systems neuroscience, introducing the concepts of tuning curves, Poisson variability, and population coding. Together, these concepts will give us a generative model for neural activity given a stimulus. We will then invert this generative model to obtain a neural likelihood function, first for a single neuron and then for a population of neurons. We link the neural likelihood function to the more abstract likelihood functions used in earlier chapters.

14.1 Generative Model of the Activity of a Single Neuron

As step 1, we start with the generative model of the spike count of a single neuron, for example, the number of spikes elicited by a flash of light that is presented for a few milliseconds. We describe the spike count of a neuron over a given time interval using a Poisson distribution. Poisson variability (figure 14.1A) produces a distribution over nonnegative integers (it can be zero). Suppose a stimulus s is presented and the mean spike count of a neuron in response to this stimulus is λ, which does not need to be an integer. Then the actual spike count will vary from trial to trial, around λ. For every possible count r, we seek its probability. A Poisson process (or in our context, a Poisson spike train) is defined as follows. Take a

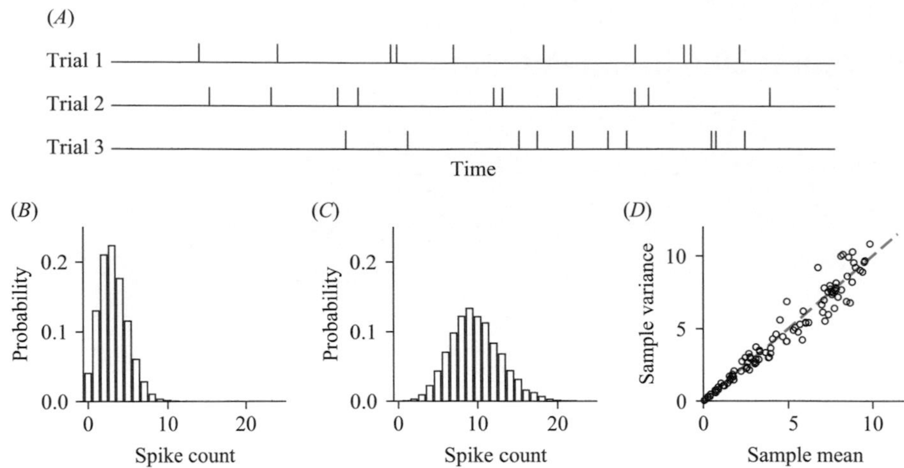

Figure 14.1
Poisson variability in a single neuron. (*A*) Hypothetical spike trains evoked in the same neuron by the same stimulus, repeated three times (trials). Not only do the spike times differ between trials; but the spike counts also differ. (*B*) Probability distribution of the spike count of a single Poisson neuron, with mean spike count $\lambda = 3.2$. Note that the x axis starts at 0, not at 1. Since the Poisson distribution is discrete, drawing it as a continuous curve would be a mistake. (*C*) Same but with $\lambda = 9.7$. (*D*) The variance of a Poisson random variable is equal to the mean. To illustrate this, we drew 100 values of the rate parameter uniformly from [0, 10]. For each value, we simulated 100 spike counts and calculated mean and variance. The Fano factor is not exactly 1 because of the finite size of the samples.

fixed time interval (for example 1 second), and divide it into small bins (e.g., 1 millisecond each). We assume that each bin can contain 0 spikes or 1 spike, and that the occurrence of a spike is independent of whether and when spikes occurred earlier (for this reason, it is sometimes said that a Poisson process "has no memory"). For a Poisson process with rate parameter λ, the probability of observing a total of r spikes on a single trial is given by the Poisson distribution,

$$p(r|s) = \frac{1}{r!} e^{-\lambda} \lambda^r, \tag{14.1}$$

where $r!$ is pronounced "r factorial" and stands for the product $1 \cdot 2 \cdot 3 \cdots r$. The Poisson distribution is shown for $\lambda = 3.2$ and $\lambda = 9.7$ in figure 14.1B–C. Keep in mind that, while r is an integer, λ can be any positive number. For large enough λ, the distribution is close to symmetrical and looks roughly Gaussian; this is not true for small λ.

An important property of the Poisson distribution is that both the mean and the variance of a Poisson-distributed variable are equal to its rate parameter λ (figure 14.1D). The ratio of the variance to mean of a neuron's spike count is called the *Fano factor*; for a Poisson neuron, the Fano factor is 1. Experiments have found the Fano factor to vary widely depending on brain area and species; therefore, Poisson variability should be considered a drastic simplification of reality.

For our generative model of neural firing, we need to specify the probability of a number of spikes r, as a function of the stimulus, s. To do this, we note that λ is a function of the stimulus: it is the height of the tuning curve (the neuron's average spike count or firing rate) at stimulus level s. Therefore, in terms of the stimulus, equation (14.1) can be written as

$$p(r|s) = \frac{1}{r!}e^{-f(s)}f(s)^r. \tag{14.2}$$

This probability distribution is sometimes called the *neural noise distribution*.

Box 14.1
Sampling Codes

Although the traditional view in systems neuroscience is that neural activity for a given stimulus is purely stochastic, other views exist. In one such view, *sampling codes*, the variability in a neuron's spike count is not random but corresponds to samples from a posterior distribution over a basic world state variable, such as the presence of a simple feature in one location in the visual field [53, 73, 78]. This view is not at odds with the approach presented here, because, no matter the origin of the variability, the neural likelihood function is still a valid concept for downstream computation.

So far, we have not specified the stimulus s. In simple models of neural firing, s is a low-dimensional variable, such as location, contrast, orientation, or speed of an object in the visual domain, and intensity, source location, or pitch of a sound in the auditory domain. Then, tuning curves $f(s)$ often admit relatively simple parametric descriptions. Such choices of s are, however, necessarily radical simplifications. For example, an image of an oriented stimulus also has contrast, size, spatial frequency, and so on. All of those are nuisance variables that would have to be taken into account in a complete description of neural tuning. Experimentally, all or nearly all nuisance parameters would have to be kept constant for a typical physiology experiment to "cover" the space.

This is not to say that the stimulus space cannot be richer. For example, one could present randomly varying images, in which the intensity of each pixel is independently drawn from a distribution, or one could present images of natural scenes. Predicting neural responses to "rich," high-dimensional stimuli s typically requires a model in which each neuron is sensitive to a small part of the stimulus and applies a linear filter to that part as the first step in the computation. In the visual domain, such models are also called *image-computable models*, since the input is a complete image. Such models also produce distribution $p(r|s)$.

In the rest of this chapter, we will focus on one-dimensional stimuli s for which $p(r|s)$ admits a compact mathematical expression.

14.2 Neural Likelihood Function for a Single Neuron

We now turn to step 2 of the Bayesian model, where the observer infers the stimulus from a specific spike count r. Given r, the neural likelihood of a hypothesized stimulus value s is the probability that r spikes were produced by that value of s. In other words, we copy equation (14.2) but consider it as a function of s rather than r:

$$\mathcal{L}(s; \mathbf{r}) = \frac{1}{r!}e^{-f(s)}f(s)^r. \tag{14.3}$$

We consider two example cases: a bell-shaped tuning curve and a monotonic tuning curve.

Box 14.2
Neural Tuning Curves

Neuroscientists have long analyzed which stimuli would activate a neuron, starting with studies by [29]. They recorded from primary visual cortex (V1) in cat, while presenting illuminated oriented bars. They found that the response of V1 neurons was systematically related to the orientation of the stimulus (figure 14.2A). There often existed one orientation at which a neuron fired most rapidly: the neuron's preferred orientation. The neuron's tuning curve is the mean firing rate (number of spikes per second) as a function of orientation. Many visual neurons have a unimodal (single-peaked) curve.

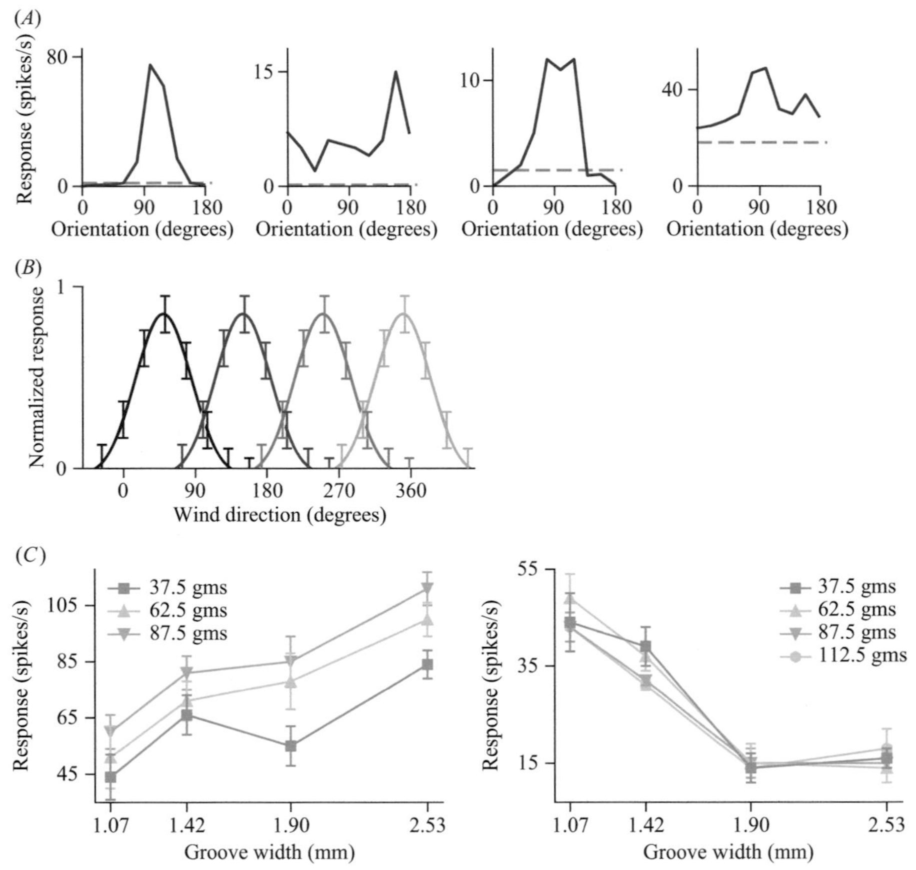

Figure 14.2
Empirical tuning curves. (*A*) Tuning curves for orientation in macaque primary visual cortex (V1). The dashed line represents the spontaneous firing rate. Reproduced from [160]. (*B*) Normalized tuning curves for the direction of air current in four interneurons in the cercal system of the cricket ("cercal" refers to the cerci, appendages covered with small hairs that respond to wind direction). Reproduced from [127]. (*C*) Tuning curves for two neurons, for the width of the groove in a tactile grating in macaque secondary somatosensory cortex (S2). Different curves are for different magnitudes of the contact force (expressed as mass). Reproduced from [147].

Tuning curves can have a wide variety of shapes, depending on the species, the brain area, and stimulus feature (figure 14.2). For example, in the motor cortex, we find that neural responses influence the direction of movement of the hand of a monkey. Instead of narrow unimodal functions, we usually find very broad tuning curves. In the auditory cortex, the frequency of the sound stimulus affects the firing rate of the neuron in a complex tuning curve. And in the hippocampus, a region of the mammalian brain involved in memory acquisition and navigation, there is a two-dimensional representation of position. Some tuning curves are not bell-shaped, but monotonic (figure 14.2C). In all these cases, reasonably simple tuning curves characterize the mapping from sensory stimuli to the activity of neurons.

14.2.1 Case 1: A Bell-Shaped Tuning Curve

Suppose that the tuning curve of the neuron has a Gaussian shape with peak location (preferred stimulus) $s_{\text{pref}} = 0$, width $\sigma_{\text{tc}} = 10$, baseline $b = 1$, and gain $g = 5$:

$$f(s) = ge^{-\frac{(s-s_{\text{pref}})^2}{2\sigma_{\text{tc}}^2}} + b. \tag{14.4}$$

This curve is depicted in figure 14.3B. A point on the curve is the mean spike count of the neuron in response to a particular stimulus. In spite of the similarity to the Gaussian probability distribution, the tuning curve is in no way a probability distribution! In particular, it is not normalized.

Suppose we are told this neuron fired four spikes in a given time interval, and we are asked what we can say about the stimulus. Based on figure 14.3B, we might say the stimulus was approximately -10 or $+10$, because then the neuron would produce the expected number of spikes. However, figure 14.3B shows us only the average spike count over many trials. The trial-to-trial response is noisy, as expressed by equation (14.2). Therefore, a total of four spikes could also have been produced by a stimulus value of say 3.7—it just so happened that on this trial, the neuron fired fewer spikes than average. Four spikes could even indicate that the stimulus was -21, although it would require that the neuron happened to fire many more spikes than its average spike count for this stimulus. Clearly, some stimulus values are

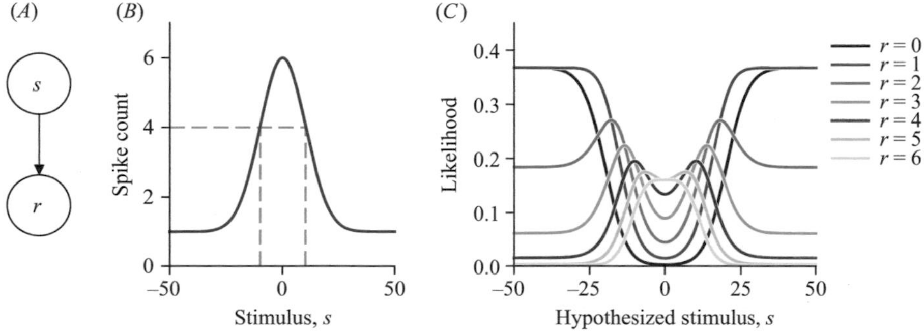

Figure 14.3
Single-neuron inference. (*A*) Generative model. (*B*) Idealized tuning curve, $f(s)$, of a single neuron with preferred stimulus 0. (*C*) Likelihood functions over the stimulus for different observed spike counts in a one-neuron brain.

more likely than others, and we can define the likelihood of a hypothesized stimulus value as the probability of observing four spikes in response to that stimulus value.

We formalize this intuition by simply substituting the expression for the tuning curve, equation (14.4) into equation (14.3) to obtain the neural likelihood function. We have plotted the resulting function for $r=4$ as well as for other values of r in figure 14.3C. These peculiar-looking functions tell us how likely each possible stimulus value is based on the observed spike count. The shape of the curve for $r=4$ confirms our intuition: values of (approximately) $+10$ and -10 are most likely, 0 is still quite likely, but -30 is very unlikely. To compute the likelihood function, we used not only the tuning curve (equation (14.4)) but also the form of neural variability (equation (14.2)). This allows us to say more about the stimulus than only that $+10$ and -10 are most likely.

These likelihood functions have several interesting properties. First, the likelihood function can have a dramatically different shape for different observations. For example, when no spikes are observed, any of the central values of the stimuli are very unlikely, so the likelihood function has an inverted U-shape. Second, unlike in earlier chapters, we do not see anything close to a Gaussian shape here. In fact, there is no tuning curve $f(s)$ we could have used to get a likelihood function that is exactly Gaussian for every r. Third, the likelihood functions are not normalized; in fact, the area under each function is infinite: the "tails" extend to arbitrarily large values.

Exercise 14.1 Use equations (14.4) and (14.3) to understand why.

In general, likelihood functions are not normalized. In the present chapter, as we are discussing neural models, the likelihood function of the stimulus is never automatically normalized.

14.2.2 Case 2: A Monotonic Tuning Curve

So far, we have considered a real-valued variable with bell-shaped tuning. We will now consider a nonnegative variable with monotonic tuning, as we encountered in figure 14.2C. An example would be a power law tuning curve with baseline:

$$f(s)=as^b+c, \tag{14.5}$$

where we enforce $a>0$ and $c\geq 0$, so that mean spike counts are guaranteed nonnegative regardless of s. Such a tuning curve would make sense for magnitude variables such as length, weight, contrast, and loudness. An example of such a tuning curve is shown in figure 14.4A.

What do we expect the likelihood to look like in this case? Suppose we observe that this neuron fires four spikes. The likelihood reflects how probable this observation is under different hypothesized values of s. From the tuning curve, we know that this neuron fires four spikes on average when the stimulus is $s=6$; therefore, we expect the probability that it will fire exactly four spikes to be quite high for a hypothesized stimulus value of 6. On the other hand, if the stimulus were 0, then the neuron's average number of spikes would be 0.5, making it improbable for the neuron to fire four spikes. Similarly, if the stimulus were 100, then the neuron's average number of spikes would be fifty-one, which would again make our observation of four spikes improbable. Therefore, we would expect the likelihood to be high for $s=6$ and drop off gradually as s moves away from 6 in either direction. Thus, we expect that the neural likelihood will be bell-shaped. This intuition is not specific to a spike count

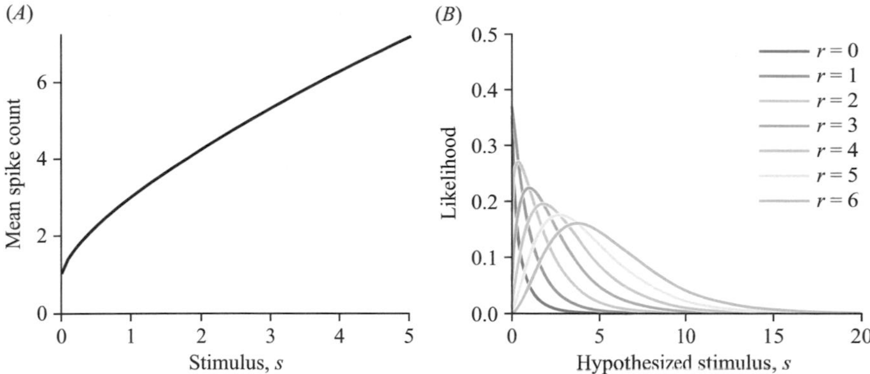

Figure 14.4
(*A*) Example of a monotonic tuning curve. We used $f(s) = as^b + c$ with $a = 2$, $b = 0.7$, and $c = 1$. (*B*) Corresponding likelihood functions, under the assumption of Poisson variability and with observations $r = 0, \ldots, 6$. Although the tuning curve is monotonic, each of the likelihood functions is bell-shaped.

of 4. In general, any particular spike count will give the highest likelihood to one stimulus value and lower likelihoods to stimuli on either side. We have plotted the likelihoods based on several different observed values of r in figure 14.4B. In line with our intuition, these likelihood functions are bell-shaped.

14.3 Neural Likelihood Function Based on a Population of Neurons

The neural likelihood functions that we have encountered so far have been very wide, indicating high uncertainty. However, when we recall that these likelihood functions were based on the firing of just a single neuron, and that this neuron was noisy (Poisson), it is in fact remarkable how much we can already say about the stimulus. Moreover, most of us have more than one neuron in our brain, and therefore the information that we have about stimuli in the world is based on the simultaneous firing of a population of neurons. The theory of population-based neural likelihood functions has been developed by many theoretical neuroscientists, including [55, 83, 113, 146, 157]; we provide only an extremely simplified version here.

We consider a population consisting of an arbitrary number of neurons; we call the number of neurons n. On a given trial, the neurons in this population will produce a set of spike counts, $r_1, \ldots, r_n$, which we will often denote in shorthand by a vector $\mathbf{r}$ and call the *(pattern of) population activity*. Mathematically, $\mathbf{r}$ is a high-dimensional vector. If 1,000 neurons were selective to s, then $\mathbf{r}$ would be a 1,000-dimensional vector. We assume that the variability in $\mathbf{r}$ across trials is independent across neurons conditioned on s:

$$p(\mathbf{r}|s) = p(r_1, \ldots, r_n|s) \tag{14.6}$$

$$= p(r_1|s) \cdots p(r_n|s) \tag{14.7}$$

$$\equiv \prod_{i=1}^{n} p(r_i|s). \tag{14.8}$$

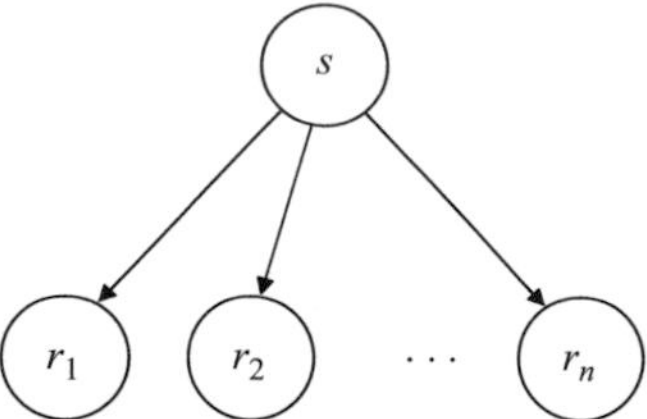

Figure 14.5
Generative model of activity in a population consisting of n neurons in response to a stimulus s.

The generative model is shown in figure 14.5. The conditional independence assumption parallels the one encountered in cue combination (figure 5.4), evidence accumulation (section 5.5), and learning (figure 6.1), but now neurons have taken the place of cues or time points.

We now turn to inference. When we observe a specific pattern of population activity $\mathbf{r}$, the neural likelihood function of s is

$$\mathcal{L}(s;\mathbf{r})=\prod_{i=1}^{n} p(r_i|s). \tag{14.9}$$

Each factor in this product can be thought of as the likelihood function based on a single neuron's spike count. Thus, the population likelihood function is the product of single-neuron likelihood functions. This product parallels the product of single-cue likelihoods in cue combination (equation (5.26)).

To make further progress, we assume that every neuron's spike count follows a Poisson distribution, with each neuron having its own rate parameter: for the ith neuron, the rate parameter is given by the ith tuning curve evaluated at s, that is, by $f_i(s)$. Then,

$$p(r_i|s)=\frac{1}{r_i!}e^{-f_i(s)}f_i(s)^{r_i}. \tag{14.10}$$

As a consequence, when we observe a specific pattern of population activity $\mathbf{r}$, the neural likelihood function of s is

$$\mathcal{L}(s;\mathbf{r})=\prod_{i=1}^{n}\frac{1}{r_i!}e^{-f_i(s)}f_i(s)^{r_i}. \tag{14.11}$$

To further evaluate the population likelihood function, we need to make assumptions about the tuning curves. We now work out an example in which the tuning curves are all bell-shaped. To model such a population, we replace the single-neuron tuning curve in equation (14.4) by a different tuning curve for each neuron:

$$f_i(s)=g_i e^{-\frac{(s-s_{\text{pref},i})^2}{2\sigma_{\text{tc},i}^2}}+b_i, \tag{14.12}$$

where g_i, $s_{\text{pref},i}$, $\sigma_{\text{tc},i}$, and b_i are the gain, preferred stimulus, width, and baseline of the tuning curve of the ith neuron. An example of such a population is shown in figure 14.6A. Amplitude, width, and baseline are highly variable, as is the case in real recordings. (However, the fact that each tuning curve is a member of the same simple parametric family of functions is not realistic.)

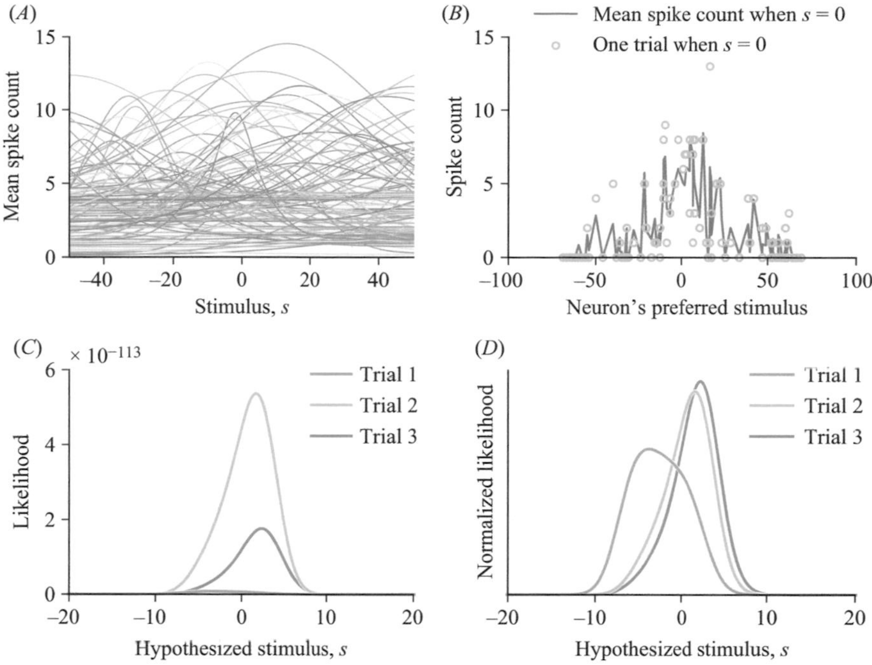

Figure 14.6
Inference based on a heterogeneous population consisting of 100 Poisson neurons with bell-shaped tuning curves. (*A*) Tuning curves. (*B*) Example pattern of activity. (*C*) Three example neural likelihood functions from this population, all obtained with $s=0$. (*D*) Normalized versions of the likelihood functions.

We simulated three patterns of activity in the population of 100 independent Poisson neurons with tuning curves as in figure 14.6A (preferred stimuli equally spaced between -60 and 60), elicited by the stimulus $s=0$. These patterns would correspond to neural recordings on three trials on which $s=0$ was presented. Such a pattern could look like figure 14.6B.

Per equation (14.9), the likelihood function of s is the product of the probabilities that neuron 1 fires two spikes, neuron 2 fires three spikes, and so on:

$$\mathscr{L}(s;\mathbf{r})=p(r_1=2|s)p(r_2=3|s)\cdots p(r_{31}=0|s). \tag{14.13}$$

The individual probabilities on the right-hand side are obtained from the Poisson equation, equation (14.10), with equation (14.12) for the tuning curves. Three resulting likelihoods are plotted together in figure 14.6C.

Several properties of these likelihood functions merit discussion. First, the smooth and structured form of the likelihood function stands in contrast to the messy and apparently structureless population pattern of activity in figure 14.6B. Second, the scale on the y axis is very small, namely of the order of 10^{-113}. This low magnitude of the likelihood functions is not a mistake: it results from the fact that the probability of multiple events (spikes in multiple neurons) is always less than the probability of any one of those events. Every time a neuron is added, the likelihood gets smaller. Third, the likelihood functions vary enormously in peak height. The one from trial 1 is so low overall that it is barely visible; its peak height is $8.2\cdot10^{-115}$. In figure 14.6D, we have normalized the same three likelihood functions for visualization purposes. This makes the trial 1 likelihood function clearly visible. The normalized likelihood function is the same as the posterior distribution when the prior is uniform. Finally, even in a large population, the likelihood function can be visibly different from Gaussian. In general, likelihood functions can skewed, multimodal, or flat-topped. That being said, the larger the population (and the higher the neurons' gains), the closer to Gaussian the likelihood function will tend to look.

The neural likelihood function contains all information that can objectively be obtained from the population activity. No more information can be obtained, and any different information would be incorrect.

Box 14.3
Representation of Uncertainty or Uncertainty Associated with a Representation?

Neural likelihood functions are sometimes described as a representation of uncertainty. However, the term "representation" in neuroscience usually refers to a world state, such as "representation of motion direction" or "representation of face identity." In this sense, uncertainty is a strange thing to represent, since uncertainty is a property of a belief of an observer instead of a world state. A more accurate formulation might be that the neural likelihood function captures the uncertainty associated with a (single-trial) representation of a world state variable.

There is a category of models in neuroscience that place uncertainty primarily in the external world. In those, uncertainty is computed purely based on the sensory input and not on an internal representation or neural activity. Subsequently, the researcher aims to find neural correlates of uncertainty, for example through neuroimaging. This approach is a descriptive model, not a process model, because it does not specify step by step the way sensory information gets processed. In such descriptive models, however, it would be more justified to speak about the representation of uncertainty. We do not consider descriptive models of this type in this book, because we take seriously the notion that the brain utilizes a generative model of its internal representations during inference.

14.4 Toy Model

Although we were able to calculate and plot population likelihood functions, we have so far not gained any intuition about how its properties depend on neural activity. To obtain such intuition, we need compact expressions that are interpretable mathematically. For that purpose, we will make additional assumptions in this section. Borrowing from the language of physics, the model in this section is a "toy model": it captures the essence of a problem without being realistic per se.

Within the class of tuning curve defined by equation (14.4), we make the additional assumption that tuning curves are "homogeneous," that is, translated versions of each other, with preferred stimuli that are equally spaced across some interval. We also assume a large number (high density) of neurons and zero baseline, $b=0$. In other words, we use equation (14.12) with $b=0$, $g_i=g$, and $\sigma_{\text{tc},i}=\sigma_{\text{tc}}$:

$$f_i(s)=ge^{-\frac{(s-s_{\text{pref},i})^2}{2\sigma_{\text{tc}}^2}}. \tag{14.14}$$

An example of such a population is depicted in figure 14.7A.

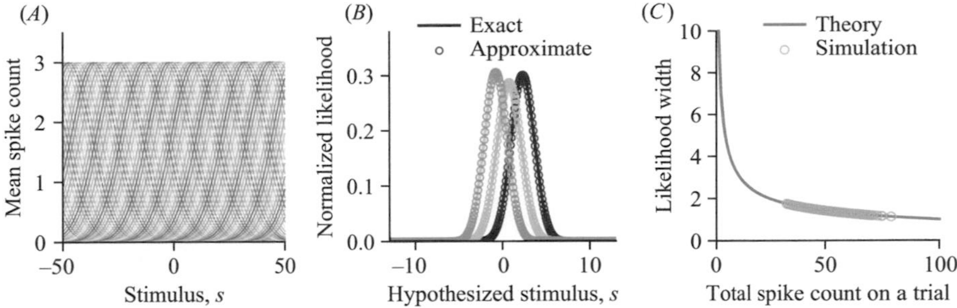

Figure 14.7
Toy model with homogeneous tuning curves. (*A*) Dense, translation-invariant tuning curves with a Gaussian shape (width 10) and baseline 0. (*B*) Three example neural likelihood functions with the constant-sum approximation overlaid. The approximation is very good. (*C*) Sensory uncertainty from a neural population: width of the neural likelihood function as a function of the total single-trial spike count in the population.

Because of the high density and the homogeneity of the neurons, the sum of the tuning curves over neurons will be approximately independent of the stimulus:

$$\sum_{i=1}^{n} f_i(s) \approx \text{constant}. \tag{14.15}$$

Since we defined tuning curves only in a limited region of space, the approximation will hold only in that region; the sum will drop to zero for values of s outside this region. Exponentiating the negative of equation (14.15), we find

$$\prod_{i=1}^{n} e^{-f_i(s)} \approx \text{constant}. \tag{14.16}$$

If one is interested only in the *shape* of the likelihood function, as is usually the case (for example to obtain an estimate of the stimulus, or a measure of uncertainty), then multiplicative constants such as $\frac{1}{r_i!}$ do not matter. Combining that fact with equation (14.16), the likelihood from equation (14.11) simplifies to

$$\mathcal{L}(s; \mathbf{r}) \underset{\sim}{\propto} \prod_{i=1} f_i(s)^{r_i}. \tag{14.17}$$

In other words, the population likelihood is approximately proportional to the product of the neurons' tuning curves raised to the powers of the corresponding spike counts. The higher a neuron's spike count, the higher the power and the more influence that neuron's tuning curve has on the likelihood function. This already gives us a bit of intuition.

Finally, we substitute equation (14.14) into equation (14.17) and simplify to find

$$\mathcal{L}(s; \mathbf{r}) \underset{\sim}{\propto} e^{-\frac{1}{2\sigma_{\text{tc}}^2} \sum_{i=1}^{n} r_i (s - s_{\text{pref},i})^2}. \tag{14.18}$$

Exercise 14.2

(a) Verify this.

(b) Why was the assumption $b = 0$ important?

We can write equation (14.18) in a simpler form:

$$\mathcal{L}(s;\mathbf{r}) \underset{s}{\propto} e^{-\frac{(s-\mu_{\text{likelihood}})^2}{2\sigma^2_{\text{likelihood}}}}, \tag{14.19}$$

where

$$\mu_{\text{likelihood}} \equiv \frac{\sum_{i=1}^{n} r_i s_{\text{pref},i}}{\sum_{i=1}^{n} r_i} \tag{14.20}$$

$$\sigma^2_{\text{likelihood}} \equiv \frac{\sigma^2_{\text{tc}}}{\sum_{i=1}^{n} r_i}. \tag{14.21}$$

Exercise 14.3 This is not obvious and deriving it requires several steps. Do the derivation.

There is a good reason to write the likelihood function like this: we recognize the form of an (unnormalized) Gaussian function in equation (14.19)! In other words, when a population of independent Poisson neurons has homogeneous Gaussian tuning curves without baselines, and we make the constant-sum approximation, then the neural likelihood function of the stimulus is approximately Gaussian. This property makes this special case valuable as a toy model. We plotted normalized likelihood functions obtained from this population using the exact equation, overlaid with the approximate expression, equation (14.19), in figure 14.7B. The approximation is indistinguishable from the complete expression.

The MLE of the stimulus is $\mu_{\text{likelihood}}$:

$$\hat{s}_{\text{ML}} = \mu_{\text{likelihood}}. \tag{14.22}$$

Thus, under the assumptions made, the MLE is a weighted average of the preferred stimuli of the neurons in the population, with weights given by the neurons' spike counts. This is also called the *center-of-mass decoder*; the equivalent for circular variables is called the *population vector decoder* (see problem 14.7).

The variance of the likelihood function is given by equation (14.21). Per section 3.4.1, we can interpret its square root as the (sensory) uncertainty that the observer has about the stimulus. First, we see that the narrower the tuning curve (smaller σ_{tc}), the narrower the likelihood function. Second, and more interestingly, the observer's uncertainty depends on the total spike count in the population, $\sum_{i=1}^{n} r_i$, and as such, varies from trial to trial even as the physical stimulus is held fixed. The higher the total spike count in the population, the narrower the likelihood function and the lower uncertainty (figure 14.7C). Since all neurons contribute to the sum in equation (14.21), we should think of uncertainty as a property of the population, not of any one neuron.

Exercise 14.4 When one neuron fires one spike, equation (14.21) states that the width of the likelihood is equal to the width of tuning curve. Explain why this is correct and why this scenario is different from the one-neuron brain discussed in section 14.2.

That we were able to write down explicit equations for the MLE and the width of the likelihood function was due to the specific assumptions we made. In general, it is rare for closed-form expressions for the maximum-likelihood estimate (MLE) and the width of the likelihood function to exist.

14.5 Relation between Abstract and Neural Concepts

So far in this chapter, we introduced a generative model for neural activity, defined the neural likelihood function, and—in a toy model—found closed-form expressions for the MLE of the stimulus and the width of the likelihood function. We will now compare and contrast these concepts with the more abstract model that we introduced in chapter 3 and have used throughout this book (see table 14.1).

The abstract model was simplified in several ways relative to the neural model: first, the likelihood function was always Gaussian, while in the neural model, it is not. Second, the MLE was identical to the observation, which is made possible by defining the observation as a measurement living in the same space as the stimulus; in the neural model, the observation lives in a completely different space (the space of n-dimensional vectors of nonnegative integers) than the stimulus, and therefore also than the MLE. Third, in the abstract model, the likelihood width was the same from trial to trial; in the neural model, since it depends on the observation, it varies from trial to trial.

We can now more fully appreciate that the concept of a measurement was an abstraction. The brain does not have scalar measurements as input to its inference; it has only neural action potentials. In fact, we could *define* a measurement in terms of the neural model (toy or not): the measurement is the MLE of the stimulus based on the neural observation, namely the population pattern of activity $\mathbf{r}$. In this view, the measurement x of a stimulus s is the value of s under which the observed neural activity $\mathbf{r}$ is most probable. We can think of the measurement x as a "processed form" of the neural activity $\mathbf{r}$. For example, if the stimulus is an orientation of a line and the neural observation is the activity of a population of V1 neurons, then the measurement would be the best guess of orientation obtained from this pattern.

Would this correspondence hold statistically across many trials? In the abstract model, for given s, the measurement has a Gaussian distribution with mean s. But is the same true for

Table 14.1
Comparison between abstract and neural concepts in Bayesian modeling of a continuous estimation task. The toy model refers to the model in section 14.4.

Concept	Abstract model	Neural model
Observation	Scalar measurement x. Possible values: real numbers.	Vector of spike counts $\mathbf{r}=(r_1,\ldots,r_n)$. Possible values: nonnegative integers.
Its distribution	$p(x\|s)$, typically Gaussian with mean s and standard deviation σ	$p(\mathbf{r}\|s)$ (n-dimensional). In the toy model: independent Poisson with Gaussian tuning curves
Likelihood of s	$\mathscr{L}(s;x)$. Gaussian if noise distribution is Gaussian	$\mathscr{L}(s;\mathbf{r})$. In the toy model: Gaussian.
MLE of s	x	In the toy model: $\frac{\sum_{i=1}^{n} r_i s_{\text{pref},i}}{\sum_{i=1}^{n} r_i}$
Width of likelihood function	σ (fixed)	Activity-dependent (variable). In the toy model: $\frac{\sigma_{\text{tc}}}{\sqrt{\sum_{i=1}^{n} r_i}}$
Distribution of MLE	Gaussian with mean s and standard deviation σ (same as noise distribution)	No analytical form; only Gaussian in the limit of many spikes

the neural model? Even in the neural *toy* model, the MLE of the stimulus is a complicated function of $\mathbf{r}$ and its distribution over many trials for given s does not have a closed-form expression. However, in the limit that gain (and thus the average spike count) is very high, the distribution of the MLE becomes indistinguishable from Gaussian. This is a property known as *asymptotic normality*. The limit is one in which the amount of information in the population is high because a sufficient number of neurons has responded with a sufficient number of spikes. Under the same conditions, the expected value of the MLE is equal to the stimulus itself. In other words, the MLE is *asymptotically unbiased*. These properties provide some justification for our assumed measurement distribution in the abstract model, but it is important to keep in mind that when one moves away from the asymptotic regime, that assumed distribution becomes an increasingly crude approximation.

14.6 Using the Neural Likelihood Function for Computation

So far in this chapter, we defined the neural generative model, which is step 1 of the Bayesian modeling recipe we outlined in chapter 3. We have described part of step 2, namely the likelihood function of the stimulus based on the activity in a typical sensory population. We know from previous chapters in which sensory noise played a role that this likelihood function can be used in a multitude of ways: in combination with prior information (chapter 3), in combination with other likelihood functions (chapter 5), or to infer a higher-level categorical variable (e.g., chapters 7, 8, 10, 11). In all these cases, each likelihood function of a stimulus is an elementary building block that is used to build the posterior distribution over the world state of interest.

The situation is exactly the same for the neural likelihood. Whereas before the present we used a stimulus likelihood function $\mathcal{L}(s;x)=p(x|s)$, we can now replace it by a neural likelihood function, $\mathcal{L}(s;\mathbf{r})=p(\mathbf{r}|s)$, and everything else would go through as before.

As an example, we now discuss the combination of a neural likelihood function with a Gaussian prior $p(s)$,

$$p(s)=\mathcal{N}(s;\mu,\sigma_s^2). \tag{14.23}$$

The posterior distribution over s is

$$p(s|\mathbf{r})\propto p(\mathbf{r}|s)p(s). \tag{14.24}$$

Under the toy model of section 14.4, but not otherwise, the likelihood is proportional to a Gaussian with mean given by equation (14.20) and standard deviation by equation (14.21). Then, we have exactly returned to the case study of chapter 3, with the only difference being the substitutions from table 14.1 for x and σ^2. Thus, we can immediately import the equation for the posterior mean estimate from chapter 3, equation (3.24), and make these substitutions, so that

$$\hat{s}_{\text{PM}}=\frac{\frac{1}{\sigma_{\text{tc}}^2}\sum_{i=1}^{n} r_i s_{\text{pref,i}}+\frac{\mu}{\sigma_s^2}}{\frac{1}{\sigma_{\text{tc}}^2}\sum_{i=1}^{n} r_i+\frac{1}{\sigma_s^2}}. \tag{14.25}$$

This equation provides the answer to the question: If the sensory input activity is $\mathbf{r}$ and the stimulus is drawn from $p(s)$, how does the brain make the best possible estimate of the stimulus? In other words, it is a neural Bayesian stimulus-response mapping. The inference problems in other chapters involving a noisy measurement x can be treated using the same substitution. These do not contribute new understanding, so we will not go into them.

Box 14.4
Artificial Neural Networks to Learn Likelihood Functions

In this chapter, we started with an idealized and mathematically beautiful generative model, which we then used to derive a likelihood function. Reality is messy, however, and in practice it can be hard to formulate a good likelihood function. In such cases, we may want to use deep learning to obtain a likelihood function. We would start with training data $(s, \mathbf{r})$ where s is the stimulus and $\mathbf{r}$ population activity. In one approach, we would assume that the posterior over s is Gaussian with an observation-dependent mean and variance. We could then estimate the mean with one network-defined function $f_\theta(\mathbf{r})$ and the standard deviation with another network-defined function $g_\theta(\mathbf{r})$. Alternatively, we could bin the stimulus and train a classification network; this would produce a discretized posterior over the stimulus [193]. In both approaches, we would convert the posterior into a likelihood by dividing out the experimentally known prior distribution. However computed, a likelihood estimated using a neural network can readily be incorporated into further Bayesian computation.

Such approaches may hold the key for generalizing analytically tractable Bayesian models to real-world behaviors. We can imagine likelihood functions where the relevant inputs are the complex stimuli people experience in their everyday lives and the estimates relate to complex choices. However, such approaches would lose much of their explanatory power relative to the relatively simple models we focused on here.

14.7 The Neural Implementation of Bayesian Computation

It is one thing to write down a formal expression for the posterior distribution of a quantity of interest, another to specify neural operations that can implement the calculation of this posterior distribution. We could approach this problem in many ways. One way is to manually construct such operations and demonstrate through math and simulations that these operations suffice. This approach, which is exemplified by *linear probabilistic population codes* [113], is theoretically elegant but the assumptions that are needed to do the math place severe limitations on the task and the form of neural variability. (The neural likelihood function represents only the most basic component of this theory.) A second method is to train an artificial neural network to learn a mapping consisting of simple building blocks [120, 140]. This approach is much more flexible but might lead to a network that is hard to interpret. Besides these two approaches, many other frameworks for the neural implementation of Bayesian computation exist (e.g., [42, 89, 107, 151, 162]). These are beyond the scope of this book.

14.8 Applications

Although the focus of this book is on how the brain performs inference based on noisy sensory information, there is a parallel literature on how an experimenter can decode brain states on a trial-by-trial basis. Traditionally, this literature has focused on point estimates;

however, in recent years, more attention has been given to decoding entire likelihood functions—thus linking to the rest of this chapter.

Functional magnetic resonance imaging (fMRI) is a method that uses big magnets and radio-frequency waves to measure the three-dimensional oxygenation of blood in the head due to neural activity. The brain is divided into voxels (like *pix*els, but *vo*lume elements—small cubes), and fMRI records "percent signal change" in each voxel in response to presented stimuli. Of natural interest is an application of "mind reading": decoding what the brain thinks or sees based on fMRI data. The logic here is a bit different from the rest of the book, since the model is not a model of the observer. After all, a human subject's observations are the activations of neurons, not voxels. Experimenters, however, use the generative model for voxel activity to decode the stimulus. Thus, the observer is the experimenter, not the subject. Just as the observer must first learn the parameters of the generative model of its observations to perform inference, the experimenter has to learn the parameters of the generative model of voxel activity. For that purpose, we use training data, in which the stimulus is considered known on every trial. Learning the parameters can be done using maximum-likelihood estimation, but it is actually better to maintain a posterior distribution over parameters, similar to the posterior over σ in section 6.4. Then, using the learned parameters, one can decode a posterior distribution over the stimulus from the voxel activity on new trials [61, 108, 185].

The calculation of posterior distributions based on neural activities for perception has a close analogue in the technical problem of decoding in the context of brain-machine interfaces. Following a spinal-cord injury that results in paralysis, a person might be outfitted with prosthetic arms. But how can these prosthetic limbs be controlled? One possibility is to use voice commands, but this procedure is cumbersome. Another possibility is to use eye movements. A third possibility, which at first seems worthy of a science fiction story, is to control the prosthetic device through the use of a brain machine interface that reads the user's intent from their neural activity. Recorded neural signals from the motor cortex, properly interpreted, can indicate the user's intended movements [189]. In such brain-machine interface scenarios, researchers may attempt to calculate the posterior distribution over the movement intent, given the recorded neural activity [24].

14.9 Summary

In this chapter, we extended the discussion of generative models to neural activity and introduced the concepts of neural likelihoods and neural uncertainty. We learned the following:

- Neural responses to stimuli can be described using tuning curves and a model for random variability around those tuning curves. A simple model is independent Poisson variability.
- By inverting the generative model so defined, we can compute a neural likelihood function of the stimulus from a given pattern of neural population activity.
- The neural likelihood function contains all information that can objectively be obtained from the population activity. No more information can be obtained, and any different information would be incorrect.

- This neural likelihood function, its mode, and its width all vary across trials, even if the stimulus is kept the same.
- The neural likelihood function tends to look Gaussian only in the limit of high information (many neurons, many spikes).
- Neural likelihood functions can be used for further Bayesian computation, in the same way as we used nonneural likelihood functions in previous chapters.
- The case of independent Poisson neurons with Gaussian tuning curves and the constant-sum approximation is a useful toy model that admits intuitive closed-form expressions for both the MLE and the likelihood width.
- The measurement used in previous chapters can approximately be identified with the MLE of the stimulus obtained from a neural population.
- Many approaches exist for implementing Bayesian computations with neural operations.
- Just as the brain could compute a neural likelihood function, experimenters can compute likelihood functions from voxel activity using an estimated generative model of that activity. This allows for the decoding of sensory (or memory) uncertainty on a trial-by-trial basis.

14.10 Suggested Readings

- Christopher R. Fetsch, Alexandre Pouget, Gregory C. DeAngelis, and Dora E. Angelaki. "Neural Correlates of Reliability-Based Cue Weighting During Multisensory Integration." *Nature Neuroscience* 15, no. 1 (2012): 146–154.
- József Fiser, Pietro Berkes, Gergő Orbán, and Máté Lengyel. "Statistically Optimal Perception and Learning: From Behavior to Neural Representations." *Trends in Cognitive Sciences* 14, no. 3 (2010): 119–130.
- Peter Földiák. "The 'Ideal Homunculus': Statistical Inference from Neural Population Responses." In *Computation and Neural Systems*, edited by Frank H. Eeckman and James M. Bower, 55–60. New York: Springer, 1993.
- Laura S. Geurts, James R. H. Cooke, Ruben S. van Bergen, and Janneke F. M. Jehee. "Subjective Confidence Reflects Representation of Bayesian Probability in Cortex." *Nature Human Behaviour* 6 (2022): 294–305.
- Mehrdad Jazayeri and J. Anthony Movshon. "Optimal Representation of Sensory Information by Neural Populations." *Nature Neuroscience* 9, no. 5 (2006): 690–696.
- Hsin-Hung Li, Thomas C. Sprague, Aspen H. Yoo, Wei Ji Ma, and Clayton E. Curtis. "Joint Representation of Working Memory and Uncertainty in Human Cortex." *Neuron* 109, no. 22 (2021): 3699–3712.
- Wei Ji Ma, Jeffrey M. Beck, Peter E. Latham, and Alexandre Pouget. "Bayesian Inference with Probabilistic Population Codes." *Nature Neuroscience* 9, no. 11 (2006): 1432–1438.
- A. Emin Orhan and Wei Ji Ma. "Efficient Probabilistic Inference in Generic Neural Networks Trained with Non-Probabilistic Feedback." *Nature Communications* 8, no. 1 (2017): 1–14.

- Ryan M. Peters, Phillip Staibano, and Daniel Goldreich. "Tactile Orientation Perception: An Ideal Observer Analysis of Human Psychophysical Performance in Relation to Macaque Area 3b Receptive Fields." *Journal of Neurophysiology* 114, no. 6 (2015): 3076–3096.
- Jonathan W. Pillow, Jonathon Shlens, Liam Paninski, Alexander Sher, Alan M. Litke, E. J. Chichilnisky, and Eero P. Sinomcelli. "Spatio-Temporal Correlations and Visual Signalling in a Complete Neuronal Population." *Nature* 454, no. 7207 (2008): 995–999.
- Alexandre Pouget, Peter Dayan, and Richard Zemel. "Information Processing with Population Codes." *Nature Reviews Neuroscience* 1, no. 2 (2000): 125–132.
- Terence David Sanger. "Probability Density Estimation for the Interpretation of Neural Population Codes." *Journal of Neurophysiology* 76, no. 4 (1996): 2790–2793.
- Ruben S. Van Bergen, Wei Ji Ma, Michael S. Pratte, and Janneke F. M. Jehee. "Sensory Uncertainty Decoded from Visual Cortex Predicts Behavior." *Nature Neuroscience* 18, no. 12 (2015): 1728–1730.
- Edgar Y. Walker, R. James Cotton, Wei Ji Ma, Andreas S. Tolias. "A Neural Basis of Probabilistic Computation in Visual Cortex." *Nature Neuroscience* 23, no. 1 (2020): 122–129.
- Richard S. Zemel, Peter Dayan, and Alexandre Pouget. "Probabilistic Interpretation of Population Codes." *Neural Computation* 10, no. 2 (1998): 403–430.

14.11 Problems

Problem 14.1 Are the following statements true or false? Explain.

(a) In general, the variance of a single neuron responding to a stimulus can be determined from the value of its tuning curve at that stimulus value.

(b) In any population of neurons, the variability of population activity is fully known if one knows the variability of each single neuron.

(c) The closer a stimulus is to the preferred stimulus of a Poisson neuron, the lower is the response variance of this neuron when the stimulus is presented repeatedly.

(d) When neurons have similar and equally spaced tuning curves, then the neural likelihood function has the same width as the tuning curve.

Problem 14.2 We assume a population of nine independent Poisson neurons with Gaussian tuning curves and preferred orientations from -40 to 40 in steps of 10. The tuning curve parameters have values $g=10$, $b=0$, and $\sigma_{\text{tc}}=20$. A stimulus $s=0$ is presented to this population. What is the probability that all neurons stay silent?

Problem 14.3 In the toy model of section 14.4, we assumed zero-baseline Gaussian tuning curves.

(a) What changes in the math if the baseline is not zero?

(b) For each of the baseline values 0, 0.25, 0.5, and 1, numerically compute and plot ten likelihood functions (assume the true stimulus is zero), and describe what you observe.

Problem 14.4 In the toy model, we assumed the same tuning width σ_{tc} for all neurons. Derive the equivalent of equations (14.20) and (14.21) for the mode and the width of the likelihood function if every neuron has its own tuning width, say $\sigma_{\mathrm{tc},i}$ for the ith neuron.

Problem 14.5 Assume a population of independent Poisson neurons responding to a nonnegative stimulus s. Each neuron has a linear tuning curve, that is,

$$f_i(s) = a_i(s). \tag{14.26}$$

Show that for any pattern of activity in this population, the normalized neural likelihood function is a gamma distribution, and find expressions for its parameters.

Problem 14.6 Here, we explore whether the width of the likelihood function might be correlated with the error of the MLE, in other words, whether uncertainty and error go hand in hand.

(a) Simulate the toy model for 10,000 trials (all at $s=0$, and gain 1) and create a scatter plot of the squared error of the MLE versus the squared width (variance) of the likelihood function. What is the correlation coefficient?

(b) Choose several different values of gain. Does the strength of the correlation depend on gain?

(c) Now divide the trials into four quartiles for the variance of the likelihood function. Compute the variance of the MLE for each of these four groups of trials. Is there a correlation?

Problem 14.7 Some stimulus variables, such as motion direction, are periodic (directional). We can think of such variables as taking values on a circle, for example from $-\pi$ to π radians. Consider a laboratory experiment in which motion direction is drawn from a Von Mises distribution with circular mean μ_s and concentration parameter κ_s:

$$p(s) \propto e^{\kappa_s \cos(s-\mu_s)}. \tag{14.27}$$

Assume that motion direction, denoted by s, is encoded in a population of n independent Poisson neurons. The tuning curve of the ith neuron has a Von Mises shape with gain g, preferred direction $s_{\mathrm{pref},i}$, and concentration parameter κ_{tc}:

$$f_i(s) = g e^{\kappa_{\mathrm{tc}} \cos(s-s_{\mathrm{pref},i})}. \tag{14.28}$$

(a) Show that the likelihood function of the stimulus based on a population pattern of activity in this population, $\mathbf{r}=(r_1,\ldots,r_n)$, is proportional to a Von Mises distribution,

$$\mathcal{L}(s;\mathbf{r}) \propto e^{\kappa_L \cos(s-\mu_L)}, \tag{14.29}$$

and find expressions for $\cos\mu_L$, $\sin\mu_L$, and κ_L, each in terms of the r_i's. The MLE μ_L is also known as the *population vector decoder*.

(b) Show that the posterior distribution is also a Von Mises distribution and calculate its circular mean.

Problem 14.8 This is a continuation of the previous problem in which we will numerically implement and compare two neural decoders of a periodic variable. It is also a neural parallel of material that we covered back in chapter 4.

(a) Draw 2,000 motion directions s from the stimulus distribution with $\mu_s = 0$ and $\kappa_s = 4$. Each drawn stimulus represents one experimental trial. For each trial, draw a pattern of population activity in response to the motion direction on that trial; assume $g = 0.2$, preferred directions at every multiple of $10°$, and $\kappa_{\text{tc}} = 1$.

(b) Then, again for each trial, compute both the MLE and the posterior mean estimate (PME). You may use the expressions obtained in the previous problem. If you did not do that problem, you may do the computations numerically. (Hint: use the inverse tangent function.)

(c) Calculate the square of the sine of the estimation error for all trials and both estimators. This quantity is a circular equivalent of the squared error for real-valued variables.

(d) Create a figure consisting of 2×2 subplots. The top left subplot should show a scatterplot of the MLE against s. The top right subplot should show a scatterplot of the PME against s. Both subplots should have both an x range and a y range from $-\pi$ to π; also draw the diagonal as a dashed black line, for reference. The bottom left subplot should show a histogram of the squared sine error of the MLE. Use twenty bins. The bottom right subplot should show the same for the PME.

(e) Does the MLE or the PME have a higher mean squared sine error? Which correlates better with the true stimulus? Are these properties expected?

15

Bayesian Models in Context

What scope and limitations do Bayesian models have within the study of behavior and of the brain?

In this concluding chapter, we describe how Bayesian models of perception and action connect to bigger themes in the field of behavioral research and beyond. We also discuss limitations of the modeling framework in this book and suggest future directions for the field.

Plan of the Chapter

We first suggest that Bayesian behaviors can plausibly be expected to emerge over lifetime and evolutionary time scales. However, it is also reasonable to expect that not all Bayesian behavior is optimal and indeed that not all behavior is Bayesian. We discuss how Bayesian models of perception and action differ from and relate to other classes of popular models within cognitive science and neuroscience. We discuss how "as if" criticisms of Bayesian models can be addressed. We conclude with open areas of research: real-world complexity, learning to be Bayesian, and approximate inference.

15.1 Bayesian versus Optimal Behavior

Animals, including humans, are faced with consequential decision problems throughout their lives. Accordingly, we expect that adaptive behaviors will tend to be selected over both lifetime and evolutionary time scales. Among the decision problems that animals encounter in the real world, a great many will involve uncertainty. For this large set of problems, the optimal solutions are Bayesian. We therefore expect that humans and other animals will tend to use Bayesian strategies and that many behaviors will tend to conform to the predictions of Bayesian models.

The use of Bayesian inference is not synonymous with optimal behavior. On the one hand, some optimization problems do not require a Bayesian approach. Such problems include finding an unbeatable way to play tic-tac-toe, a fast way to sort a list, or the way to pack the largest number of apples into a crate. However, such problems are often relatively artificial computer science problems without great ecological relevance to organisms. In most ecologically important problems, uncertainty about world states is not far away.

On the other hand, a decision maker may be fully Bayesian but suboptimal. For example, when faced with an unfamiliar task, they might apply incorrect priors or likelihood calculations, leading to model mismatch (see section 3.5). In such cases, the decision maker's behavior will typically be suboptimal with respect to the generative model characterizing the task at hand. It would still be optimal with respect to the counterfactual assumed generative model. A Bayesian model is still useful in this case. It can serve to attribute the suboptimality to a specific wrong belief, for example about a wrong prior based on past experience. It can also happen that behavior is not optimal with respect to the generative model of a laboratory task, but it is with respect to a generative model in the natural world. In those cases, it would be misleading to call the behavior suboptimal.

Exercise 15.1 Take three of your favorite papers in the broader space of Bayesian models of perception and action and clarify for yourself which exactly are its assumptions. What is assumed optimal? How is the model justified?

15.2 Overly Strong Claims of Optimality

To claim that a behavior is close to optimal is often attractive, because this provides a crisp, principled account of the behavior, whereas dissecting suboptimality tends to be a messier business. Thus, it is perhaps not surprising that authors occasionally play somewhat loose with the notion of optimality. This could take the form of adding ad hoc mechanisms in order to "salvage" optimality or of only performing a qualitative comparison between data and model. Bayesian models are inherently quantitative and every effort should be made to make detailed quantitative comparisons. A special kind of questionable optimality claim arises when the prior is given excessive flexibility instead of grounding it in either natural statistics or experimental statistics that the participants have been provided the opportunity to learn.

The specification of the task should also include a meaningful cost function. How constrained the modeler is in the choice of cost function depends on the task. In models of categorical perceptual tasks without an external reward structure, the 0-1 cost function, which corresponds to maximizing accuracy, is the most natural choice (section 13.5). In continuous perceptual tasks, a squared error cost function is often assumed for convenience; however, there is room for alternatives. When external rewards are provided, a complication is a potential nonlinearity between reward and subjective utility (section 13.7.2). In the most challenging cases, the cost function is an unknown function of multiple variables; for example, what is the cost of moving your body in a certain way when catching a ball, or the cost of misunderstanding someone in a conversation? Generally speaking, we believe that researchers should be prudent in allowing flexibility in any component of a Bayesian model, and every form of flexibility should be carefully justified.

Given occasional overly strong claims of optimality, it is not surprising that there has been some skepticism of such claims and, by extension, of Bayesian models [22, 86]. Some critiques are of the type "Bayesian studies can make everything sound optimal." While this is true about Bayesian models without any constraints, it simply is not true for Bayesian models that derive likelihoods, priors, and cost function from the task specification as much as possible. Many Bayesian models have only a handful of (well-justified) free parameters.

Exercise 15.2 Do you know of any Bayesian papers that you feel make excessive claims of optimality? How would their message be affected if those claims were removed?

15.3 Understanding Why Some Behaviors Are Optimal and Others Are Not

If behavior is close to optimal in some cases and not in others, what distinguishes these situations? A common notion is that behavior in perceptual and motor tasks tends to be optimal, whereas behavior in cognitive tasks tends to be suboptimal. Arguments for the suboptimality of cognition typically revolve around the length of the list of cognitive biases and fallacies that humans exhibit [87, 88], from the gambler's fallacy to the anchoring fallacy to confirmation bias to the availability heuristic. On very general grounds, such a distinction between perception and cognition seems plausible. After all, perceptual systems are evolutionarily older and therefore have had more time to become optimized. On closer look, however, the notion of optimality in these cognitive effects is very different in many ways than ones on the perceptual and motor side. For example, the cognitive biases typically involve explicit manipulations of probabilities. Replacing probabilities by frequencies reduces or eliminates some biases; in addition, the specific wording sometimes matters a lot. Also, cognitive fallacies are only crudely measured at the individual-subject level: a subject gives either the "biased" or the "unbiased" answer, yielding no insight into the individual's decision-making process. By contrast, using parametric measurements in more robust experimental paradigms, human reasoning and learning are often found to be well accounted for by Bayesian models at least at a qualitative level (e.g., [11, 34, 48, 68, 71, 134, 210]). Optimality, however, is much harder to establish in the context of high-level cognitive tasks.

An interesting approach that deserves more attention is to package what is essentially the same task in both a perceptual and a cognitive framing, and compare human behavior between them. People were close to optimal in the rapid movement task that we discussed in section 13.8.2. However, this task was designed to be the movement equivalent of a lottery task in which people exhibit forms of suboptimality [182]. Similarly, Stephanie Chen and colleagues found that in a task in which optimal behavior required marginalizing over categories, people would fail to do so when giving verbal reports but not when catching an object [35].

15.4 Bayesian Models Are Not Mechanistic Models

Bayesian models have been criticized for being "as if" models, in the sense that people might follow something close to the Bayesian decision rule, but the Bayesian computations leading up to that rule might not have any grounding in reality. In one reading of this criticism, grounding in reality is achieved only by causal-mechanistic models. Indeed, Bayesian models of perception and action are functional, not causal-mechanistic explanations, in the sense that they do not specify the neural parts and neural operations that implement the Bayesian computations [40, 54]. That being said, there is an abundance of work on the neural implementation of Bayesian models (see chapter 14). We are faced with the problem of *multiple realizability*, in the sense that the same Bayesian computation can be implemented by neurons in multiple (and potentially infinitely many) different ways. Thus, the Bayesian model cannot make any fine-grained predictions for neuroscientists interested in mechanism. In our view, this problem is not specific to Bayesian models; it is shared by nearly all models of behavior. A more positive spin on the problem is that the field is blessed with a form of *separation of scales*: the macro scale (the inference computation) can be modeled separately from the micro scale (the neural implementation).

One place where Bayesian models might intersect with mechanism is in *resource-rational models* [109]. This is a category of models that deviate from purely Bayesian models in that the cost of representation or computations is taken into account. In some cases, this cost is biologically motivated, for example as corresponding to the number of neurons devoted to a computation or to the total activity of a population of neurons.

15.5 Bayesian Transfer

Another way to interpret the "as if" criticism is not in terms of neural implementation but in terms of the meaning of the internal constructs of the Bayesian model. In a Bayesian model, likelihoods, priors, and cost functions determine which behaviors will be successful. But are these constructs meaningful and actually used by the individual? This question is related to the one in the Introduction about whether Bayesian models are process models.

A strategy for demonstrating that the internal constructs of a Bayesian model are meaningful is what Maloney and Mamassian call *Bayesian transfer* [121]. The idea is that if prior, likelihood, and cost function are meaningful, they should be flexible and generalize across tasks or conditions. For example, if a participant is near optimal with the combination of likelihood 1 and prior 1, as well as with the combination of likelihood 2 and prior 2, then the generalization test would consist of presenting a condition that calls for the use of likelihood 1 and prior 2, or likelihood 2 and prior 1. In such a new condition, relearning a rule through trial and error would be slow and would require many trials, whereas recombining "computational modules" should be fast. Of course, it is possible that some constructs of Bayesian computations are flexible and generalizable with respect to variations of one component but not others. Ma and Jazayeri propose a hierarchy of Bayesian flexibility [114]. Bayesian flexibility is the main way of obtaining evidence about the degree to which priors and likelihoods can be combined in arbitrary ways.

Taken to the extreme, the adjustment in a Bayesian transfer experiment should be instantaneous: varying the prior and likelihood on a trial-by-trial basis should produce near optimal behavior. In other words, we could use every trial as a generalization test from other trials. In practice, few studies vary both prior and likelihood at the same time, but varying the likelihood from trial to trial is very common. For instance, in cue combination studies, the reliability of at least one sensory cue may be varied from trial to trial.[1] To strengthen the test, trial-by-trial feedback should be withheld, so as to make trial-and-error rule learning nearly impossible. There could be initial training only at the start of the session or of the entire experiment, but perhaps not even for all likelihood conditions—forcing the participants to generalize from the limited training conditions combined with preexperiment experience with similar stimuli.

It is less common to test the generalizations of priors or cost functions than of likelihoods, but a number of relevant studies have been conducted. For example, the light-from-above

1. We saw in chapter 14 that the likelihood might vary from trial to trial even when the stimulus is held fixed; however, experimenters cannot control this variation so they have to resort to changing the stimulus.

prior [5] and a prior over faces [77] seem to be task-general. Acerbi and colleagues [2] varied the prior from trial to trial, and Whiteley and Sahani [201] varied the cost function from trial to trial (section 13.8.1). In all these situations, it seems that people are capable of Bayesian transfer in those situations as well.

Altogether, we think that "as if" criticisms have been and are being adequately addressed empirically, and we do not consider them major challenges to the Bayesian modeling framework of perception and action.

15.6 Probabilistic Computation and Hybrid Models

Bayesian transfer studies of the sensory likelihood function have shown that uncertainty is computed by the brain and taken into account in decisions. To distinguish this from Bayesian inference more generally, we have previously named such computation *probabilistic computation* [112]. Combining the need to compare Bayesian models against suboptimal alternatives with the notion of probabilistic computation, we arrive at a category of hybrid models, namely models of suboptimal probabilistic computation. These are models in which stimulus uncertainty is represented by the brain on a trial-by-trial basis, but then used suboptimally in subsequent computations. For example, in section 8.6, we derived a Bayesian decision rule for a categorization rule; this rule, one form of which is equation (8.38), depends on the level of sensory uncertainty σ. However, it is possible that the brain takes uncertainty into account but according to a different rule. In several visual decision-making paradigms, evidence for suboptimal probabilistic computation has been found. The suboptimality may or may not be Bayesian under a different generative model than the one dictated by the task structure—this is usually difficult to establish. As long as it is not established that the decision rule is Bayesian, the model should be thought of as hybrid, with a Bayesian front end and a non-Bayesian back end. In neural studies, one could in principle trace the propagation of the likelihood function across brain regions involved in different parts of a computation, but to our knowledge this has not been done.

15.7 Real-World Complexity

A main challenge to Bayesian models of perception and action consists of scaling up to complex and/or naturalistic problems. The vast majority of tasks studied in this book and in the field are simple, involving one or a few variables, each of which is binary or one-dimensional. This matches well with most laboratory tasks but poorly with naturalistic tasks, in which there are more variables, those variables are higher-dimensional, and they can take on more values. We saw a glimpse of this in section 12.2.2, where we calculated a likelihood function over a change-point vector that could take 2^T values (also see problem 11.7). Even for modest T, this yields a very large number of hypotheses to keep in mind. Another example is top-level nuisance parameters, as discussed in chapter 9. In a real scene, there are many such parameters for every object: viewing angle, viewing distance, lighting, and so on. Moreover, these nuisance parameters themselves have strong spatial structure, with objects also playing a special role. Properly marginalizing over all these parameters, as we did in our simple examples in chapter 9, is a daunting task. In fact, such problems may easily be computationally prohibitive [188].

In addition, real-world behavioral output is nothing like the response options in laboratory tasks. For example, when hiking in the forest, a behavior to predict is where one is going to move one's foot next, at which angle, at what time, and with what force. Another example of real-world behavior is conducting a conversation. There, the behavior to be predicted is not a simple binary choice or one-dimensional estimate, but natural language (even apart from body posture, eye movements, etc.). In addition, the cost function is complex, likely involving multiple terms such as the accuracy of one's understanding, the time a conversation takes, and the impression one makes on the other. Consequently, Bayesian modelers have not tried to tackle this problem.

It is easy to imagine that if the environment or the task becomes too complex, humans may cease to follow the predictions of Bayesian models. However, it is exactly under such circumstances that it is not even clear how to formulate a Bayesian model—for example because we have no handle on the distributions involved in the generative model. We could call this the *Bayesian tragedy*: we can specify Bayesian models in regimes where they are likely to work, and we have a hard time specifying them in regimes where they may be unlikely to work.

Box 15.1
Approximate Inference

While in this book we have mostly studied exact Bayesian inference, there is a whole world out there of approximate Bayesian computation [19]. Some approximate algorithms have made inroads in cognitive science and neuroscience.

Belief propagation. In this class of algorithms, we start with some variables, propagate messages through the system, and apply potential corrections. We often assume that the relevant variables have distributions for which the integrals can be analytically solved or approximated. We also assume that the relations between variables are such that they can be described as a (generally) sparse graph. In certain kinds of graphs, belief propagation provides an efficient way of solving many Bayesian problems. Sometimes it is approximately correct even when the assumptions are not satisfied.

Variational Bayesian approaches. For complex probability distributions, belief propagation cannot work in a fully analytical fashion. In fact, the probability distribution of each unobserved variable given the observed variables may be arbitrarily complicated. Variational Bayes uses the following trick. We approximate the probability distribution of p by a probability distribution (q_θ) that depends on a number of parameters θ. Subsequently, we optimize the parameters of q_θ so that q_θ becomes as similar as possible to the real probability distribution p. Making them the same is not generally possible, so variational methods usually optimize an approximation, the so-called *evidence lower bound* (ELBo).

Markov chain approaches. A different approach toward the solution of similar problems is Markov chain methods. These initialize a vector of all variables to be inferred and change it iteratively to produce a sequence of vectors. This is set up in such a way that, over time, the expected number of times of being in a state is proportional to the actual probability of the state given the observed variables. These methods can deal with arbitrary probability distributions. They are particularly useful if the probability distribution is concentrated in small areas of high probability. However, all the proofs for correctness hold only in the limit of large, often astronomical, numbers of iterations. In practice, Markov chain methods are usually limited by the difficulty of problems they can solve by available runtime. In cognitive science, it has been proposed that human decision makers behave like Markov chains with a limited number of iterations [192].

15.8 Learning to Be Bayesian

In this book, we have modeled the brain as a system that implements Bayesian inference. However, how does the brain get there? In section 6.4, we addressed a narrow form of this question, describing the Bayesian algorithm for learning a parameter of a generative model. A much broader question is how the ability to accurately invert a generative model is acquired. Are we born with a specialized mechanism for doing Bayesian inference? Or do we learn through experience that the Bayesian approach provides the best solution to real-world problems that involve uncertainty? In the realms of perception and action, these are open questions. In favor of an account centered on learning, young children seem to be farther from optimality than adults in combining sensory information with a prior [31] and in combining multisensory cues [69, 133, 144].

Although Bayesian models are not mechanistic, the field more broadly has a tendency to try to answer questions about learning in mechanistic terms. This is greatly facilitated by artificial neural networks, general-purpose learners that can serve as a playground for testing ideas (box 6.4 and box 15.2). For problems that involve uncertainty, the Bayesian solution is always best. The solutions to the Bayesian problems in this book (and in the experimental literature) can be learned relatively easily by a neural network [140]. In fact, this finding could make one wonder why very young children are not already proficient Bayesians! However, the Bayesian problems studied in the laboratory tend to be extremely simplified versions of real-world inference problems. Moreover, these neural networks were provided with trial-to-trial feedback on the state of the world. Such feedback is unrealistic for human learners. A particularly pessimistic view is that the whole Bayesian approach must be rejected because the world does not provide observers with sufficient ground-truth information to build accurate generative models on which to base inferences [148]. A more constructive view would be to look for general-purpose learning mechanisms that combine limited ground-truth feedback (labeled data) with unsupervised learning [17, 120] and priors [175]. An additional constraint is to use biologically plausible learning rules, which usually means that the update of a parameter in the network is only affected by activity of "nearby" units in the network; gradient descent is not biologically plausible in this sense.

Box 15.2
Artificial Neural Networks versus Bayesian Inference

The general approach in this book has been to define a relatively simple generative model, $p(x|s)$, and invert it to obtain an expression for the posterior, $p(s|x)$. However, the world is sufficiently complex that we may not be able to specify a satisfactory generative model for a particular task, and even if we are, $p(s|x)$ may be very difficult to compute. The field of artificial neural networks (ANNs) follows a different approach. Instead of starting with a human-defined generative model of the world, ANNs use a general-purpose function $f_\theta(x)$ with many free parameters θ (see box 6.4) to approximate the posterior. The parameters are optimized to obtain the best possible results. The focus is thus not on inverting a generative model but on finding a model that works. That being said, a trained ANN might for all practical purposes be Bayesian. In other words, it is conceivable that a system that is nowhere overtly Bayesian passes, after training, every test for Bayesian decision making that a researcher would apply to a human subject.

15.9 Suggested Readings

- Wendy J. Adams. "A Common Light-Prior for Visual Search, Shape, and Reflectance Judgments." *Journal of Vision* 7, no. 11 (2007).
- John R. Anderson. "The Adaptive Nature of Human Categorization." *Psychological Review* 98, no. 3 (1991): 409–429.
- John R. Anderson. *The Adaptive Character of Thought*. New York: Psychology Press, 2013.
- Jeffrey S. Bowers and Colin J. Davis. "Bayesian Just-so Stories in Psychology and Neuroscience." *Psychological Bulletin* 138, no. 3 (2012): 389–414.
- Nick Chater, Noah Goodman, Thomas L. Griffiths, Charles Kemp, Mike Oaksford, and Joshua B. Tenenbaum. "The Imaginary Fundamentalists: The Unshocking Truth about *Bayesian Cognitive Science*." Behavioral and Brain Sciences 34, no. 4 (2011): 194–196.
- Nick Chater and Mike Oaksford, eds. *The Probabilistic Mind: Prospects for Bayesian Cognitive Science*. New York: Oxford University Press, 2008.
- Stephanie Y. Chen, Brian H. Ross, and Gregory L. Murphy. "Implicit and Explicit Processes in Category-Based Induction: Is Induction Best When We Don't Think?" *Journal of Experimental Psychology: General* 143, no. 1 (2014): 227–246.
- Jerry A. Fodor. *Psychological Explanation: An Introduction to the Philosophy of Psychology*. New York: Random House, 1968.
- Monica Gori, Michele Del Viva, Giulio Sandini, and David C. Burr. "Young Children Do Not Integrate Visual and Haptic Form Information." *Current Biology* 18, no. 9 (2008): 694–698.
- Gerd Gigerenzer. "On Narrow Norms and Vague Heuristics: A Reply to Kahneman and Tversky." *Psychological Review* 103, no. 3 (1996): 592–596.
- Matt Jones and Bradley C. Love. "Bayesian Fundamentalism or Enlightenment? On the Explanatory Status and Theoretical Contributions of Bayesian Models of Cognition." *Behavioral and Brain Sciences* 34, no. 4 (2011): 169–188.
- Daniel Kahneman. *Thinking, Fast and Slow*. New York: Macmillan, 2011.
- Charles Kemp, Andrew Perfors, and Joshua B. Tenenbaum. "Learning Overhypotheses with Hierarchical Bayesian Models." *Developmental Science* 10, no. 3 (2007): 307–321.
- Falk Lieder and Thomas L. Griffiths. "Resource-Rational Analysis: Understanding Human Cognition as the Optimal Use of Limited Computational Resources." *Behavioral and Brain Sciences* 43, no. e1 (2020): 1–60.
- Wei Ji Ma. "Organizing Probabilistic Models of Perception." *Trends in Cognitive Sciences* 16, no. 10 (2012): 511–518.
- Wei Ji Ma and Mehrdad Jazayeri. "Neural Coding of Uncertainty and Probability." *Annual Review of Neuroscience* 37 (2014): 205–220.
- Laurence T. Maloney and Pascal Mamassian. "Bayesian Decision Theory as a Model of Human Visual Perception: Testing Bayesian Transfer." *Visual Neuroscience* 26, no. 1 (2009): 147–155.

- Barbara Mellers, Ralph Hertwig, and Daniel Kahneman. "Do Frequency Representations Eliminate Conjunction Effects? An Exercise in Adversarial Collaboration." *Psychological Science* 12, no. 4 (2001): 269–275.
- Daniel J. Navarro and Amy F. Perfors. "Similarity, Feature Discovery, and the Size Principle." *Acta Psychologica* 133, no. 3 (2010): 256–268.
- Dobromir Rahnev and Rachel N. Denison. "Suboptimality in Perceptual Decision Making." *Behavioral and Brain Sciences* 41 (2018): e223
- Julia Trommershäuser, Laurence T. Maloney, and Michael S. Landy. "Decision Making, Movement Planning and Statistical Decision Theory." *Trends in Cognitive Sciences* 12, no. 8 (2008): 291–297.
- Edward Vul, Noah Goodman, Thomas L. Griffiths, and Joshua B. Tenenbaum. "One and Done? Optimal Decisions from Very Few Samples." *Cognitive Science* 38, no. 4 (2014): 599–637.
- Louise Whiteley and Maneesh Sahani. "Implicit Knowledge of Visual Uncertainty Guides Decisions with Asymmetric Outcomes." *Journal of Vision* 8, no. 3 (2008): 1–15.
- Fei Xu and Joshua B. Tenenbaum. "Word Learning as Bayesian Inference." *Psychological Review* 114, no. 2 (2007): 245–272.

Appendices

Appendix A: Notation

Probability Distributions

Notation for probabilities can be confusing. At the core of this confusion lies the (sensible) convention of using the same letter p for any probability distribution, even when it is associated with different random variables. To distinguish distributions, mathematicians use a subscript to the p to denote the random variable that the probability distribution belongs to. For example, in $p_X(x)$, X is the random variable and x is a value that it takes. This notation is used for both probability mass functions and probability density functions, although some texts use f instead of p for density functions.

Unfortunately, the subscripts quickly become cumbersome or redundant. For example, the "correct" way to write Bayes' rule would be

$$p_{X|Y}(x|y) = \frac{p_{Y|X}(y|x)p_X(x)}{p_Y(y)}, \tag{A.1}$$

which feels redundant. Therefore, we use our own conventions, except in Chapter 3 and Appendix B, where we use the formal notation. When we describe the probability distribution of random variables with generic values (as opposed to specific numbers), we leave out the subscript and let the argument of the distribution indicate the variable. Thus, we write $p(x)$, $p(x,y)$, and $p(x|y)$ instead of $p_X(x)$, $p_{X,Y}(x,y)$, and $p_{X|Y}(x|y)$.

If the value is specific rather than generic, then we do have to specify both the random variable and its value. For discrete random variables, we use intuitive expressions such as $p(X=3)$, $p(X=3,\ Y=2)$, and $p(X=3|Y=2)$ instead of $p_X(3)$, $p_{X,Y}(3,2)$, and $p_{X|Y}(3|2)$. (Some other texts use a capital P or "Pr" for the probability of an event or assertion, such as $P(X=3)$ or $\Pr(X=3)$. We do not.)

For continuous random variables, using such equalities is problematic. For a continuous random variable X, we cannot replace $p_X(3)$ by $p(X=3)$, because p_X is a probability density function and not a probability mass function and, therefore, $p(X=a)$ would equal 0 for any a (see appendix section B.5.4). In such cases, we keep the subscript: we write $p_X(3)$, $p_{X,Y}(3,y)$, and $p_{X|Y}(3|y)$, where X is a continuous and Y any random variable. If the specific value of a continuous variable is "conditioned on," however, we *can* use an equality after the "given" sign, for example, we write $p(x|Y=2)$ instead of $p_{X|Y}(x|2)$, regardless of the types of X and Y.

Overall, we hope that these conventions strike the right balance between readability and mathematical precision.

Expected Values

We now discuss expected values.

1. If $f(X)$ is a function of only one random variable, X, then we write $\mathbb{E}[f(X)]$ to denote the expected value or mean of $f(X)$ with respect to $p(X)$, that is, $\sum_{i=1}^{n} f(x_i)p(x_i)$ (discrete) or $\int f(x)p(x)dx$ (continuous).
2. If $f(X, Y, \ldots)$ is a function of two or more random variables, then we use subscripts to indicate which variable(s) the expected value is taken over. For example, in the continuous case, the following expected values are different:

$$\mathbb{E}_X[f(X,Y)] \equiv \int f(x,y)p(x)dx; \tag{A.2}$$

$$\mathbb{E}_{X,Y}[f(X,Y)] \equiv \int f(x,y)p(x,y)dxdy. \tag{A.3}$$

3. When using a conditional distribution to calculate the expectation, we write the random variable being conditioned on in the subscript, and the value of that variable inside the square brackets. For example, in the continuous case,

$$\mathbb{E}_{X|Y}[f(X)|y] \equiv \int f(x)p_{X|Y}(x|y)dx, \tag{A.4}$$

where y could be a generic or a specific value. If y is a generic value, we may drop the subscript and write $\mathbb{E}[f(x)|y]$ instead of $\mathbb{E}_{X|Y}[f(X)|y]$.
4. Conventions for variance and standard deviation are like those for expected values. We denote them by Var[·] and Std[·], respectively.

Gaussian Distribution

For the Gaussian, or normal, distribution, we use the notation $\mathcal{N}$ with parameters mean and variance:

$$\mathcal{N}(y;\mu,\sigma^2) \equiv \frac{1}{\sqrt{2\pi\sigma^2}} e^{-\frac{(y-\mu)^2}{2\sigma^2}}. \tag{A.5}$$

A word of caution: many software packages use standard deviation, not variance, as a parameter of their built-in normal distribution.

Terminology Note

Even though mean, variance, and standard deviation are properties of random variables, it can be a handful to say things like “the variance of X under its posterior distribution.” Instead, we often use the common shorter expression “the variance of the posterior distribution” when there is no ambiguity about the variable.

Appendix B: Basics of Probability Theory

Calculus helps you understand math, but probability helps you understand life. Unfortunately, probability theory is not a standard component of most undergraduate science curricula. We hope that this will change, but in the meantime, this appendix provides some basics of probability theory. It is by no means an exhaustive introduction as one would find in a textbook on probability theory. Instead, it is a tutorial that focuses only on the concepts and calculations used in the book.

B.1 Objective and Subjective Probability

Probability is the degree of possibility. In its most restrictive sense, probability can be defined as the expected frequency of an outcome of a repeatable event, such as the probability that a coin will come up heads or the probability that someone rolls a 5 on a die. These events can be repeated an arbitrarily large number of times, and the long-run outcome frequencies can be tallied. If the proportion of tosses on which a coin lands heads converges to 0.5 as the number of tosses approaches infinity, we can state that the coin has a 0.5 probability of landing heads. This type of probability is sometimes called *objective probability*, and it is the only valid type of probability according to a strict frequentist view of probability.

A much broader—and, we believe, more useful—conceptualization of probability is as the degree of belief in a possibility. This is sometimes called *subjective probability*. The everyday terms *confidence* and *uncertainty* refer to subjective probabilities. If I know that a die has $\frac{1}{6}$ chance of landing 5, then my confidence in the proposition that it will land 5 is $\frac{1}{6}$. This particular example is trivial, because it involves simply converting an objective probability (a long-run outcome frequency) into a belief statement. However, the vastly wider applicability of the subjective conceptualization becomes clear when we consider degrees of belief in outcomes that cannot be repeated, for example, the probability that candidate A will beat candidate B in an upcoming election. This is not a probability that can be obtained by repeating the same event many times, but we may nevertheless have a strong prediction regarding the outcome. Indeed, examples of subjective probabilities that cannot be phrased as long-run outcome frequencies abound in daily life: What is the probability that it will rain today? What is the probability that I will enjoy a class? Many scientific questions also can be phrased only in terms of subjective and not in terms of objective probabilities: What is the probability that Saturn's mass lies between 10^{25} and 10^{26} kg? What is the probability that disease X is caused by a virus?

The distinction between objective and subjective probability has a long history [41] and is not always clear. For example, to determine the probability that it will rain today, a forecaster might run a large number of simulations starting from the current state of the atmosphere, each with a different instantiation of the stochastic factors in the model, and record the frequency of rain among these runs. While the resulting probability is subjective, it has been obtained in an "objective" way, namely by counting. Similarly, if I observe dark clouds in the sky and express my opinion that there is a high probability of rain, I am expressing a subjective probability judgment, but I am basing this judgment on a large number of previously observed, similar (though not identical) skies.

Bayesian inference treats subjective and objective probabilities in the same way: the same mathematical relationships (Bayes' rule, marginalization, etc.) apply identically to both types of probability. Bayesian inference is therefore extremely widely applicable. Bayesian models of perception, however, are grounded fundamentally in subjective probability: there is only one true world state, but from the point of view of an organism trying to infer it, there are many possibilities, and degrees of belief can be assigned to these possibilities.

B.2 The Intuitive Notion of Probability

We call the set of all possibilities under consideration the *sample space*. An *event* or *hypothesis* is a subset of the sample space. The term "event" is commonly used when discussing objective probability, and the term "hypothesis" when discussing subjective probability. The sample space could be "all possible numbers I can roll on a die" or "all possible weather patterns that can occur today." Given the former sample space, an event could be "I will roll an even number." Given the second sample space, a hypothesis could be "It will rain today." The probability of an event or hypothesis is a real number between 0 and 1, indicating the degree of possibility of the event or hypothesis. An event that is certain has a probability of 1, and an impossible event has a probability of 0. For events, one can think of probability as the frequency that the event happens among a very large number of random samples from the sample space. For instance, the probability that I will roll an even number on a six-sided die is 0.5. As explained above, this can also be conceptualized as a degree of belief. For a hypothesis, the frequency concept does not generally apply but probability still represents a degree of belief; for example, the degree of belief that it will rain today could be 0.35. We will denote the probability of an event or hypothesis X by $p(X)$; you will also find the notation $p(X)$ in the literature. For example, $p(\text{coin will come up heads}) = 0.50$.

B.3 Complementary Event

Given an event or hypothesis, its complementary event or hypothesis is that the first event or hypothesis does not occur or is false. For example, the complementary event to "rolling a 1 on a die" is "rolling a 2, 3, 4, 5, or 6 on a die." If the event or hypothesis is denoted X, then its complementary event or hypothesis, or complement, is denoted $\neg X$ (read "not X"). Here, $\neg$ is the symbol for a logical negation. The probability of the complementary event or hypothesis is 1 minus the probability of the event or hypothesis:

$$p(\neg X) = 1 - p(X). \tag{B.1}$$

In some situations, it is easier to calculate the probability of the complement of an event or hypothesis than of an event or hypothesis itself. For example, if you are asked to calculate the probability that the sum of the eyes on two dice is at least 3, it is easiest to first calculate the probability that the sum is lower than 3, and subtract that from 1 (the answer is $\frac{35}{36}$).

B.4 Venn Diagram Representation

Events and hypotheses can be represented graphically through *Venn diagrams* (figure B.1). (Strictly speaking, the areal diagrams that we show in this book are *Euler diagrams*, as they represent only relationships that actually occur; Venn diagrams, strictly defined, represent all possible logical relationships between sets. However, we follow the prevalent convention of naming any areal probability diagram a Venn diagram.) First draw a large rectangle whose interior represents all possible outcomes, that is, the sample space. Assign the area of this rectangle a value of 1, representing a total probability of 1. Then draw inside this rectangle a circle that represents all outcomes consistent with a particular event or hypothesis.

For example, the rectangle could represent all people in a group, and the circle all students among them. The area enclosed by the circle is a fraction of the area enclosed by the rectangle; this fraction represents the probability of the event or hypothesis—in our example, the probability that a randomly selected person is a student. The complement of the event or hypothesis is represented by the points that are inside the rectangle but outside the circle. Its area divided by the area of the rectangle represents the probability that a randomly selected person in the group is not a student.

B.5 Random Variables and Their Distributions

A random variable is a variable whose values cannot be known with certainty. Examples include the number rolled on a die, the date of birth of a person, the shoe size of a random person on your street, the time it takes to travel from home to work, the number of voters who

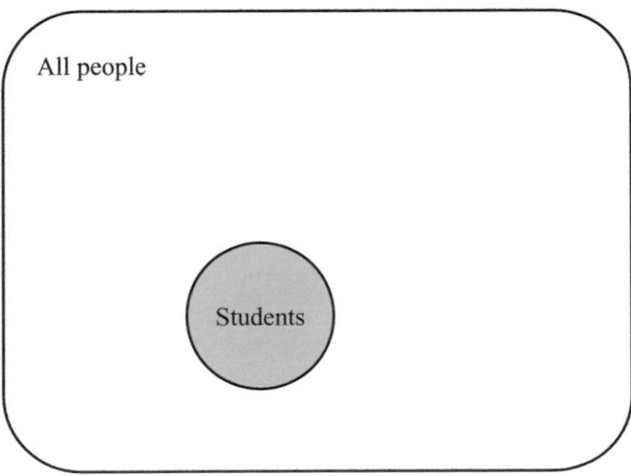

Figure B.1
Example of Venn diagram used to represent probabilities.

will participate in an upcoming election, or the price of a stock tomorrow. The opposite of a random variable is a variable whose value is known with certainty. Examples of nonrandom variables are the number of planets between us and the Sun (two), the number of days in a week (seven), the ratio of the circumference to the diameter of a circle (π), and the distance between two adjacent cm marks on a ruler (1 cm).

This is not an airtight distinction. Variables that appear nonrandom might be subject to measurement or production noise, which makes them random. For example, the distance between two adjacent cm marks on a ruler might vary, since the machine that produced the rulers was probably programmed by a computer that set the centimeter marks. However, computer-generated numbers have only a finite number of decimals, perhaps ten. As a consequence, the centimeter marks will never reach femtometer precision. In addition, the paint used for the marks will not attach itself to the surface in an identical way every time a ruler is produced. Therefore, one can think of the distance between two adjacent marks as a random variable. For reasons such as these, it might be useful to think of all variables as random, just with some having very low uncertainty.

Randomness, also called variability, noise, or stochasticity, is often a consequence of a lack of knowledge. When I roll a die, if you could somehow record exactly the position, direction, and speed with which the die left my hand, and you were able to simulate exactly the interactions the die had with air and table, then you would be able to predict with certainty the outcome of the roll. Since nobody knows the values of all these variables, the outcome of the die roll is considered to be random. Whether true randomness exists is a philosophical question that is beyond the scope of this book.

B.5.1 Discrete versus Continuous Random Variables

Random variables can be distinguished based on the values they can take. The most important distinction is between discrete and continuous random variables.

A *discrete random variable* takes on a set of values that can be counted, even though there might be infinitely many. Examples are the number of children in a household, the number of action potentials fired by a neuron, the number of moves in a chess game, the age of a person when counted in whole years, and the identity of a spoken word. A discrete random variable that takes on only two possible values is called *binary*.

A *continuous random variable* takes on values on a continuum. Examples are the length of a line segment, the direction one can walk in an open field, the waiting time in front of a red light, the speed of a car, and the frequency of a musical note. One can think of a continuous variable as discrete but with values that come in infinitesimally small increments. For example, distance is a continuous variable, but when it is measured in whole millimeters, it is a discrete variable. In a computer program, truly continuous variables do not exist; they must always be discretized.

B.5.2 Total Probability = 1

The total probability of all possible values of a random variable equals 1. This total is a reflection of the fact that the possibilities are mutually exclusive. If one were to increase the probability of one value, the probability of at least one other value has to decrease.

B.5.3 Discrete Probability Distributions

Discrete probability distributions are functions that assign a probability to each possible value of a discrete random variable. The probability distribution over a discrete random variable is also called a *probability mass function*. A discrete random variable X taking a particular value x is an event, denoted $p(X=x)$. As we now vary x over all its possible values, we obtain a function of x. This is the probability mass function, which we will denote by $p_X(x)$, or by $p(x)$ if there is no ambiguity about the identity of the random variable.

$$p_X(x)=p(X=x). \tag{B.2}$$

This means that the probability mass function evaluated at x is equal to the probability that the random variable takes this value. We use a subscript X (uppercase) to refer to a random variable, and the argument x (lowercase) to refer to a specific value of this random variable. The term "mass" is borrowed from physics. Roughly speaking, it is based on using matter as a metaphor for a possibility (a point in the sample space). The larger the probability of an event or hypothesis, the larger the mass of the piece of matter in the metaphor.

For example, the random variable X is the number rolled on a die. Its possible values x are 1 to 6. If the die is fair, the probability of each of these values is $\frac{1}{6}$, that is, $p(X=x)=\frac{1}{6}$ for all x. This is an example of a discrete uniform distribution. If the die is not fair, the probability of at least two of the values will differ from $\frac{1}{6}$ and the distribution will no longer be uniform.

For discrete random variables, total probability is computed by summing over all possible values; it should return 1. This is denoted as follows:

$$\sum_x p(x)=1. \tag{B.3}$$

In a specific case where the possible values of X are given, we can put those above and below the $\sum$ sign. For example, the total probability of a die roll would be

$$\sum_{x=1}^{6} p(x)=\frac{1}{6}+\frac{1}{6}+\frac{1}{6}+\frac{1}{6}+\frac{1}{6}+\frac{1}{6}. \tag{B.4}$$

Binary random variables are a special case of discrete random variables. Suppose a binary random variable X can take values x_1 and x_2. We know that total probability equals 1. Therefore, the probability of x_2 is 1 minus the probability of x_1, that is, $p(X=x_2)=1-p(X=x_1)$.

B.5.4 Continuous Probability Distributions

What is the probability that someone is exactly 160 cm tall? It is zero since "exactly" means that the length is accurate to an infinite number of decimal places. This problem is characteristic of continuous random variables and illustrates that probability mass functions, which worked well for discrete distributions, have to be replaced by a different concept in order to accommodate continuous variables.

Suppose we are interested in the probability distribution of the height of an adult (figure B.2). As a first approximation, we could consider possible heights in bins of 10 cm increments: between 120 and 130 cm, between 130 and 140 cm, and so on. Each bin has an

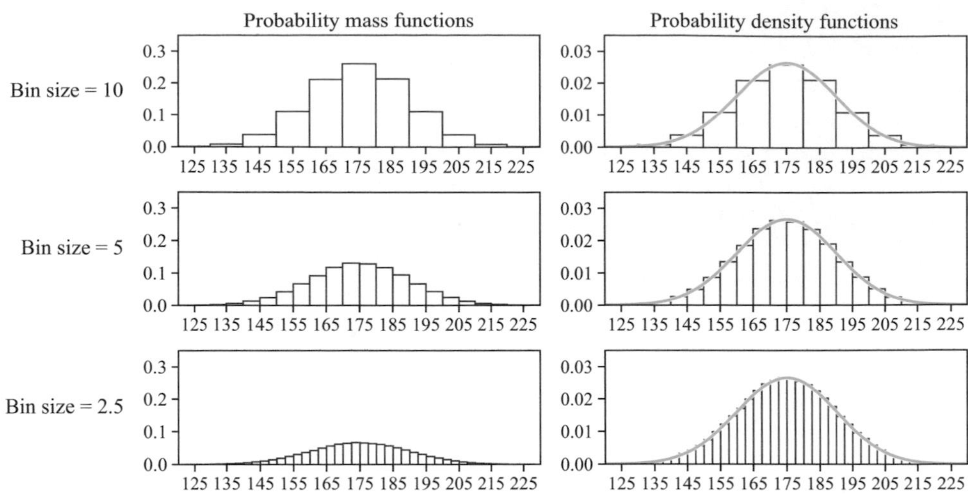

Figure B.2
When a random variable can take on a continuum of values (x axis), a probability mass function is only defined when outcomes are binned. The values of the function will decrease with decreasing bin size (left column). Probability density functions are obtained by dividing the probability mass values by bin size. This yields values that are independent of the bin size. The process of making the bin size smaller can be continued until the bins are infinitesimally small. This produces a continuous probability density function, overlaid in green.

associated probability, and, in this way, we can build up a probability mass function. However, we might want to describe height more finely, say in bins of 5 cm: between 122.5 and 127.5 cm, and so on. Each original bin is thus replaced by two new bins, each of which has on average half the probability mass of the original bin. Thus, the new probability mass function is scaled to about half the height of the original one (figure B.2). As we keep decreasing bin width in order to increase precision, the probability mass per bin keeps decreasing as well—it can become arbitrarily small. This is not very satisfactory. Is there a way to prevent the probability mass function from "disappearing"? Yes, this can be accomplished by dividing the probability mass in a bin by the width of the bin. By doing so, the function does not change much as we decrease bin width—it only becomes more precise. The *probability density function* is the result of this process as the bin width approaches zero (green curve). Again, there is an analogy with physics: if the probability in a bin is regarded as mass, then dividing this probability by bin width is analogous to computing a linear density: mass per unit length of the x axis.

The similarity in notation between the probability mass function for discrete variables and the probability density function for continuous variables, both written as $p_X(x)$ or $p(x)$, is misleading as there are some important conceptual differences between the two. For a discrete distribution, the probability mass of a single point never exceeds 1, since the probability mass values have to sum to 1. For a continuous distribution, the probability density at a single point is meaningless and can take arbitrarily large values. Consider, for example, a uniform distribution on the interval $[0, 0.01]$. It will have a probability density of 100 at every point. Only the integral over an interval will always be less than or equal to 1. Stated differently, for a discrete distribution, the probability $p(X=x)$ is a meaningful number that can take any value between 0 and 1, and is in fact identical to $p(x)$. For a continuous distribution, $p(X=x)$ is always 0, and only probabilities of the form $p(a \leq X \leq b)$, with a and b arbitrary numbers, are meaningful. For example, the probability that you hit the exact center point of a dartboard with an (infinitely) sharp dart is 0.

We will use the terms *probability distribution function* (PDF), *probability distribution*, or simply *distribution* to refer to the probability mass function of a discrete random variable or the probability density function of a continuous random variable.

Just as for discrete variables, the total probability of all values of a continuous variable equals 1. Total probability for a continuous variable is computed not as a sum, but as an *integral*. The integral of a continuous probability density function, as defined above, is the width of a bin multiplied by the function value in that bin, summed over all bins, in the limit that bin width approaches zero. Calculus provides recipes to compute integrals of certain functions. In this chapter, we familiarize ourselves with various integrals over probability density functions, especially because they directly parallel expressions with sums over probability mass functions; however, we will not evaluate these integrals, so no calculus is needed. The rule of total probability for a continuous variable X is written as

$$\int p(x)dx = 1. \tag{B.5}$$

The dx is in essence the width of a very small bin, and the integral sign $\int$ is a stretched-out S for "sum."

The most important continuous distribution is the normal distribution, which we discuss in detail below. Another important one is the uniform distribution. The uniform distribution on an interval $[a, b]$ has a constant PDF

$$p(x) = \frac{1}{b-a}. \tag{B.6}$$

The following continuous distributions are also common in applications of probability theory. The exponential distribution is given by $p(x) = \lambda e^{-\lambda x}$, with λ a constant and x defined on the positive real line. The power law distribution is given by $p(x) \propto x^{-a}$, with a a constant and x again defined on the positive real line.

B.5.5 Formal Definition of the Probability Density Function

Consider a continuous random variable X, such as the waiting time in a queue. The probability that the value of X is less than or equal to x is denoted by $p(X \leq x)$. This is the *cumulative distribution function* (CDF) of X at x, denoted $P_X(x)$:

$$P_X(x) = p(X \leq x). \tag{B.7}$$

By definition, this is a monotonically increasing function that takes values between 0 (as x approaches ∞) and 1 (as x approaches ∞). The PDF of X is now the derivative of this function:

$$p_X(x) = \frac{dP_X}{dx}. \tag{B.8}$$

Figure B.3 shows an example of a CDF and a PDF. For discrete random variables, the CDF can be defined in the same way, but doing so is not a necessary step in defining the probability mass function.

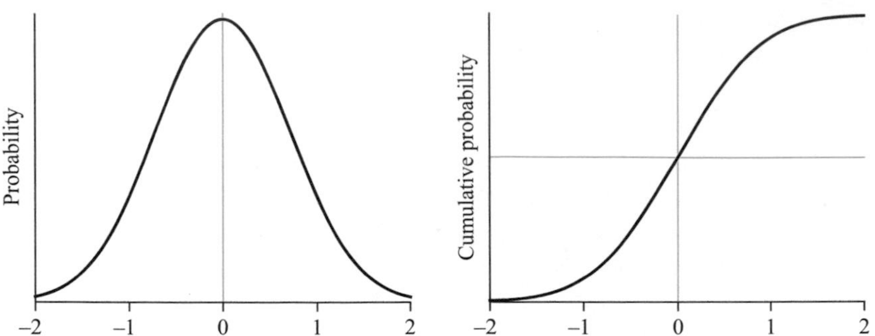

Figure B.3
A probability density function and a cumulative distribution function.

To go back from the PDF to the CDF, we integrate:

$$P_X(x) = p(X \leq x) = \int_{-\infty}^{x} p_X(y)dy. \tag{B.9}$$

(The notation $p_X(y)$ is not a typo. It means the probability density of the random variable X, evaluated at a value y of the dummy variable.) The physics equivalent of this statement is that the integral over a density is a mass. It immediately follows that the probability that X takes values in an interval (x_1, x_2) can be obtained by integrating $p_X(x)$ between x_1 and x_2:

$$p(x_1 \leq X \leq x_2) = \int_{x_1}^{x_2} p_X(x)dx. \tag{B.10}$$

It also follows from the definition that $\int_{-\infty}^{\infty} p(x)dx = 1$.

B.5.6 Normalization

A function can be made into a probability distribution by dividing each value by the total value on the entire domain, provided that this total value is finite. As a result, the probability distribution will integrate (or sum) to 1. This process is called *normalization*. If the total value on the entire domain is infinite, normalization is not possible.

Exercise B.1 Prove that the exponential distribution, $p(x) = \lambda e^{-\lambda x}$, is normalized.

Exercise B.2 Try to normalize the power law distribution, $p(x) \propto x^{-a}$, and find a condition on a for which normalization is possible.

B.6 Mean, Variance, and Expected Value

For a discrete random variable X, the mean or expected value of X is

$$\mathbb{E}[X] = \sum_x x p(x). \tag{B.11}$$

The variance, which is a measure of the spread around the mean, is defined as

$$\text{Var}[X] = \sum_x (x - \mathbb{E}[X])^2 p(x). \tag{B.12}$$

The standard deviation is the square root of the variance. Mean and variance are special cases of the *expected value* of any function of a random variable. If we denote the function by f, then the expected value of f is

$$\mathbb{E}[f(X)] = \sum_x f(x)p(x). \tag{B.13}$$

Thus, the mean is the expected value of X and the variance is $\mathrm{Var}[X] = \mathbb{E}[(x - \mathbb{E}[X])^2]$.

For a continuous random variable X with probability density $p(x)$, the analogous expressions are obtained by replacing sums with integrals:

$$\mathbb{E}[X] = \int_{-\infty}^{\infty} xp(x)dx; \tag{B.14}$$

$$\mathrm{Var}[X] = \int_{-\infty}^{\infty} (x - \mathbb{E}[X])^2 p(x); \tag{B.15}$$

$$\mathbb{E}[f(X)] = \int_{-\infty}^{\infty} f(x)p(x)dx. \tag{B.16}$$

Both for discrete and continuous variables, there is a common, equivalent expression for the variance, namely the difference between the mean of the squares and the square of the mean:

$$\mathrm{Var}[X] = \mathbb{E}[X^2] - \mathbb{E}[X]^2. \tag{B.17}$$

Exercise B.3 Prove this.

B.7 The Normal Distribution

B.7.1 Definition

The most important continuous distribution in applications of probability theory is the normal or Gaussian distribution. Its PDF is

$$p(x) = \frac{1}{\sqrt{2\pi\sigma^2}} e^{-\frac{(x-\mu)^2}{2\sigma^2}}. \tag{B.18}$$

We sometimes use $p(x) = \mathcal{N}(x; \mu, \sigma^2)$ as shorthand notation. The parameters μ and σ^2 do not have an a priori meaning (they are just denoted suggestively), but they turn out to be equal to the mean and variance of the distribution, respectively (see problem B.9. The factor $\frac{1}{\sqrt{2\pi\sigma^2}}$ is needed for normalization. The *standard normal* distribution is a normal distribution with mean 0 and standard deviation 1.

B.7.2 Central Limit Theorem

The importance of the normal distribution derives mainly from the central limit theorem. Roughly, the central limit theorem states that the mean of a large number of independent random variables with identical probability distributions will follow an approximately normal distribution, regardless of the distribution of the original variables. This theorem is most powerful because of its last part: the distribution of the original variables is irrelevant. The theorem can be relaxed to allow for independent, but not identically distributed variables.

In mathematical models of perception, the central limit theorem always plays a role in the background: whenever we assume that the noise corrupting a stimulus is normally distributed, we are essentially motivating this using the central limit theorem. The random variable describing the noise corrupting a stimulus might be the sum of a large number of independent noise processes.

B.7.3 Multiplying Two Normal Distributions

Let's consider the product of two Gaussian probability distributions over the same random variable X. One has mean μ_1 and variance σ_1^2, and the other mean μ_2 and variance σ_2^2. We will multiply these two distributions just as we would multiply regular functions, and then we will normalize the result (since the product is not automatically normalized). What is the resulting probability distribution? We first write down the expressions for the two probability density functions:

$$p_1(x) = \frac{1}{\sqrt{2\pi\sigma_1^2}} e^{-\frac{(x-\mu_1)^2}{2\sigma_1^2}}; \tag{B.19}$$

$$p_2(x) = \frac{1}{\sqrt{2\pi\sigma_2^2}} e^{-\frac{(x-\mu_1)^2}{2\sigma_2^2}}. \tag{B.20}$$

Multiplying these two functions comes down to summing the exponents. We will do that first:

$$\text{sum of exponents} = -\frac{(x-\mu_1)^2}{2\sigma_1^2} - \frac{(x-\mu_2)^2}{2\sigma_2^2} \tag{B.21}$$

$$= -\frac{1}{2}\left(\frac{x^2 - 2\mu_1 x + \mu_1^2}{\sigma_1^2} + \frac{x^2 - 2\mu_2 x + \mu_2^2}{\sigma_2^2}\right). \tag{B.22}$$

We reorganize by collecting all terms containing x^2, and all containing x:

$$\text{sum of exponents} = -\frac{1}{2}\left(\left(\frac{1}{\sigma_1^2} + \frac{1}{\sigma_2^2}\right)x^2 - 2x\left(\frac{\mu_1}{\sigma_1^2} + \frac{\mu_2}{\sigma_2^2}\right) + \cdots\right) \tag{B.23}$$

$$= -\frac{1}{2}\left((J_1 + J_2)x^2 - 2x(J_1\mu_1 + J_2\mu_2) + \cdots\right), \tag{B.24}$$

where we used *precision* notation,

$$J_1 \equiv \frac{1}{\sigma_1^2}; \tag{B.25}$$

$$J_2 \equiv \frac{1}{\sigma_2^2}. \tag{B.26}$$

Moreover, here and in the following, the dots represent all terms that do not depend on x. When exponentiated, these terms become a multiplicative constant that is independent of x. Since the resulting product of distributions must be normalized at the end of the calculation anyhow, any multiplicative constant that we insert or leave out until that point is irrelevant.

The sum of exponents can be written as

$$\text{sum of exponents} = -\frac{1}{2(J_1+J_2)^{-1}}\left(x^2 - 2x\frac{J_1\mu_1+J_2\mu_2}{J_1+J_2} + \cdots\right) \quad \text{(B.27)}$$

$$= -\frac{1}{2(J_1+J_2)^{-1}}\left(x - \frac{J_1\mu_1+J_2\mu_2}{J_1+J_2}\right)^2 + \cdots. \quad \text{(B.28)}$$

Thus, the product of the distributions in equation (B.20) is

$$p_1(x)p_2(x) \propto e^{-\frac{1}{2(J_1+J_2)^{-1}}\left(x^2 - 2x\frac{J_1\mu_1+J_2\mu_2}{J_1+J_2} + \cdots\right)} \quad \text{(B.29)}$$

$$= e^{-\frac{1}{2(J_1+J_2)^{-1}}\left(x - \frac{J_1\mu_1+J_2\mu_2}{J_1+J_2}\right)^2}, \quad \text{(B.30)}$$

where the proportionality sign absorbs all factors that are independent of x. We recognize this as another normal distribution, now with mean $\frac{J_1\mu_1+J_2\mu_2}{J_1+J_2}$ and variance $\frac{1}{J_1+J_2}$.

Exercise B.4 What is the correct normalization constant in equation (B.30)?

B.7.4 Multiplying Multiple Normal Distributions

We now generalize the previous section to N normal distributions. This is used in problem 10.7. Consider a set of N normal distributions over the same variable x The ith distribution has mean μ_i and variance σ_i^2. The (unnormalized) product of these distributions is equal to

$$\sqrt{\prod_i \frac{J_i}{2\pi}}\, e^{-\frac{\sum_i J_i}{2}\left(\mu - \frac{\sum_i J_i x_i}{\sum_i J_i}\right)^2} e^{-\frac{1}{2}\sum_i J_i x_i^2} e^{-\frac{(\sum_i J_i x_i)^2}{2\sum_i J_i}}. \quad \text{(B.31)}$$

B.7.5 The Cumulative Normal Distribution

The cumulative distribution function of a Gaussian distribution is not an elementary function (i.e. one built from exponentials, logarithms, and powers using addition, subtraction, multiplication, and division). However, it has a dedicated notation simply because it occurs often. We define the cumulative distribution of the standard normal density as:

$$\Phi_{\text{standard}}(y) \equiv \frac{1}{\sqrt{2\pi}}\int_{\infty}^{y} e^{-\frac{x^2}{2}}\,dx. \quad \text{(B.32)}$$

This function takes values between 0 and 1.

Exercise B.5 Show that $\Phi_{\text{standard}}(0) = 0.5$.

Exercise B.6 Show that the integral of a Gaussian distribution can be expressed in terms of the cumulative normal distribution as follows:

$$\int_{-\infty}^{x} \mathcal{N}(y;\mu,\sigma^2)dy = \Phi_{\text{standard}}\left(\frac{x-\mu}{\sigma}\right). \quad \text{(B.33)}$$

This is possible because any normal distribution can be shifted (by $-\mu$) and scaled (by dividing by σ) to obtain a standard normal distribution.

Two other integrals of the Gaussian distribution are useful:

$$\int_{-\infty}^{\infty} e^{-\frac{(x-\mu)^2}{2\sigma^2}}\,dx = \sqrt{2\pi\sigma^2}, \quad \text{(B.34)}$$

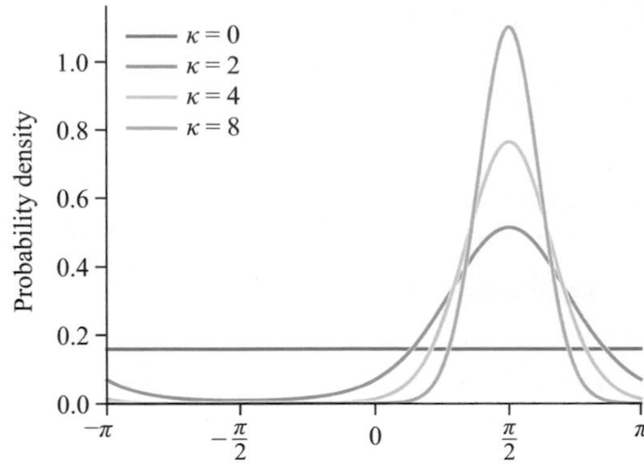

Figure B.4
Examples of Von Mises distributions. The domain of the random variable is periodic.

$$\int_a^b \mathcal{N}(x;\mu,\sigma^2)dx = \Phi_{\text{standard}}\left(\frac{b-\mu}{\sigma}\right) - \Phi_{\text{standard}}\left(\frac{a-\mu}{\sigma}\right). \tag{B.35}$$

B.7.6 The Von Mises Distribution

Some variables, such as orientation and motion direction naturally have a circular (periodic) domain. For such variables, a Gaussian distribution does not make sense. One solution is to choose a Von Mises distribution (see figure B.4). This can be regarded as the circular analog of a Gaussian distribution. It has two parameters: a circular mean, and a concentration parameter, which is similar to the reciprocal of the variance of a Gaussian. A Von Mises distribution over a circular variable s with domain $[0, 2\pi)$, circular mean μ, and concentration parameter κ is

$$p(s) = \frac{1}{2\pi I_0(\kappa)} e^{\kappa\cos(s-\mu)}. \tag{B.36}$$

Here, I_0 is the modified Bessel function of the first kind of order 0. This is a so-called *special function*, a function that is defined in terms of an integral or infinite series. Its precise definition is not important here; for us, all that matters is that $2\pi I_0(\kappa)$ normalizes the Von Mises distribution. When $\kappa = 0$, $I_0(\kappa) = 1$, and the Von Mises distribution becomes a uniform distribution on the circle. The higher κ, the more similar the Von Mises distribution becomes to a normal distribution. This is illustrated in figure B.4.

Exercise B.7 We said that μ is the circular mean, but how would you define the mean of a circular variable?

Exercise B.8 Show analytically that in the limit of large κ, the Von Mises distribution becomes the normal distribution. (Hint: use the Taylor expansion of the cosine.) Moreover, show that the precision of that normal distribution is κ.

B.8 The Delta Function

A special type of random variable that we will encounter quite often is one that takes only one possible value. There is a special notation for the probability distribution of such a random variable. If X is a continuous random variable that always takes the value $X = a$, then we write for its distribution,

$$p(x) = \delta(x - a). \tag{B.37}$$

Here, δ is the *Dirac delta function*. It returns 0 unless the argument equals 0, in which case it returns infinity. Of course, infinity is not a number and therefore the Dirac delta function is strictly speaking not an ordinary function. This is not of practical concern, since the only place in which we will use the delta function is inside an integral. There, the following property holds for any function $f(x)$:

$$\int \delta(x - a) f(x) dx = f(a), \tag{B.38}$$

where we assumed that the region of integration contains a. In fact, equation (B.38) is the defining property of the Dirac delta. The delta function has the effect of evaluating the function f inside the integral at a single point, a.

We find it convenient to use the same notation for discrete as for continuous variables, that is, write $\delta(x - a)$ instead of δ_{xa}. Then, the discrete analog of equation (B.38) is

$$\sum_x \delta(x - a) f(x) = f(a). \tag{B.39}$$

B.9 The Poisson Distribution

A discrete probability distribution that we use to describe neural activity (section 14.1) is the Poisson distribution. The possible values of a Poisson random variable are $0, 1, 2, 3, \ldots$ (there is no upper limit). The Poisson distribution has a free parameter, which we will call λ. The probability distribution of X is given by

$$p(x) = \frac{e^{-\lambda} \lambda^x}{x!}, \tag{B.40}$$

where $x! = 1 \cdot 2 \cdot 3 \cdots x$ is thc factorial operation. The Poisson distribution is discrete and only defined on the nonnegative integers. The factor $e^{-\lambda}$ acts as a normalization factor. Unlike x, λ does not have to be an integer. The mean and the variance of a Poisson-distributed variable are both equal to λ.

B.10 Drawing from a Probability Distribution

In probabilistic modeling, we often have to draw random numbers according to a specified probability distribution. These draws are also called samples. Drawing random numbers is by no means trivial, but fortunately, most software packages have built-in random number generators for the most common probability distributions. We can then use these functions to custom-write code for drawing from probability distributions that are not preprogrammed.

B.11 Distributions Involving Multiple Variables

Random variables can depend on each other in interesting ways. This is formalized in joint and conditional probability distributions, and in Bayes' rule, which we derive here formally. The concepts discussed in this section apply to both continuous and discrete variables. Thus, probability or p can refer to either probability mass or probability density. Since we consider multiple variables at the same time, we will generally use a subscript on p to denote which random variable(s) the probability distribution belongs to.

B.11.1 Joint Probability

The *joint probability distribution* of random variables X and Y is denoted $p_{X,Y}(x,y)$, or in shorthand, $p(x,y)$. It is the probability of the values x and y as a pair. Summing over both x and y gives 1:

$$\sum_x \sum_y p(x,y) = 1. \tag{B.41}$$

For continuous variables, the integral over both variables is 1:

$$\iint p(x,y)dxdy = 1. \tag{B.42}$$

Joint probability is symmetric:

$$p(x,y) = p(y,x). \tag{B.43}$$

If X and Y represent events, the joint probability of X and Y is the probability that both occur, denoted $p(X, Y)$ or $p(X \cap Y)$. In the Venn diagram representation (figure B.5), we represent Y by another circle, intersecting the first one. The joint probability of X and Y is equal to the area of the intersection. It is always less than or equal to the area of each individual circle. This expresses the relations $p(X, Y) \leq p(X)$ and $p(X, Y) \leq p(Y)$. For example, the probability that it rains on a given day and you will be at work on time is smaller than the probability that it rains. These relations only hold for discrete variables.

B.11.2 Marginalization

Imagine you have a cat and a dog. You carefully track what the probabilities are during the day that only your cat is present in the living room, only your dog, neither, or both. These probabilities are shown in table B.1; this table, called a *contingency table*, represents the

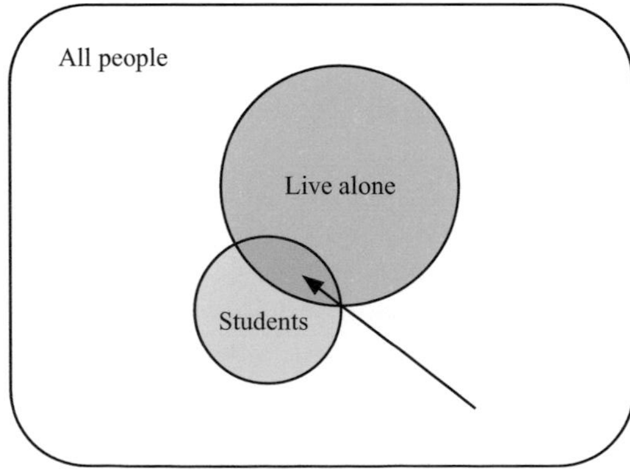

Figure B.5
The joint probability of the events "being a university student" and "living alone" is represented by the area of the intersection, indicated by the arrow.

Table B.1
Frequencies of combinations of two random variables.

	cat absent	cat present	total
dog absent	0.40	0.05	0.45
dog present	0.30	0.25	0.55
total	0.70	0.30	1

probabilities of joint outcomes. The *marginal probabilities* are the probabilities that the cat is present or absent regardless of the dog, and the probabilities that the dog is present or absent regardless of the cat.

Marginalization is the operation of obtaining from a joint distribution over multiple variables the distribution over a subset of those variables. For example, if $p(x, y)$ is the joint distribution of X and Y, then summing over Y produces the distribution of X alone:

$$\sum_y p(x, y) = p(x). \tag{B.44}$$

The continuous analogue is obtained by replacing the sum by an integral:

$$\int p(x, y) dy = p(x). \tag{B.45}$$

This summation or integration is called "marginalization" because $p(x)$ and $p(y)$ are called the marginals of $p(x, y)$. If you think of (x,y) as a point in two-dimensional space, and the joint distribution providing z values in this space, then the marginals are the distributions obtained by summing in either dimension (figure B.6). This results in two one-dimensional distributions that live in the "margins" of the original two-dimensional distribution.

B.11.3 Conditional Probability

For events X and Y, the conditional probability $p(X|Y)$ is the answer to the question, "Of all outcomes that are consistent with event Y, what fraction is also consistent with event X?" The conditional probability of an event always lies between 0 and 1. In the Venn diagram representation (figure B.9), $p(X|Y)$ is equal to the area of the intersection divided by the area of the second circle. Similarly, the probability that Y occurs given that X occurs is equal to the area of the intersection divided by the area of the first circle.

Three examples of conditional probability:

- If the probability that it rains today and you arrive at work on time is equal to 0.4, and the probability that it rains today is 0.5, then the probability that you arrive at work on time given that it rains is $\frac{0.4}{0.5} = 0.8$.
- The probability that I roll a 6 on a die given that I roll an even number is equal to $\frac{1}{6}/\frac{1}{2} = \frac{1}{3}$.
- In a given country, each state has a different proportion of taxi drivers. The probability that a person randomly selected from a particular state is a taxi driver, $p(x|y)$, is equal to the proportion of people in the country who live in that state *and* are taxi drivers, $p(x, y)$, divided by the proportion of people living in that state, $p(y)$.

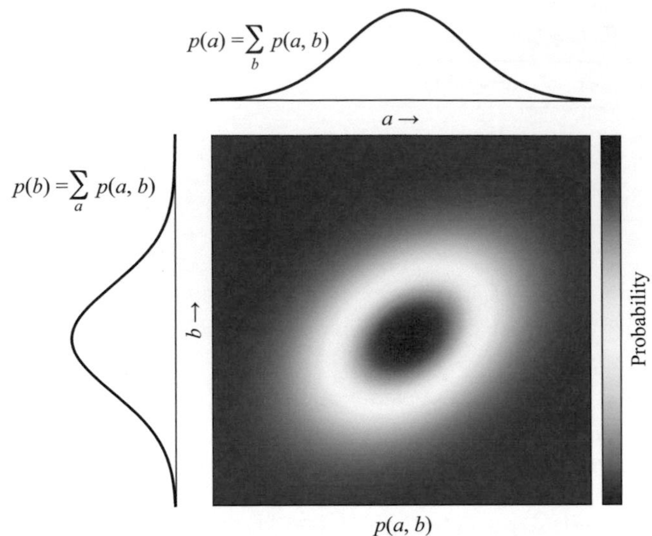

Figure B.6
The color plot represents the joint probability distribution of two random variables A and B. The black curves represent the marginals, obtained by summing the joint across one of the two variables.

- From the contingency table in table B.1, conditional probabilities can be computed. For example, the probability that the cat is present given that the dog is present is 0.25 (cat and dog both present) divided by $0.30+0.25=0.55$ (dog present).

We extensively discussed in chapter 2 that $p(X|Y)$ is not equal to $p(Y|X)$.

If X and Y are random variables, the probability distribution of X given Y is denoted by $p_{X|Y}(x|y)$, or by $p(x|y)$ if there is no ambiguity about the identities of the variables. The "|" sign is read as "given" or "conditioned on." It is defined as the probability of x and y as a pair, divided by the probability of y:

$$p(x|y)=\frac{p(x,y)}{p(y)}. \tag{B.46}$$

Exercise B.9

(a) Show formally that $p(x|y)$ is normalized as a function of x.

(b) Give an example that shows that $p(x|y)$ is not normalized as a function of y.

We will now combine the notion of marginalization with the definition of conditional probability.

Exercise B.10 Show that:

$$p(x)=\sum_y p(x|y)p(y). \tag{B.47}$$

The integral form of this equation is

$$p(x)=\int p(x|y)p(y)dy. \tag{B.48}$$

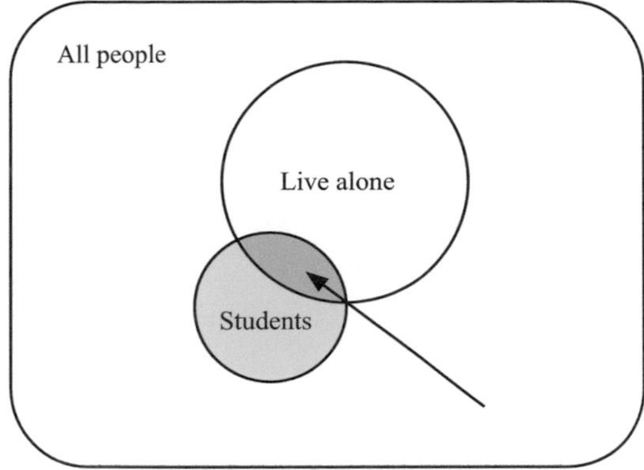

Figure B.7
The conditional probability of "living alone" given "being a university student" is represented by the area of the intersection divided by the area of the blue disc.

Equations (B.47) and (B.48) are rules that we use throughout the book. Continuing on the taxi driver example: suppose you are interested in the probability that a randomly selected citizen is a taxi driver. You know for each state the proportion of taxi drivers. I also know the proportion of all citizens living in each state. To obtain my answer, you multiply those two proportions for every state and then sum over all provinces.

We can condition every probability in equations (B.44) and (B.47) on a third random variable, z (this can be done with any rule in probability calculus). Then we obtain

$$p(x|z) = \sum_y p(x,y|z) \tag{B.49}$$

$$= \sum_y p(x|y,z)p(y|z), \tag{B.50}$$

or in its integral form,

$$p(x|z) = \int p(x|y,z)p(y|z)dy. \tag{B.51}$$

We will prove this in problem B.5.

Conditional distributions are not limited to a single random variable before and after the given sign. For example, one could consider the distribution of X and Y given Z and W, denoted by $p(x,y|z,w)$.

B.11.4 Independence

Two random variables X and Y are called independent if their joint distribution factorizes into the marginals, that is, if

$$p(x,y) = p(x)p(y) \tag{B.52}$$

for all x and y. For example, the probability that I roll a 6 on a die and toss heads on a coin is the product of both events taken separately. Independence can be depicted graphically as in figure B.8: one can reconstruct the joint distribution by multiplying the marginals. The

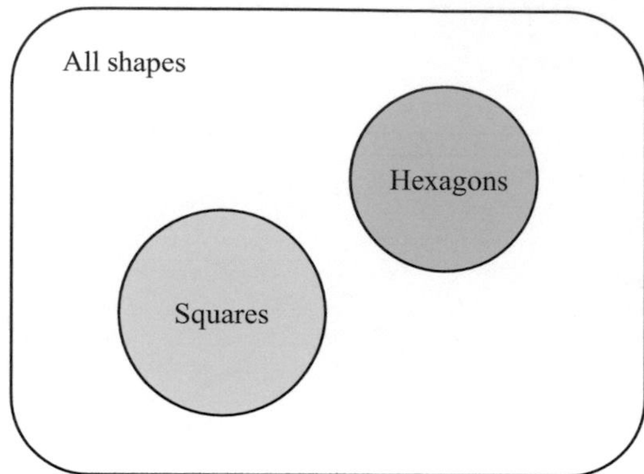

Figure B.8
Venn diagram depiction of the independence of two random variables.

notion of independence is closely related to that of correlation: two independent random variables are also uncorrelated. The opposite is not true.

Exercise B.11 Why not?

Exercise B.12 If X and Y are independent, what can one say about the conditional distributions $p(x|y)$ and $p(y|x)$?

If X, Y, and Z denote three random variables, then X and Y are *conditionally independent* given Z if

$$p(x,y|z) = p(x|z)p(y|z) \tag{B.53}$$

for any values x, y, and z. We use the notion of conditional independence for the first time in chapter 5. Be careful to not confuse conditional independence with independence!

B.11.5 Bayes' Rule

Bayes' rule relates the conditional probabilities $p(x|y)$ and $p(y|x)$ to each other:

$$p(x|y) = \frac{p(y|x)p(x)}{p(y)}. \tag{B.54}$$

Here, $p(y|x)$ as a function of x is the *likelihood function* over x, $p(x)$ is the *prior distribution* over x, and $p(x|y)$ is the *posterior distribution* over x.

Exercise B.13 Before reading on, try to prove Bayes' rule using the equations in the preceding sections.

Here is how the proof goes. From equation (B.46), we know that $p(x,y) = p(x|y)p(y)$. By renaming x and y, we also obtain $p(y,x) = p(y|x)p(x)$. Joint probability is symmetric, $p(x,y) = p(y,x)$. From these three equations, it follows that $p(x|y)p(y) = p(y|x)p(x)$. Dividing both sides by $p(y)$ gives Bayes' rule.

Exercise B.14 Prove that the right-hand side of equation (B.54) is normalized over x.

The Venn diagram interpretation of Bayes' rule for events X and Y is that the area of overlap can be calculated in two ways (figure B.9): as a fraction of the X circle area times

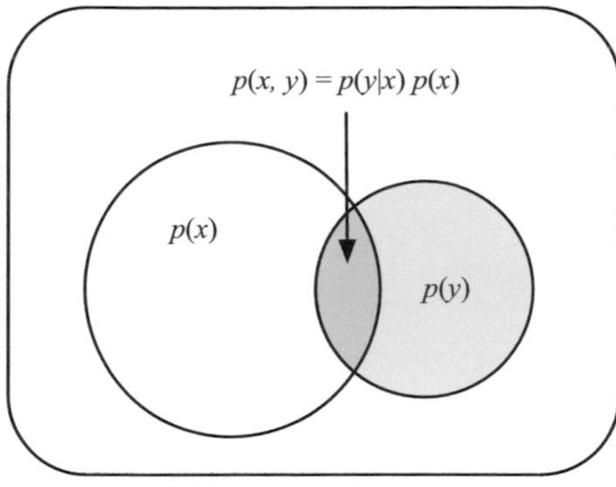

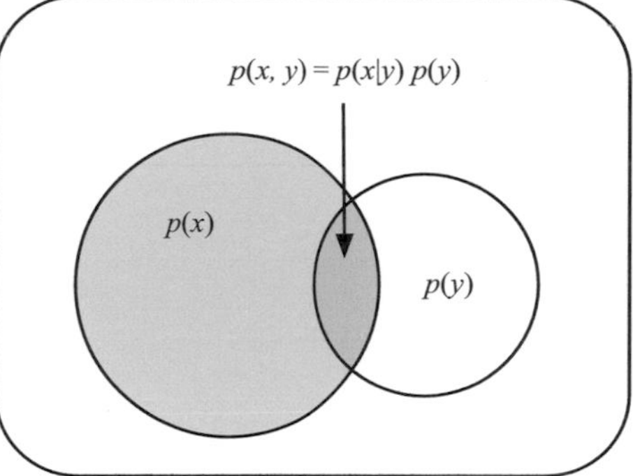

Figure B.9
Bayes' rule is obtained by writing the intersection area in two different ways and equating the two expressions.

the X circle area, or as a fraction of the Y circle area times the Y circle area. Since the outcomes should be identical, this means that the two fractions can be expressed in terms of each other if one knows the ratio of areas of the X and Y circles.

Suppose that 1 in 100,000 people is a professional basketball player, that 1 in 100 people is tall, and that 95 percent of basketball pros are tall. What is the probability that a tall person is a professional basketball player? We solve this problem using a direct application of Bayes' rule: If X is "being a basketball pro" and Y is "being tall," then $p(X) = 0.00001$, $p(Y) = 0.01$, and $p(Y|X) = 0.95$. It follows that $p(X|Y) = \frac{0.95 \cdot 0.00001}{0.01} = 0.0095$, or about 1 in 1000.

Exercise B.15 Prove a different form of Bayes' rule:

$$p(x|y) = \frac{p(y|x)p(x)}{\sum_x p(y|x)p(x)}. \tag{B.55}$$

A final way of writing Bayes' rule is with a proportionality sign (see box 3.6):

$$p(x|y) \propto p(y|x)p(x). \tag{B.56}$$

B.12 Functions of Random Variables

B.12.1 Functions of One Variable: Changing Variables

In this section, we discuss the frequently occurring problem of transforming the distribution of a continuous random variable. The question is as follows. If X is a random variable with probability distribution $p_X(x)$ and $Y = f(X)$ is a new random variable obtained by applying the function or transformation f to X, what is the distribution of Y? In this section, we will use subscripts, such as X in $p_X(x)$, to avoid confusion, since there are multiple random variables.

An example: X is a random variable following a uniform distribution on [0, 1]. $Y = X^2$ is a new random variable obtained by squaring outcomes of X. What is the distribution of Y? An easy but wrong answer would be that because X follows a uniform distribution, Y does as well. It can be understood intuitively that this answer is wrong. When a number x lies between 0 and 1, squaring it will always make it smaller. Thus, even though the values of Y will also lie between 0 and 1, lower values in this range will have greater probability density than higher values do. The question can be answered correctly by considering the cumulative distribution functions of X and Y, which we will denote $P_X(x)$ and $P_Y(y)$, respectively:

$$P_Y(y) = p(Y \le y) \tag{B.57}$$

$$= p(X^2 \le y) \tag{B.58}$$

$$= p(X \le \sqrt{y}) \tag{B.59}$$

$$= P_X(\sqrt{y}). \tag{B.60}$$

Now, we use the fact that the probability density function is the derivative of the CDF, equation (B.8), to find the PDF of y, denoted $p_Y(y)$:

$$p_Y(y) = \frac{dP_Y}{dy} \tag{B.61}$$

$$= \frac{d}{dy} P_X(\sqrt{y}) \tag{B.62}$$

$$= \left. \frac{dP_X}{dx} \right|_{x=\sqrt{y}} \frac{d}{dy}\sqrt{y} \quad \text{(chain rule)} \tag{B.63}$$

$$= p_X(\sqrt{y}) \frac{1}{2\sqrt{y}} \tag{B.64}$$

$$= 1 \cdot \frac{1}{2\sqrt{y}} \tag{B.65}$$

$$= \frac{1}{2\sqrt{y}}. \tag{B.66}$$

The resulting distribution, $p_Y(y)$, is normalized (verify this) and conforms to our intuition: the probability density is higher for lower values of y. We can verify the result through simulation: draw many random numbers from a uniform distribution between 0 and 1, square them, plot a finely binned, normalized histogram of the squares, and plot the function $\frac{1}{2\sqrt{y}}$ on top of it.

We could have stated the same problem with $p_X(x)$ being any distribution instead of a uniform distribution. The calculation is then identical except for the last step. Then, we find

$$p_Y(y) = p_X\left(\sqrt{y}\right) \frac{1}{2\sqrt{y}}. \tag{B.67}$$

Thus, the distribution of the squared variable is a product of the original distribution evaluated at the value of x that maps to y, $p_X(\sqrt{y})$, and an extra factor. The extra factor, called the *Jacobian*, is equal to the derivative of the mapping from y to x. It would be wrong to leave out the Jacobian and to assert that $p_Y(y) = p_X(\sqrt{y})$.

The Jacobian appears not just in this example of squaring a random variable, but in our original, general problem. Suppose X is a random variable with probability distribution $p_X(x)$, and $Y = f(X)$, where f is a monotonically increasing function. What is the distribution of Y? We first define the inverse function f^{-1} as the function of y that "undoes" the effect of f, in other words, $f^{-1}(f(x)) = x$. This inverse function is well defined because we assumed that f is monotonically increasing. Tempting but incorrect ways to obtain the distribution of Y would be to substitute the inverse function into $p(X)$, $p_Y(y) = p(X)(f^{-1}(y))$, or to assume that an operation applied to a distribution is the same as the distribution applied to the operation, $p_Y(y) = f(p_X(y))$. The correct approach is again to calculate the cumulative distribution of Y,

$$P_Y(y) = p(Y \leq y) \tag{B.68}$$

$$= p(f(X) \leq y) \tag{B.69}$$

$$= p(X \leq f^{-1}(y)) \tag{B.70}$$

$$= P_X(f^{-1}(y)) \tag{B.71}$$

and from that the probability density function of y,

$$p_Y(y) = \frac{dP_Y}{dy} \tag{B.72}$$

$$= \frac{d}{dy} P_X(f^{-1}(y)) \tag{B.73}$$

$$= \left.\frac{dP_X}{dx}\right|_{x=f^{-1}(y)} \frac{df^{-1}}{dy} \quad \text{(chain rule)} \tag{B.74}$$

$$= p_X(f^{-1}(y)) \frac{df^{-1}}{dy}. \tag{B.75}$$

So far, we considered a monotonically increasing function f. When f is instead monotonically decreasing, the final expression for $p_Y(y)$ acquires an extra minus sign.

Exercise B.16 Show this.

We can summarize both cases—monotonically increasing and decreasing—in a single equation:

$$p_Y(y) = p_X(f^{-1}(y)) \left| \frac{df^{-1}}{dy} \right|. \tag{B.76}$$

B.12.2 Apple Example

As a final illustration of the change-of-variables procedure, let us suppose that you are going to visit an apple orchard. You know very little about how fast apples grow or the duration of the growing season in the area, and you do not know the type of apples in the orchard. If a friend asked you what you thought the size of the apples in the orchard was going to be, you might initially respond that you have no idea. On more careful consideration, drawing upon your limited knowledge of apples in general, suppose you state that you have a uniform prior density over the diameter of apples in the orchard, from 3 to 13 cm. What, then, is your prior density over apple *volume*?

Before we derive the answer, let us appreciate the problem. Your uniform prior over apple diameter means that, for example, you consider it equally probable that an apple's diameter will lie between 5 and 6 cm as between 10 and 11 cm. If we approximate apples as spheres,

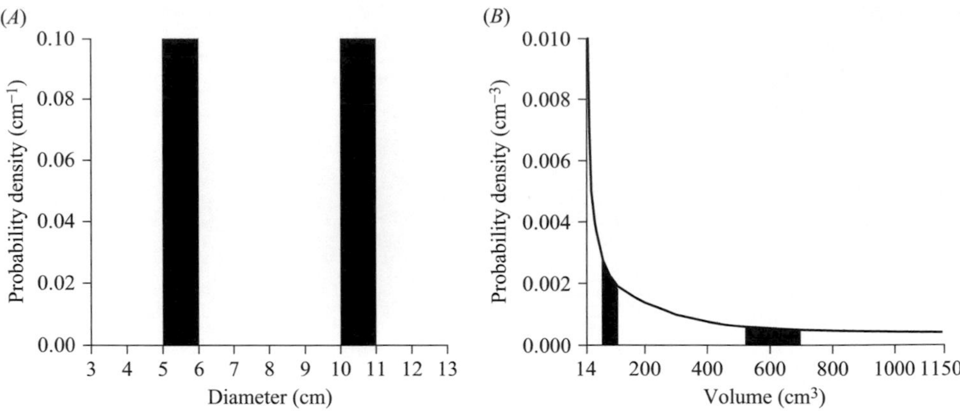

Figure B.10
Change of variables. (*A*) A uniform prior over apple diameter, from 3 to 13 cm (a range of 10 cm). The prior density is a line at height 0.1 cm^{-1} because the total area under the density must equal 1. The probability that the diameter lies between 5 and 6 cm is 0.1, as is the probability the diameter lies between 10 and 11 cm (filled rectangular areas). (*B*) The prior over apple volume. Each filled rectangular area is the probability that apple volume lies in the range corresponding to the apple diameters covered by the filled rectangles in A. Again, the total area under the density is 1, and the area of each filled rectangle is 0.1. (Note the differences in *y* axis scales).

then the volume of an apple is

$$v = \frac{4\pi r^3}{3} = \frac{\pi w^3}{6}, \tag{B.77}$$

where r is the radius and $w = 2r$ is the diameter of the apple. The volumes corresponding to diameters of 5, 6, 10, and 11 cm are therefore (to the nearest integer) 65, 113, 524, and 697 cm^3, respectively. This means that you believe it is equally probable (10 percent probable, to be exact) for the volume of an apple to lie between 65 and 113 cm^3—a range of 48 cm^3—as it is to lie between 524 and 697 cm^3—a range of 173 cm^3. Your prior density over apple volume, then, is clearly not flat. Rather, the density will be higher at smaller volumes (see figure B.10).

To derive the probability density over volume, we note that:

$$w = \sqrt[3]{\frac{6v}{\pi}}. \tag{B.78}$$

For the derivative, we find that:

$$\frac{dw}{dv} = \sqrt[3]{\frac{2}{9\pi v^2}}. \tag{B.79}$$

Therefore,

$$p_V(v) = p_W(w)\left|\frac{dw}{dv}\right| \tag{B.80}$$

$$= p_W(w)\sqrt[3]{\frac{2}{9\pi v^2}} \tag{B.81}$$

$$= \frac{1}{10\,\text{cm}}\sqrt[3]{\frac{2}{9\pi v^2}}. \tag{B.82}$$

This is the curve plotted in figure B.10B.

B.12.3 The Commitment of Ignorance

An interesting consequence of the change-of-variables procedures, illustrated by the apple orchard example, is that it is not possible to be ignorant about every feature of a problem. For instance, it is not possible to be fully ignorant about apple size, generally defined. As we have just seen, if we are ignorant about apple diameter, in the sense that we consider a wide range of diameters to be equally probable, then we are consequently not ignorant about apple volume! When doing Bayesian statistical analysis, a researcher may want to incorporate as little prior opinion as possible into an analysis about which they feel they have almost no relevant background knowledge. How can they best do this, if by specifying their ignorance about a parameter they are consequently specifying knowledge about related parameters? For instance, if a researcher has "no knowledge" of the standard deviation, σ, of a random variable, they may choose to use a flat prior over a very wide range of σ, but then they are implicitly specifying a nonuniform prior over the variance, σ^2. The search to develop appropriate default or reference priors for such situations is an interesting topic in the field of Bayesian statistical analysis. In Bayesian models of perception and action, this is usually not an issue, since the prior is assumed to be derived either from the experimental statistics or from natural statistics.

B.12.4 Marginalization Formulation

It is instructive to phrase the problem of transforming the distribution of a random variable as a formal problem of marginalization. This formulation is equivalent but gives more insight in some ways. We assume again that X is a random variable with probability distribution $p_X(x)$, and $Y=f(X)$, where f is a monotonically increasing function. As we discussed in section B.8, a deterministic mapping such as f can be expressed as a delta distribution. Here, this distribution would take the form

$$p_{Y|X}(y|x)=\delta(y-f(x)). \tag{B.83}$$

Now we can compute the probability density at y formally using the marginalization identity from equation (B.47):

$$p_Y(y)=\int_{-\infty}^{\infty} p_{Y|X}(y|x)p_X(x)dx \tag{B.84}$$

$$=\int_{-\infty}^{\infty} \delta(y-f(x))p_X(x)dx. \tag{B.85}$$

In words, the probability density of y is the total probability of all values of x that get mapped to y by f. We can evaluate this expression by making a transformation of variables: $x=f^{-1}(t)$, so that $dx=\frac{df^{-1}}{dt}dt$. Substituting, we find

$$p_Y(y)=\int_{-\infty}^{\infty} \delta(y-t)p_X(f^{-1}(t))\frac{df^{-1}}{dt}dt. \tag{B.86}$$

We can now use equation (B.38) to evaluate the integral:

$$p_Y(y)=p_X(f^{-1}(y))\frac{df^{-1}}{dy}dy, \tag{B.87}$$

which is the same as equation (B.75). Again, when f is monotonically decreasing instead of increasing, we obtain the same result but with a minus sign.

Exercise B.17 Where does the minus sign come from in this formulation?

The advantage of this integral formulation is that the first equality in equation (B.84) is general and not limited to deterministic mappings from X to Y. Thus, the problem of transforming a random variable is simply a special case of a probabilistic mapping from Y to X, and the first equality in equation (B.84) can be applied for *any* conditional distribution $p(y|x)$.

A second advantage is that the expected value of any function $g(Y)$ of a random variable $Y = f(X)$ is now easy to transform:

$$\mathbb{E}[g(Y)] = \int g(y) p_Y(y) dy \tag{B.88}$$

$$= \int g(y) \left(\int_{-\infty}^{\infty} \delta(y - f(x)) p_X(x) dx \right) dy \tag{B.89}$$

$$= \int \left(\int_{-\infty}^{\infty} g(y) \delta(y - f(x)) dy \right) p_X(x) dx \quad \text{(swapped order of integration)} \tag{B.90}$$

$$= \int g(f(x)) p_X(x) dx \tag{B.91}$$

$$= \mathbb{E}[g(f(X))]. \tag{B.92}$$

In other words, the combination $p_Y(y)dy$ inside an integral is identical to $p_X(x)dx$, as long as $y = f(x)$ is substituted elsewhere in the integral. (The integration limits might also change accordingly.)

Exercise B.18 Use this result to show that the mean of $aX + b$ is $a\mathbb{E}[X] + b$, and that its variance is $a^2 \mathrm{Var}[X]$.

A third advantage of the marginalization formulation is that it directly generalizes to functions of multiple variables, as we now examine.

B.12.5 Functions of Multiple Variables

Suppose you roll two fair dice and add the outcomes. What is the probability distribution of the sum? Simple counting gives the answer: outcome 2 can be reached in only one way $(1 + 1)$ and therefore has probability $\frac{1}{36}$. Outcome 3 can be reached in two ways $(1 + 2$ and $2 + 1)$ and therefore has probability of $\frac{2}{36}$, and so on. This results in the probability distribution shown in figure B.11. How do we calculate this distribution formally?

We call the random variables corresponding to both die rolls X and Y. Their sum is a new random variable, $Z = X + Y$. In other words,

$$p_{Z|X,Y}(z|x, y) = \delta(z - x - y). \tag{B.93}$$

To calculate the distribution of Z, denoted $p_Z(z)$, we apply the discrete analog of equation (B.84):

$$p_Z(z) = \sum_{x=1}^{6} \sum_{y=1}^{6} p_{Z|X,Y}(z|x, y) p_{X,Y}(x, y) \tag{B.94}$$

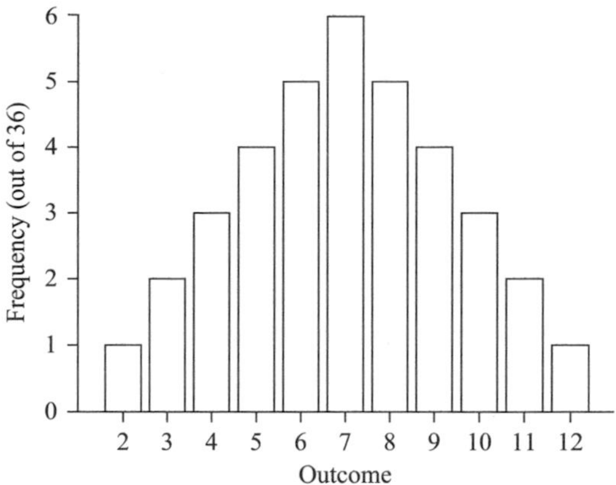

Figure B.11
The probability distribution of the sum of two die rolls.

$$\sum_{x=1}^{6}\sum_{y=1}^{6} p_{Z|X,Y}(z|x,y)p_X(x)p_Y(y) \tag{B.95}$$

$$\sum_{x=1}^{6} p_X(x)\sum_{y=1}^{6} \delta(z-x-y)p_Y(y). \tag{B.96}$$

We now use the property of the delta function, equation (B.39), as well as the condition that for $p_Y(y)$ to be nonzero, we must have $1 \leq y \leq 6$, therefore $1 \leq z-x \leq 6$, and therefore $z-6 \leq x \leq z-1$. Then,

$$p_Z(z) \sum_{x=\max(1,z-6)}^{\min(z-1,6)} p_X(x)p_Y(z-x) \tag{B.97}$$

$$\sum_{x=\max(1,z-6)}^{\min(z-1,6)} \frac{1}{6}\cdot\frac{1}{6} \tag{B.98}$$

$$\frac{1}{36}(\min(z-1,6)-\max(1,z-6)+1). \tag{B.99}$$

The same logic can be applied to a continuous distribution. Let X and Y be independent continuous variables with respective PDFs $p_X(x)$ and $p_Y(y)$. We define a new variable $Z = f(X, Y)$, with f any function, and denote by f_x^{-1} the inverse function of f for given x: $Y = f_x^{-1}(Z)$. (Such an inverse function does not always exist, but in the examples in this book, it does.) Then the distribution of Z is

$$p_Z(z) = \iint p_{Z|X,Y}(z|x,y)p_X(x)p_Y(y)dxdy \tag{B.100}$$

$$= \iint \delta(z-f(x,y))p_X(x)p_Y(y)dxdy \tag{B.101}$$

$$= \int dx p_X(x) \left(\delta(z-t) p_Y(f_x^{-1}(t)) \left| \frac{df_x^{-1}}{dt} \right| dt \right) \tag{B.102}$$

$$= \int dx p_X(x) p_Y(f_x^{-1}(z)) \left| \frac{df_x^{-1}}{dz} \right|, \tag{B.103}$$

where from the second to the third line we have made the transformation of variables $y = f_x^{-1}(t)$. One can think of the delta function as selecting a region of N-dimensional space—namely all points that map onto y—and of the integral as the total probability under $p(X)$ in that region.

Exercise B.19 If X and Y are independent and have a uniform distribution on $[0, 1]$, compute the distribution of $Z = X + Y$. The answer is a special case of the *Irwin-Hall distribution*.

Exercise B.20 If X and Y are independent and have normal distributions, prove that $Z = X + Y$ also has a normal distribution.

So far, we have computed the distribution of a sum random variable. We can also use equation (B.103) to compute the distribution of nonlinear combinations of random variables, such as a product or quotient. The distribution of the product (or quotient) of two variables is not equal to the product (or quotient) of their distributions, and is often very different. We will examine this in problem B.14.

B.13 Problems

Problem B.1 In a flower garden, 20 percent of flowers are tulips. Of those, one-quarter are red. What is the probability that a random flower in this garden is a tulip but not red? Although it is easily possible to solve this problem using intuition, we would like you to formally apply the rules of probability.

Problem B.2 Four players, sitting around a table, are about to play a game. To determine who starts, one person throws two dice. The sum of the two numbers determines who starts, with the counting going clockwise starting with the roller being 1 (so with a sum of 5 or 9, the roller starts). What is the probability of starting for each player?

Problem B.3 You and I each roll a die once.

(a) What is the probability that one of us rolls a 6 and the other an odd number?
(b) What is the probability that at least one of us rolls a 6?
(c) What is the probability that you roll higher than me?
(d) What is the probability that our total is higher than 8?

Problem B.4 You are a student in a class of thirty.

(a) What is the probability that a particular classmate shares your birthday?
(b) What is the probability that any classmate shares your birthday?
(c) What is the probability that any two students share the same birthday?

Problem B.5 Let x, y, and z be discrete random variables. Prove the following.

(a) Conditional marginal:

$$p(x|z) = \sum_y p(x|y,z)p(y|z). \tag{B.104}$$

(b) Conditional Bayes' rule:

$$p(x|y,z) = \frac{p(y|x,z)p(x|z)}{p(y|z)}. \tag{B.105}$$

Problem B.6 You and I alternately toss a coin. You start. If you toss heads, you win instantly. If you toss tails, it is my turn. If I toss tails, I win instantly. If I toss heads, it is your turn again. This repeats until one of us has won. What is your probability of winning this game?

Problem B.7 Email programs automatically classify emails as spam or not spam based on the words in the email. To do this, they use a Bayesian algorithm that works very much like the Bayesian inference that we discussed for medical diagnosis. Assume that 70 percent of all email is spam. Suppose that if an email is not spam, it has a 0.1 percent probability of including the word "bargain." Suppose further that if an email is spam, it has a 1 percent probability of including the word "bargain."

(a) Draw the generative model diagram.

(b) Fill in the associated probabilities:

$$p(\text{spam}) = \ldots \tag{B.106}$$

$$p(\text{no spam}) = \ldots \tag{B.107}$$

$$p(\text{contains "bargain"}|\text{spam}) = \ldots \tag{B.108}$$

$$p(\text{does not contain "bargain"}|\text{spam}) = \ldots \tag{B.109}$$

$$p(\text{contains "bargain"}|\text{not spam}) = \ldots \tag{B.110}$$

$$p(\text{does not contain "bargain"}|\text{not spam}) = \ldots \tag{B.111}$$

(c) What is the prior probability that a random email is spam? What is the prior probability that a random email is not spam?

(d) Suppose that a particular email contains the word "bargain." What is the likelihood that it is spam? What is the likelihood that it is not spam?

(e) Multiply the prior over "spam" by the likelihood of "spam."

(f) Multiply the prior over "not spam" by the likelihood of "not spam."

(g) Now you have what we called the "protoposterior." Divide each of the answers to (e) and (f) by the sum of both answers. What is the posterior probability that this particular email is spam?

Problem B.8 This problem is about the famous Monty Hall problem [155]. You are on a game show. The host shows you three doors. Behind one of them, a prize is hidden. You choose one door. The host, who knows behind which door the prize lies, opens a remaining door that does not contain the prize. The host then gives you the opportunity to switch your choice to the remaining unopened door or to stay with your original choice. Your door of choice gets opened and you receive the prize if it is there.

(a) To maximize the probability of receiving the prize, what should you do?

(b) If there are N doors and the host opens m of them (where $m < n - 1$, what is the probability of receiving the prize under the best strategy?

(c) Would the answer to (a) change if the host did not know which of the two remaining doors contained a prize, but the one he opens just happens not to contain the prize? Explain.

(d) Speculate on why most people believe it does not matter whether you stay or switch.

Problem B.9 The probability density function of a Gaussian random variable X is given by equation (B.18). Show that the mean and variance of this random variable are equal to μ and σ^2, respectively.

Problem B.10 If X is an exponentially distributed random variable where X has the positive real line as domain, what are the domain and distribution of $Y = e^X$?

Problem B.11

(a) Someone tosses a fair coin three times. You observe the outcome of one toss, which is heads. What is the probability that heads are more common than tails among all three tosses?

(b) Someone tosses a fair coin N times. You observe the outcome of one toss, which is heads. What is the probability that heads are more common than tails among all N tosses? (For even N, interpret “more” as “strictly more.”)

Problem B.12 Use equation (B.76) to prove that if X is normally distributed with mean μ and variance σ^2, that $aX + b$ is normally distributed with mean $a\mu$ and variance $a^2\sigma^2$.

Problem B.13 If X and Y are independent standard normal variables, show that the quotient random variable $Z = \frac{Y}{X}$ has a Cauchy distribution, that is,

$$p_Z(z) = \frac{1}{\pi(1+z^2)}. \tag{B.112}$$

Problem B.14 If X and Y are independent and have a uniform distribution on $[0, 1]$ $(0 <= x <= 1)$, show that the product random variable $Z = XY$ has distribution $p_Z(z) = -\log z$. Verify that this distribution is normalized even though the density at 0 is infinity. This example illustrates how the distribution of a product can be wildly different from the distributions of each of the factors.

Problem B.15 Consider two Von Mises distributions for the same random variable, one with mean μ_1 and concentration parameter κ_1, the other with mean μ_2 and concentration parameter κ_2. Show that the normalized product of these distributions is again a Von Mises distribution, and compute its mean and concentration parameter.

Appendix C: Model Fitting and Model Comparison

In this appendix, we describe how to fit a model to data. This is shorthand for "fitting the parameters of a model to data," in other words, adjusting the model's unknown parameters (such as σ) so that the data are accounted for as well as possible. The most popular specification of parameter fitting is *maximum-likelihood estimation* (MLE), which is the idea that we should prefer parameters that make the observed data most probable. The methods in this appendix do not apply exclusively to the Bayesian formalism. *Any* mathematical model of observer behavior can be fitted to data.

C.1 What Is a Model?

Here, we mean this question in the most practical sense possible. For a Bayesian model, step 3 outputs a *response distribution*. This is a probabilistic mapping from stimuli to responses: $p(\text{subject response} \mid \text{stimuli})$. Such a distribution characterizes not only a Bayesian model but any model of behavior. Indeed, we can *define* a model of behavior as a response distribution $p(\text{subject response} \mid \text{stimuli})$.

C.2 Free Parameters

Models have *free parameters* (or simply parameters): variables of unknown value that are assumed to be constant throughout an experiment. In Bayesian models, free parameters are typically used for properties or beliefs that may differ between subjects. Examples include the sensory noise level σ with which a particular stimulus type is measured, the assumed prior mean $\mu_{s,\text{assumed}}$ in section 3.5.1, and the inverse temperature parameter β in equation (13.34). We will denote the set of free parameters of a model collectively by θ and the model itself by M. We will make the dependence of the response distribution on the parameters and the model explicit by writing $p(\text{subject response} \mid \text{stimuli}; \theta, M)$. The semicolon serves to separate the variables that vary from trial to trial (response and stimuli) from the model identity and the model parameters, which do *not* vary from trial to trial. When it is unambiguous which model we are talking about, we will leave out M.

C.3 The Parameter Likelihood

Free parameters have to be adjusted to best describe the data; this is called *parameter fitting*, or *model fitting*. The dominant approach to parameter fitting is *maximum-likelihood estimation*. The *likelihood* of a parameter combination θ under a model M is defined as the probability of the observed subject data given the stimuli experienced by the subject, the parameter combination θ, and the model:

$$\mathcal{L}_M(\theta; \text{data}) = p(\text{subject responses across all trials} \mid \text{stimuli across all trials}; \theta, M) \quad \text{(C.1)}$$

In other words, the likelihood of a parameter combination is high when the model with that parameter combination, applied to the stimuli experienced by the subject, would relatively often produce the subject responses.

To enable fitting, behavioral scientists almost always make a conditional independence assumption: on a given trial, the probability of a particular subject response depends only on the stimuli in that trial, the parameter combination, and the model. It does not depend on the subject responses in previous trials or on the stimuli in previous trials. As sequential dependencies between trials are well documented in psychophysics, this assumption is often violated. However, relaxing the assumption of conditional independence requires a model for sequential dependencies, which is beyond the scope of this book. The conditional independence assumption can be formulated as

$$\mathcal{L}_M(\theta; \text{data}) = \prod_{i=1}^{n_{\text{trials}}} p(\text{subject response on trial } i \mid \text{stimuli on trial } i; \theta, M), \quad \text{(C.2)}$$

where n_{trials} is the number of trials. The conditional probability in equation (C.2) is, for general responses and stimuli, exactly the specification of the predictions of a model; for example, in a Bayesian model, it would be the response distribution produced by step 3. In equation (C.2), however, we substitute the *actual* subject response and the *actual* stimuli on the ith trial.

C.4 Maximum-Likelihood Estimation

Maximum-likelihood estimation of the parameters θ means finding the values of θ such that $\mathcal{L}_M(\theta; \text{data})$ is highest. This is equivalent to maximizing $\log \mathcal{L}_M(\theta; \text{data})$, because the logarithm is a monotonically increasing function. It is often more convenient to maximize the log likelihood than the likelihood itself. The log likelihood is

$$\log \mathcal{L}_M(\theta; \text{data}) = \sum_{i=1}^{n_{\text{trials}}} \log p(\text{subject response on trial } i \mid \text{stimuli on trial } i; \theta, M). \quad \text{(C.3)}$$

It bears emphasizing that the parameter likelihood maximized in model fitting is conceptually similar to but different from the likelihood functions in the Bayesian observer model that this book is really about. It is over the parameters of the model (which are unknown to the experimenter), not over a relevant state of the world (which is typically known to the experimenter but unknown to the subject). This distinction is illustrated in figure C.1. Throughout this appendix and in Bayesian modeling practice in general, it is important to

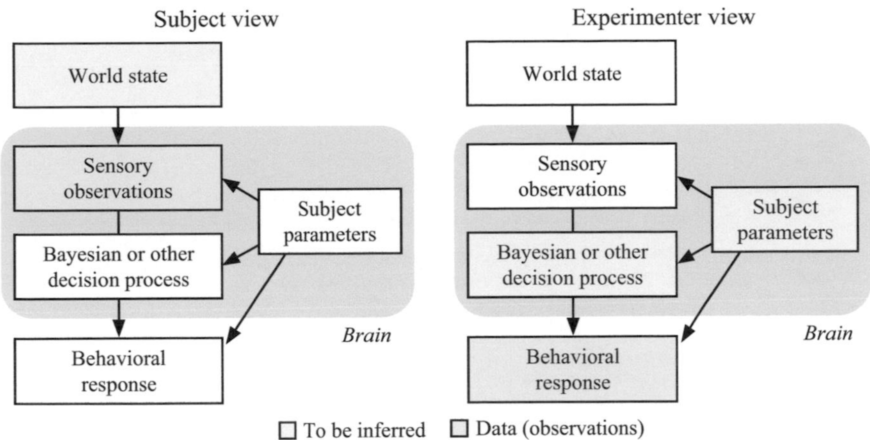

Figure C.1
Subject view versus experimenter view. A subject infers a world state from sensory observations, while an experimenter infers subject parameters and decision strategy from subject responses.

distinguish the observer's decision process (step 2) and the experimenter's model of the observer (step 4).

We note that the use of maximum-likelihood estimation for parameter fitting is equivalent to Bayesian estimation in which the investigator begins with a uniform prior distribution over model parameters. Sometimes, nonuniform priors over parameters are used in model fitting. To use such priors we need to have a meaningful justification based in our understanding of the aspects of the world that they represent. The use of priors is often unnecessary for fitting models, as long as the data set is sufficiently large (many trials) and there are sufficiently few parameters. We will thus not discuss priors over parameters here.

If we wanted to be fully Bayesian when model fitting, we would calculate full probability distributions. There is an extensive literature about fully Bayesian approaches to model fitting. Potential techniques involve Markov chain Monte Carlo and variational approaches. When these details of the model fitting process are important, we are generally in the domain where posterior uncertainty is large. We do not cover such approaches in this book; we refer the interested reader to [59, 103, 164].

In the following sections, we will revisit several models from previous chapters and perform maximum-likelihood estimation of the model's parameters based on hypothetical data sets. In these examples, as in most of this book, the subject's response is equal to their estimate of the world state of interest.

C.5 Fitting Data from an Estimation Task

The first step in fitting a model is to specify the nature of your data. Suppose you perform an estimation experiment like the one in chapter 3, where you draw a stimulus from a distribution $p(s)$ and the observer estimates the stimulus. Over fifteen trials, you collect the following data (table C.1):

These data are shown in figure C.2.

Table C.1
Example data from a continuous estimation experiment. s: stimulus; $\hat{s}$: estimate.

Trial	1	2	3	4	5	6	7	8	9	10	11	12	13	14	15
s	1.61	5.50	−6.78	2.59	0.96	−3.92	−1.30	1.03	10.74	8.31	−4.05	9.10	2.18	−0.19	2.14
$\hat{s}$	0.37	1.62	−1.17	1.66	1.17	−0.79	−1.14	0.76	4.31	2.86	−0.61	3.25	0.48	0.12	0.18

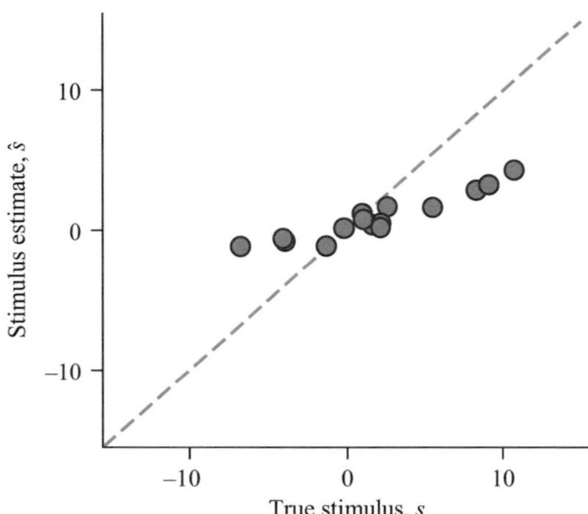

Figure C.2
Visualization of the data in table C.1.

C.5.1 Simple Model

Even based on this small data set, it is already possible to fit a model. Let us first consider a model in which the observer does not use a prior and simply reports the measurement. Then, $\hat{s}=x$ and the distribution of $\hat{s}$ given s is

$$p(\hat{s}|s,\sigma)=\frac{1}{\sqrt{2\pi\sigma^2}}e^{-\frac{(\hat{s}-s)^2}{2\sigma^2}}. \tag{C.4}$$

This fully specifies the model. We now want to estimate the parameter σ. The log likelihood of σ is, from equation (C.3),

$$\log\mathscr{L}(\sigma;\text{data})=\sum_{i=1}^{n_{\text{trials}}}\log p(\hat{s}_i|s_i,\sigma) \tag{C.5}$$

$$=\sum_{i=1}^{n_{\text{trials}}}\left[-\frac{1}{2}\log(2\pi\sigma^2)-\frac{(\hat{s}_i-s_i)^2}{2\sigma^2}\right]. \tag{C.6}$$

Our goal is to find the maximum likelihood of σ, which we will denote by $\hat{\sigma}$; this is the value of σ for which the log likelihood (and hence the likelihood) is highest.

Method 1: Analytical calculation

In this case, we can maximize the log likelihood analytically, by setting the derivative of $\log\mathscr{L}$ with respect to σ to 0:

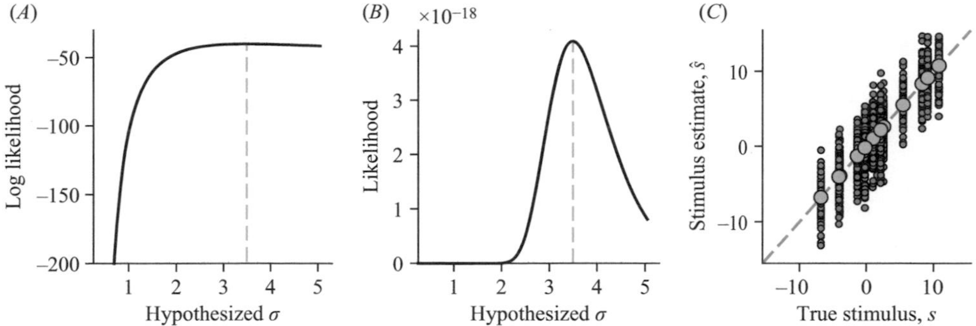

Figure C.3
Fitting the simple model in equation (C.4) to the data in table C.1. (*A*) Log likelihood and (*B*) likelihood of σ. The dashed lines mark the MLE of σ, which is $\hat{\sigma} = 3.49$. The value of the log likelihood at the maximum is $\log \mathscr{L}^* = -40.0$. (*C*) Model checking: prediction of the simple model after fitting σ. Black dots: individual simulations. Green circles: average across simulations for the same stimulus. Comparing with figure C.2, we see that the model does not fit the data well.

$$0 = \frac{d}{d\sigma} \log \mathscr{L}(\sigma; \text{data}) \tag{C.7}$$

$$= -\frac{n_{\text{trials}}}{\sigma} + \frac{1}{\sigma^3} \sum_{i=1}^{n_{\text{trials}}} (\hat{s}_i - s_i)^2. \tag{C.8}$$

Solving for σ, we find the maximum-likelihood estimate (MLE) of σ:

$$\hat{\sigma} = \sqrt{\frac{1}{n_{\text{trials}}} \sum_{i=1}^{n_{\text{trials}}} (\hat{s}_i - s_i)^2}. \tag{C.9}$$

Since $\hat{s}_i - s_i$ is the observer's estimation error on the ith trial, the right-hand side is the *root-mean-square error* (RMSE). This answer is not surprising: under a model in which estimates are normally distributed around the true value, as in equation (C.4), the standard deviation of that normal distribution is estimated as the empirical standard deviation. In the data set above, the answer is $\hat{\sigma} = 3.49$.

There are no disadvantages to deriving the minimum analytically, but the option is rarely available: setting the derivative to 0 will yield a solvable equation in only very few simple cases. In all other cases, the log likelihood in equation (C.3) has to be maximized numerically. We will now explore methods for numerical maximum-likelihood estimation.

Method 2: Grid search

The simplest numerical method is to define a grid of possible values of σ, for example from 0.5 to 10 in steps of 0.01, and simply evaluate the log likelihood on this grid. This has the advantage that we can plot the log likelihood function (or the likelihood function), as we did in figure C.3A–B.

Exercise C.1 Choosing a suitable grid of hypothesized values is a matter of common sense and experience, and, of course, you can always modify it.

(a) What happens if we include values lower than 0.5 in the grid?

(b) Why would it be pointless to extend the range beyond 10?

In figure C.3, we first observe that the likelihood function is clearly not Gaussian, or equivalently, the log likelihood function is clearly not a parabola. In the observer models of previous chapters, the likelihood of a continuous variable was usually Gaussian. However, it is not surprising that this cannot be the case here: since σ can take only positive values, a Gaussian likelihood (which stretches to negative infinity) would be meaningless.

Second, we note that the log likelihood takes values in the negative hundreds. In behavioral experiments, it is common for log likelihoods to be in the negative hundreds or thousands. This happens because probability mass functions always, and probability density functions often, take values smaller than 1. That means that each individual probability is smaller than 1, and its log is negative. The sum in equation (C.3) will on average be proportional to the number of trials.

We can now simply find the value on the grid for which the log likelihood is maximal. This returns an estimate of 3.49, consistent with our analytical calculation.

It is useful to reflect on the advantages and disadvantages of a grid search. The advantages are: (1) transparency: in a grid search, you know exactly what you are doing; (2) visualization: plotting the likelihood is useful so that you are more confident that it does not have a strange shape, and so that you know its width. The disadvantages are: (1) one is limited to the values on the grid and cannot find values in-between; and (2) the method becomes unwieldy with increasing numbers of parameters. For example, 1,000 values per parameter for four parameters would be 10^{12} parameter combinations, which would take a long time to compute. These two problems interact: if you use a finer grid to counter the first problem, you make the second problem worse.

Method 3: Numerical optimization algorithm

Because of the aforementioned combinatorial problem, it is often necessary to look beyond grid search. All ecosystems for scientific computation, including Python, Matlab, R, and Mathematica, have algorithms for minimizing arbitrary functions. Optimization algorithms are usually *iterative*: starting from an initial point, a particular routine is executed repeatedly until a termination criterion is satisfied. Many optimization algorithms are based on some form of gradient descent where, from the current point, the next point is chosen nearby, in the direction of the steepest slope (see section 6.4). Others are *global optimization algorithms*, which do not always choose the next point to be nearby and sometimes try to estimate the large-scale structure of the function to be optimized. Many optimizers that work well in that space are themselves Bayesian. We recommend trying a range of algorithms to be sure—such optimization problems are hard and the specific algorithm used can make a difference (see box C.1). You can feel more confident that you found the true maximum if different methods return the same result, and you will in the long run develop a large toolkit.

Optimization algorithms do not always find the global optimum. This could be because there is a “ridge” of parameter combinations at which the log likelihood is close to the maximum log likelihood, or because the log likelihood landscape is irregular with many local maxima. A common way to alleviate this problem is “multistart,” that is, to initialize the optimizer with many different parameter combinations (chosen either randomly or systematically but sufficiently far apart), pick the result of the best run, and use agreement across runs as a diagnostic for convergence.

Box C.1
Optimization Algorithms

In fitting behavioral data, many parameter optimization algorithms are useful. Here is a practical overview.

- If the function is smooth, deterministic, and generally well behaved, using gradient-based methods is typically preferable. An example of such an approach is the popular Broyden-Fletcher-Goldfarb-Shanno (BFGS) algorithm. There are standard implementations of BFGS (or variants thereof, such as limited-memory BFGS) in most programming languages. With these methods, we can provide an analytically calculated gradient of the target or an automatically calculated one (autograd) to the optimizer, making the code run much faster. If the gradient is not provided explicitly, the method will approximate the gradient numerically using finite differences.
- If the function is complex and rough or stochastic (for example because we use sampling to estimate a posterior), using covariance matrix adaptation evolution strategy (CMA-ES) is often a good idea. CMA-ES is freely available in most popular programming languages. CMA-ES often shines when it can run for a large number of function evaluations (tens or hundred of thousands), which might not be feasible or might take quite a long time if the target function is expensive to evaluate.
- If the function is somewhat expensive to evaluate and possibly rough or stochastic, then Bayesian optimization is often useful. These techniques try to estimate, in a Bayesian way, the distribution of target function compatible with the existing measurements. These methods are under rapid development at this time but often offer efficient estimation in cases where target functions are complicated. We want to mention one specific algorithm, Bayesian Adaptive Direct Search (BADS, [1]), which has been specifically designed for use in scenarios like the ones we cover in this book.
- Finally, if the function is high-dimensional and stochastic, but you have the gradient, then you can use stochastic gradient optimization methods such as ADAM, commonly used in machine learning. However, if your function is low-dimensional, the methods listed above might work better.

Model checking

So, now you have found parameter estimates using one of the three methods. In modeling behavioral data, a common mistake is to take estimates of parameters seriously without verifying at least superficially that the model actually fits the data well. Perhaps the root of this problem is a confusion of "best" with "good": even though maximum-likelihood estimation provides the *highest* likelihood within the context of the model that is being fitted, that highest likelihood might still be *low*. This type of mistake has consequences. Parameter estimates of a poorly fitting model are meaningless: it is like estimating the size of the earth while assuming that the earth is flat.

The easiest (and in many cases sufficient) way to check whether a model fits well, is to plot the data along with the corresponding representation of the fitted model. One way to do that in our current example is to draw stimulus estimates from the normal distribution in equation (C.4), using $\hat{\sigma} = 3.49$ in place of σ. In figure C.3C, we did that for 500 repetitions of each presented stimulus. Clearly, the trend in the model is very different from the data, and we have to look for a new model.

In short, the act of fitting a model does not make the model good! It is necessary to always verify that a model fits well before attaching any importance to its parameter estimates. If a model does not fit well, there should be ways of finding a better model.

C.5.2 A Better Model

For the data in figure C.2, we can easily come up with a better model: it looks like the observer's estimates are affected by a prior centered at 0. Therefore, we also fit the full Bayesian model from chapter 3: the observer uses a Gaussian prior with mean 0 and standard deviation σ_s. Then, the distribution of $\hat{s}$ given s is, from equation (4.5),

$$p(\hat{s}|s;\sigma,\sigma_s) = \mathcal{N}\left(\hat{s}; \frac{\frac{s}{\sigma^2}+\frac{\mu}{\sigma_s^2}}{\frac{1}{\sigma^2}+\frac{1}{\sigma_s^2}}, \frac{\frac{1}{\sigma^2}}{\left(\frac{1}{\sigma^2}+\frac{1}{\sigma_s^2}\right)^2}\right). \tag{C.10}$$

We will not assume that we know σ_s. Then, this model has two parameters, σ and σ_s. The log likelihood of both parameters is:

$$\log \mathcal{L}(\sigma,\sigma_s;\text{data}) = \sum_{i=1}^{n_{\text{trials}}} \log p(\hat{s}_i|s_i;\sigma,\sigma_s). \tag{C.11}$$

An analytical approach is not feasible in this case. Therefore, the MLEs of σ and σ_s have to be calculated numerically. Here, we do this using method 2, grid search. We define a grid for σ and one for σ_s. We choose both vectors to range from 0.02 to 10 in 100 steps. Then we can loop over both vectors (double for-loop), and for each combination of σ and σ_s on the grid evaluate the log likelihood from equation (C.11). After completing this process, we plot the log likelihood as a function of σ and σ_s; since there are two independent variables, a heat map suggests itself (figure C.4A). We can find the maximum numerically: $\hat{\sigma} = 1.76$ and $\hat{\sigma_s} = 1.24$; these values are close to the ones we used to generate the data, which were 1.5 and 1, respectively. (Because of noise in the data, we never expect to find the exact values that generated the data.) The value of the maximum is $\log \mathcal{L}^* = -13.2$, whereas in the simple model, it was much worse, −40.0.

To check the model, we again plot its prediction for the scatterplot of the stimulus estimate against the true stimulus (figure C.4B). The prediction is now much better than in figure C.2B. We now witness the consequences of blindly believing parameter estimates in poorly fitting models: if we had believed the parameter estimate $\hat{\sigma} = 3.49$ from section C.5, we would have been completely off from the true value of 1.5. This confirms the importance of checking a model and, if it fits poorly, rejecting it and looking for a better one.

C.5.3 Model Comparison

In the simple model, the log likelihood at the maximum was roughly $\log \mathcal{L}^* = -40.0$. In the more complex model, this was roughly $\log \mathcal{L}^* = -13.2$. Is a difference of 26.8 big enough that we can reject the simple model? The more complex model has one parameter extra, which makes it fit better, but is the gain in maximum log likelihood "worth it"? This is an

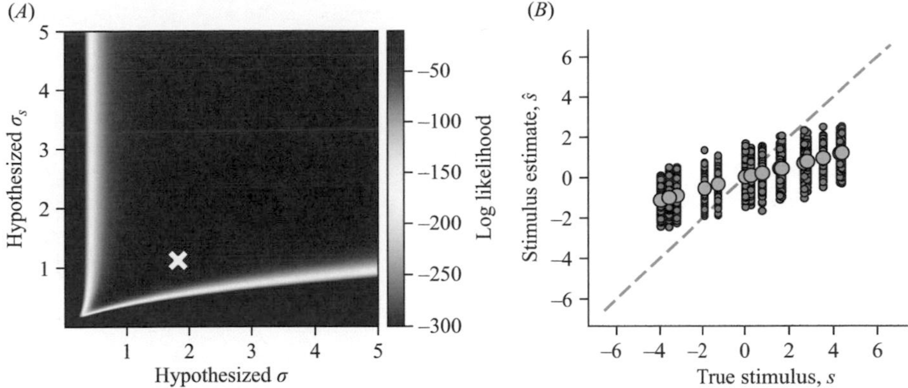

Figure C.4
A better model of the same data (in figure C.2). (*A*) Log likelihood landscape as a function of σ and σ_s for the data in figure C.2A. A redder color represents a higher log likelihood. Log likelihood values smaller than −300 were set to −300, because otherwise, the differences in log likelihood near the peak would not be visible. The maximum is shown as a white X; the value at the maximum is $\log \mathcal{L}^* = -10.2$. (*B*) Model checking: predictions of the model with the parameter estimates from part (*A*). Black dots: individual simulations. Green circles: average across simulations for the same stimulus. This model is visually much more similar to the data than the simple model in figure C.3.

important and very common question in model fitting: how to compare models after fitting? Two common methods for answering these questions are the Akaike Information Criterion (AIC) [8] and the Bayesian Information Criterion (BIC) [159]. Both take the maximum log likelihood, $\log \mathcal{L}^*$, as the starting point, but then penalize the model for the number of free parameters. The nature of the penalty differs between AIC and BIC. In AIC, the number of free parameters gets subtracted from $\log \mathcal{L}^*$, then the result gets multiplied by −2. The factor does not do anything meaningful, it is just a convention so that for a Gaussian model, the leading term of AIC and BIC is the sum of squared errors. In BIC, the number of free parameters multiplied by half the logarithm of the number of trials gets subtracted from $\log \mathcal{L}^*$, then the result gets multiplied by −2. In equations,

$$\mathrm{AIC} = -2\left(\log \mathcal{L}^* - n_{\mathrm{parameters}}\right) \tag{C.12}$$

$$\mathrm{BIC} = -2\left(\log \mathcal{L}^* - \frac{\log n_{\mathrm{trials}}}{2} n_{\mathrm{parameters}}\right). \tag{C.13}$$

Higher AIC or BIC is then interpreted as evidence that a model is *worse*. However, both AIC and BIC derive from strong assumptions about the involved data or models and care is thus recommended in the interpretation of their results. AIC and BIC are, arguably, problematic instruments for evaluating fit quality in the context of the models represented in this book. Cross-validation (see below) may be seen as conceptually superior as it truly reflects the model's predictive abilities, naturally ruling out overly complex models.

C.5.4 Likelihood Ratio Test

The likelihood ratio test is a method that is a bit different in nature from AIC and BIC. This method can be applied when one model is a subset of another model. For example, a model with a guessing rate can reduce to a model with that guessing rate set to 0. To compare two such models, one can use a *likelihood ratio test*. When M_1 is the more specific (restricted; null) model and M_2 is the more general (unrestricted) one, the test statistic is

$$D = 2\left(\log \mathscr{L}^*_{M_2} - \log \mathscr{L}^*_{M_1}\right). \tag{C.14}$$

If the true model were M_1 (null model), then D would approximately follow a chi-squared distribution with number of degrees of freedom equal to the difference in the numbers of free parameters of M_2 and M_1. The p-value of the test is the probability that D under the null model is greater than the observed value of D. When one model is not "nested" inside another model, the likelihood ratio cannot be used.

C.5.5 Cross-Validation

While AIC, BIC, and likelihood ratio tests are based on a relatively coherent set of ideas, cross-validation is quite different. The idea of cross-validation is that a good model should assign high probability to not-yet-seen data. In K-fold cross-validation, the data are divided into K parts. These parts have to be statistically equivalent—for example, it cannot be the case that all stimuli of one type are in the same part. In turn, one of the parts is left out of the data; the models are then fitted to the remaining data. The goodness of fit of the model is evaluated by computing the log likelihood of the fitted parameters on the left-out part of the data. By cycling through the K parts, K values are obtained, which are then averaged; the result is called the cross-validated log likelihood. It can be used directly to compare models. Good values for K are 5 or 10; if K is too low, the result will too much depend on the random composition of the parts, whereas if K is too high, there will be too few trials in each part. The philosophy of cross-validation is that a good model should be able to make good predictions for unseen data. The underlying assumptions of cross-validation are thus comparably simple.

C.5.6 Comparing Model Comparison Methods

The origin of these metrics is beyond the scope of this book, but it is worth pointing out that they have different roots: AIC and cross-validation are meant to measure a model's (in)ability to *predict* new data, BIC is meant to *explain* existing data—a subtle but important conceptual difference. In practice, the most important difference between AIC and BIC is that the BIC penalty for an extra parameter is generally greater than the AIC penalty. However, conceptually, cross-validation is simpler and, arguably, requires weaker assumptions than AIC and BIC.

It is believed that AIC might underpenalize free parameters whereas BIC might overpenalize free parameters, but these issues are subtle and subject to debate. Another debate is whether cross-validation is universally preferable over using either AIC and BIC.

Pragmatically, we recommend calculating AIC, BIC, potentially likelihood ratios, as well as the cross-validated log likelihood, and being confident in the results only if they are highly consistent across the metrics. Drawing strong conclusions when their opposite would be obtained by changing the model comparison metric is problematic.

Besides consistency, magnitude matters. How big a difference in model comparison metric is large? Of course, any classification is arbitrary, but that has not stopped authors from postulating them anyhow. For example, Jeffreys came up with a narrative classification that we can apply to a difference in AIC, a difference in BIC, or to -2 times a difference in cross-validated log likelihood. In this classification, a difference greater than 4.6 is "strong," between 2.3 and 4.6 is "substantial," and below 2.3 "barely worth mentioning." Such labels implicitly refer to the distribution of strength of evidence found in experiments in a particular field, but this distribution is hard to establish and field-dependent. Playing along, in the context of psychology and neuroscience, we prefer to be more cautious and call differences greater than approximately 10 "substantial" and larger than approximately 20 "large." Better than playing along would be to completely extract oneself from narrative labels, but that is difficult for human scientists.

C.5.7 Parameter Recovery and Model Recovery

In practice, parameter estimation and model comparison can get complicated when there are several models, each having multiple parameters, and the parameter likelihood is not easily computed. Therefore, programming mistakes are easily made. An important debugging check is to create "fake data sets" from each of the models under consideration, by simulating measurements and observer decisions according to the model, as we have done several times in previous chapters, and subsequently run the maximum-likelihood fitting and Bayesian model comparison codes on these fake data sets. On each data set, the ML fitting should return parameter values close to the ones with which the data set was created. To verify this properly, it is best not to create a single data set per model, but multiple, with different parameter combinations. Moreover, on each data set, Bayesian model comparison should show the true underlying model as the winner. These checks are contingent on the models being sufficiently distinguishable and the fake data sets being large enough. However, if systematically model B wins on data sets generated from model A, then you know for sure something is wrong in your model comparison code. Besides debugging, this model recovery process also serves to determine whether two models are in practice distinguishable.

C.5.8 Limitations of Model Comparison

Different scientists want to answer distinct questions, and hence model comparison does not have a simple correct answer. For some questions, some scientists might look only for the best-fitting model. For other questions, scientists may want them to encapsulate concepts that make sense as descriptions of perceptual, cognitive, or motor processes. Just adding parameters to a model to achieve a better model comparison metric does not make that model or those parameters meaningful. Many scientists try to make every component of the model justified by one of the following:

- an understanding of the ecological niche or the problem;
- previous models that have that parameter or independent experiments
- a narrative hypothesis about a psychological process.

For example, when we work toward normative models, we may want our models to start with a description of the problem, maybe uncertainty in perception and a loss function. Or if we work with reductionist models, we may want to describe behavior in terms of what neurons do. As such, for the normative models, we would like them to be cast in terms of variables describing the problem and for reductionist models we may want them to be cast in terms of variables describing parts of the nervous system.

Bayesian models usually have strong justification in terms of a narrative hypothesis: the brain has evolved to perform certain tasks close to optimally, apart from noise. But also in Bayesian models it is possible to make arbitrary, poorly justified assumptions just to fit the data. Any inclinations to do so should be avoided if at all possible.

C.6 Absolute Goodness of Fit

A technical limitation of model comparison is that the best model can still be a bad model. It would be useful to understand how good a model is in an absolute sense, that is, compared to the best possible model. The best possible model of the data is not the data themselves, in view of intrinsic stochasticity (noise) in the data. For example, on a given trial, you might predict that a subject chooses option A with probability p and option B with probability $1-p$, but you will not be able to predict the choice on that trial perfectly. The best you can do is to make p match the empirical probability of A responses across trials of this type.

The log likelihood of such an ideal model contributed by one trial of that type will then be $\log p$ when the response is A, that is, with probability p, and $\log(1-p)$ when the response is B, that is, with probability $1-p$. Combining the two, the expected contribution of this trial to the log likelihood of this ideal model will be

$$\log \mathscr{L}(\text{ideal model; data}) = p \log p + (1-p)\log(1-p). \tag{C.15}$$

This is exactly the *negative entropy* of the data. The entropy of a (discrete) probability distribution is a measure of uncertainty: a deterministic mapping has the lowest entropy (0) and a uniform distribution has the highest possible entropy $\left(\frac{1}{N}\right.$, where N is the number of alternatives$\left.\right)$.

The relationship between model log likelihood and negative entropy turns out to be true in general:

$$\log \mathscr{L}(\text{any model; data}) \leq -\text{Entropy(data)}. \tag{C.16}$$

In other words, the negative entropy of the data provides an upper bound on the goodness of fit of any model; for that reason, it is also called the *noise ceiling*. The noisier the data, the higher the entropy, and the lower the upper bound on the log likelihood of any model. Machine learning methods can be used to estimate the noise ceiling [7, 64].

C.7 Fitting Data from a Discrimination Task

We conclude this appendix with two more model-fitting examples. We first consider the following hypothetical data from a discrimination task, as in chapter 7, with $\Delta s = 1$. The data can be fully summarized as in table C.2. The two stimulus values are equally frequent (in this case, occurring on 200 trials each).

C.7.1 Simple Model

Our goal is again to fit a noise parameter σ. In this experiment, since both s and $\hat{s}$ can take on only two values, the model prediction $p(\hat{s}|s,\sigma)$ consists of just four numbers, namely $p(\hat{s}=s_+|s=s_+,\sigma)$, $p(\hat{s}=s_-|s=s_+,\sigma)$, $p(\hat{s}=s_+|s=s_-,\sigma)$, and $p(\hat{s}=s_-|s=s_-,\sigma)$. If the observer is optimal and therefore knows that the two stimuli are equally frequent, we use

Table C.2
Example data from a discrimination task.

True stimulus	Observer responds s_+	Observer responds s_-
s_+	138 trials	62 trials
s_-	90 trials	110 trials

equations (7.20)–(7.21) to evaluate these four probabilities as

$$p(\hat{s}=s_+|s=s_+,\sigma)=p(\hat{s}=s_-|s=s_-,\sigma)=\Phi_{\text{standard}}\left(\frac{\Delta s}{2\sigma}\right); \tag{C.17}$$

$$p(\hat{s}=s_-|s=s_+,\sigma)=p(\hat{s}=s_+|s=s_-,\sigma)=1-\Phi_{\text{standard}}\left(\frac{\Delta s}{2\sigma}\right). \tag{C.18}$$

The only free parameter is σ. The log likelihood of σ is, from equation (C.3),

$$\log \mathscr{L}(\sigma;\text{data})=\sum_{i=1}^{n_{\text{trials}}}\log p(\hat{s}_i|s_i,\sigma). \tag{C.19}$$

The trials can be grouped together into the four combinations of stimulus and response. Then, the sum in equation (C.19) becomes a sum of four terms:

$$\log \mathscr{L}(\sigma;\text{data})=n_{++}\log p(\hat{s}_i=s_+|s_i=s_+,\sigma)+n_{+-}\log p(\hat{s}_i=s_-|s_i=s_+,\sigma) \tag{C.20}$$

$$+n_{-+}\log p(\hat{s}_i=s_+|s_i=s_-,\sigma)+n_{--}\log p(\hat{s}_i=s_-|s_i=s_-,\sigma), \tag{C.21}$$

where we introduced the notation

n_{++}	number of trials where $s=s_+$ and $\hat{s}=s_+$
n_{+-}	number of trials where $s=s_+$ and $\hat{s}=s_-$
n_{-+}	number of trials where $s=s_-$ and $\hat{s}=s_+$
n_{--}	number of trials where $s=s_-$ and $\hat{s}=s_+$

Both the log likelihood and the likelihood are plotted as a function of σ in figure C.5. The likelihood was obtained by exponentiating the log likelihood.

To find the MLE of σ, the analytical method (method 1) is feasible here. We take the derivative of equation (C.21), set it to 0, and solve for σ. The answer is

$$\hat{\sigma}=\frac{\Delta}{2\Phi^{-1}_{\text{standard}}\left(\frac{n_{\text{correct}}}{n_{\text{correct}}+n_{\text{incorrect}}}\right)}, \tag{C.22}$$

where $n_{\text{correct}}=n_{++}+n_{--}$ and $n_{\text{incorrect}}=n_{+-}+n_{-+}$ are the number of correct and incorrect trials, respectively. This expression depends on the fact that in this particular model, $p(\hat{s}_i=s_+|s_i=s_+,\sigma)$ and $p(\hat{s}_i=s_-|s_i=s_-,\sigma)$ are equal. In the example given, equation (C.22) returns $\hat{\sigma}=1.64$. The corresponding maximum log likelihood is log $\mathscr{L}^*=-265.6$.

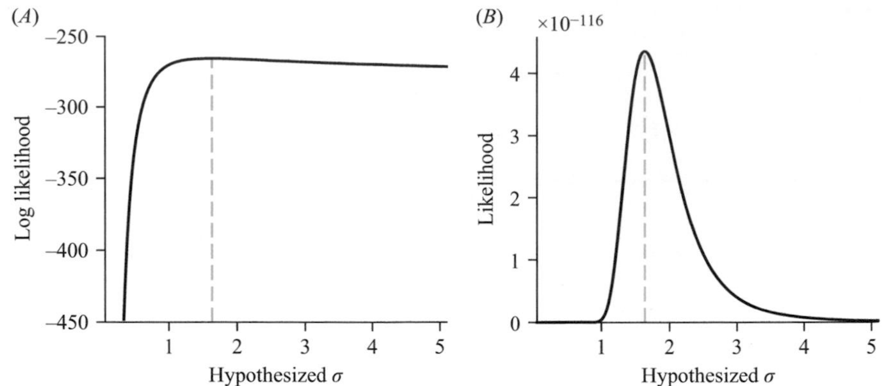

Figure C.5
Log likelihood function and likelihood function over σ obtained from the simple model in section C.7.1.

Table C.3
Checking the simple model.

True stimulus	Model responds s_+	Model responds s_-
s_+	124	76
s_-	76	124

As in section C.5.1, it is important to check the model. The probability correct is 0.62 according to the model, and the predicted numbers of responses are as in table C.3. Thus, this model does not account for the unequal proportions correct between s_+ and s_- seen in the data (table C.2). To try to account for those unequal proportions, we consider a more flexible model.

C.7.2 A Better Model?

In chapter 7, we also considered a more flexible model, in which the observer's prior is not equal to 0.5. Then, the model has two parameters, σ and that prior. Since our data in this very simple experiment also amount to two numbers (proportion correct when the stimulus is s_+ and when the stimulus is s_-), we would be fitting two data points with two parameters. Without even fitting those parameters, we know that this can be done perfectly (see problem C.2). That does not mean that every trial can be predicted perfectly, just that the predicted *probabilities* of the responses exactly match the empirical probabilities; in other words, this is an "ideal model" as described in section C.6.

Per equation (C.21), the log likelihood again consists of four terms. The maximum log likelihood of this two-parameter model is (see problem C.1)

$$\log \mathscr{L}^* = n_{++} \log \frac{n_{++}}{n_{++}+n_{+-}} + n_{+-} \log \frac{n_{+-}}{n_{++}+n_{+-}} + n_{-+} \log \frac{n_{-+}}{n_{-+}+n_{--}} + n_{--} \log \frac{n_{--}}{n_{-+}+n_{--}}. \tag{C.23}$$

This evaluates to -261.4, which is 4.2 higher than in the simple model of section C.7.1. This difference is large enough that this model wins in both AIC and BIC. Nevertheless, this is not a particularly insightful model. The fact that the winning model is equivalent to a mere description of each condition is a sign that the data in this experiment are too simple to draw strong conclusions.

C.8 Fitting Data from a Classification Task

In chapter 8, we considered binary classification tasks. Data from such a task are richer than the discrimination task in section C.7. Example data are shown in table C.4.

The model in this task is a probability distribution $p(\hat{C}|s,\theta)$, where θ represents all parameters. Thus, the log likelihood function takes the form

$$\log \mathcal{L}(\theta; \text{data}) = \sum_{i=1}^{n_{\text{trials}}} \log p(\hat{C}_i|s_i,\theta). \tag{C.24}$$

We work this out in problem 8.11. In some cases, the large sum over trials can be split up in a sum over unique stimulus-response combinations. Then, per combination, we have to multiply the summand by the number of instances of each combination, similar to equation (C.21).

C.9 Good Experimental Design for Bayesian Modeling

To allow you to successfully build a Bayesian model of your behavioral data, the first requirement is to design an experiment well. There are well-known general guidelines for this [202], and some additional tips are specific to Bayesian models. In general, one would like to control as many of the parameters that are not of interest to the scientific question. For example, to study how humans perform discrimination, we want to present stimuli for a short time (a few tens of milliseconds), to avoid complications associated with eye movements, the time course of attention, and the integration of information over time, all of which can effect the quality of encoding (i.e., the standard deviation of the noise distribution) in a potentially complex way. Response-time experiments are typically more complex to model

Table C.4
Example data from the classification task.

Trial number	Stimulus s	Subject response, $\hat{s}$
1	-4.3	-1
2	-1.0	1
3	2.1	1
4	0.7	-1
5	4.5	1
6	-3.5	-1
...	...	...

than accuracy experiments with short presentation times. Therefore, if your scientific question allows you to do an accuracy version of the same experiment, it will likely save you work and computation time during modeling.

Similarly, we want to keep attributes of the stimuli that are not of interest as much the same between stimuli. Specifically, make sure to carefully control the reliability or noise level of the stimuli. For example, if you arrange multiple items in a display, arranging them in a circle around the fixation point instead of in a rectangular grid ensures that the eccentricity (distance from the fixation point) is the same and therefore encoding precision is at least approximately the same, allowing you to model reliability with a single parameter.

Furthermore, one needs to be well aware of domain-specific effects that can influence performance in the task. For example, when two stimuli are brought close together, an effect known as crowding can occur, in which the internal representations of both stimuli influence each other. If crowding is not of primary interest, it is best to minimize it by placing the stimuli sufficiently far apart from each other. Specific to Bayesian modeling, it is often useful to use stimuli whose feature of interest is one-dimensional or at most two-dimensional. For example, when studying cue combination, it is easier to model a flash and a beep presented on a horizontal line than to model the integration of the auditory and visual information in speech perception. In the perceptual experiments discussed in this book, we use stimuli that are as simple as possible: they have only a single relevant dimension, for example orientation. Stimuli like letters, line drawings, images of objects, or natural scenes are much more difficult to cast into a model because they have many features and in some cases it is not even clear what the relevant features (perceptual building blocks) are. Moreover, a large number of features translates into a large number of dimensions, and noise models in high numbers of dimensions have even higher numbers of free parameters. This is not to say that studying complex stimuli is not interesting, on the contrary. In a way, we are merely pointing out the limitations of Bayesian modeling.

Finally, we recommend that anyone interested in building a Bayesian model of their task write out the model and simulate it before even starting to collect data. For Bayesian models, since they are based on principles of optimality, this is always possible. This process can highlight potential problems in the experimental design.

C.10 Suggested Readings

- Hirotugu Akaike. "A New Look at the Statistical Model Identification." *IEEE Transactions on Automatic Control* 19, no. 6 (1974): 716–723.
- Luigi Acerbi and Wei Ji Ma. "Practical Bayesian Optimization for Model Fitting with Bayesian Adaptive Direct Search." In *Advances in Neural Information Processing Systems*, edited by Ulrike von Luxburg, Isabelle Guyon, Samy Bengio, Hann Wallach, and Rob Fergus, 1834–1844. Red Hook, NY: Curran, 2017.
- Sylvain Arlot and Alain Celisse. "A Survey of Cross-Validation Procedures for Model Selection." *Statistics Surveys* 4 (2010): 40–79.
- David J. C. MacKay. *Information Theory, Inference and Learning Algorithms*. Cambridge: Cambridge University Press, 2003.

- In Jae Myung. "Tutorial on Maximum Likelihood Estimation." *Journal of Mathematical Psychology* 47, no. 1 (2003): 90–100.
- Gideon Schwarz. "Estimating the Dimension of a Model." *Annals of Statistics* 6, no. 2 (1978): 461–464.
- Lionel Rigoux, Klaas E. Stephan, Karl J. Friston, and Jean Daunizeau. "Bayesian Model Selection for Group Studies—Revisited." *Neuroimage* 84 (2014): 971–985.
- Bas van Opheusden, Luigi Acerbi, and Wei Ji Ma. "Unbiased and Efficient Log-Likelihood Estimation with Inverse Binomial Sampling." *PLoS Computational Biology* 16, no. 12 (2020): e1008483.
- Scott I. Vrieze. "Model Selection and Psychological Theory: A Discussion of the Differences between the Akaike Information Criterion (AIC) and the Bayesian Information Criterion (BIC)." *Psychological Methods* 17, no. 2 (2012): 228–243.
- Robert C. Wilson and Anne G. E. Collins. "Ten Simple Rules for the Computational Modeling of Behavioral Data." *Elife* 8 (2019): e49547.

C.11 Problems

Problem C.1 Derive equation (C.23).

Problem C.2 Equations (7.22)–(7.23) specified the response probabilities in a Bayesian model of discrimination:

$$p(\hat{s}=s_+|s=s_+)=\Phi_{\text{standard}}\left(\frac{\Delta s}{2\sigma}+\frac{\sigma}{\Delta s}\log\frac{p_+}{1-p_+}\right); \tag{C.25}$$

$$p(\hat{s}=s_+|s=s_-)=\Phi_{\text{standard}}\left(-\frac{\Delta s}{2\sigma}+\frac{\sigma}{\Delta s}\log\frac{p_+}{1-p_+}\right), \tag{C.26}$$

where $p_+=p(s=s_+)$ is the prior probability of s_+. As the experimenter, you know the value of Δs. The subject's responses give you the two proportions on the left-hand sides. Solve for σ and p_+. This shows that using these two parameters, the responses proportions in this task can always be exactly predicted.

Problem C.3 This problem is about the two models in section C.7.

(a) Calculate the AIC values of both models. Which model is better according to AIC?

(b) Calculate the BIC values of both models. Which model is better according to BIC?

(c) Calculate the ten fold cross-validated log likelihood of both models. Which model is better according to the cross-validated log likelihood?

(d) Perform a likelihood ratio test. Is the more complex model significantly better than the simpler model?

Bibliography

[1] Luigi Acerbi and Wei Ji Ma. "Practical Bayesian Optimization for Model Fitting with Bayesian Adaptive Direct Search". *Advances in Neural Information Processing Systems,* edited by Ulrike von Luxburg, Isabelle Guyon, Samy Bengio, Hann Wallach, and Rob Fergus, 1834–1844. Red Hook, NY: Curran, 2017. (cited on page 349)

[2] Luigi Acerbi, Sethu Vijayakumar, and Daniel M. Wolpert. "On the Origins of Suboptimality in Human Probabilistic Inference." *PLoS Computational Biology* 10, no. 6 (2014): e1003661. (cited on page 305)

[3] Daniel E. Acuña and Paul Schrater. "Structure Learning in Human Sequential Decision-Making." *PLoS Computational Biology* 6, no. 12 (2010): e1001003. (cited on page 142)

[4] Ryan Prescott Adams and David J. C. MacKay. "Bayesian Online Changepoint Detection." *arXiv preprint arXiv:0710.3742* (2007). (cited on pages 253, 254)

[5] Wendy J. Adams. "A Common Light-Prior for Visual Search, Shape, and Reflectance Judgments." *Journal of Vision* 7, no. 11 (2007) . (cited on pages 305, 308)

[6] Wendy J. Adams, Erich W. Graf, and Marc O. Ernst. "Experience Can Change the 'Light-from-above' Prior." *Nature Neuroscience* 7, no. 10 (2004): 1057–1058. (cited on page 77)

[7] Mayank Agrawal, Joshua C. Peterson, and Thomas L. Griffiths. "Scaling up Psychology via Scientific Regret Minimization." *Proceedings of the National Academy of Sciences* 117, no. 16 (2020): 8825–8835. (cited on page 354)

[8] Hirotugu Akaike. "A New Look at the Statistical Model Identification." *IEEE Transactions on Automatic Control* 19, no. 6 (1974): 716–723. (cited on pages 351, 358)

[9] David Alais and David Burr. "The Ventriloquist Effect Results from Near-Optimal Bimodal Integration." *Current Biology* 14, no. 3 (2004): 257–262. (cited on pages 117, 119)

[10] Alhacen. *Alhacen's Theory of Visual Perception: A Critical Edition, with English Translation and Commentary, of the First Three Books of Alhacen's "De Aspectibus," the Medieval Latin Version of Ibn Al-Haytham's "Kitāb Al-Manāẓir".* Edited by A. Mark Smith. Vol. 1. Philadelphia, PA: American Philosophical Society, 2001. (cited on pages 24, 26)

[11] John R. Anderson. "The Adaptive Nature of Human Categorization." *Psychological Review* 98, no. 3 (1991): 409–429. (cited on pages 4, 303, 308)

[12] John R. Anderson. *The Adaptive Character of Thought.* New York: Psychology Press, 2013. (cited on page 308)

[13] J. Yu Angela. "Adaptive Behavior: Humans Act as Bayesian Learners." *Current Biology* 17, no. 22 (2007): R977–R980. (cited on page 254)

[14] Sylvain Arlot and Alain Celisse. "A Survey of Cross-Validation Procedures for Model Selection." *Statistics Surveys* 4 (2010): 40–79. (cited on page 358)

[15] Bahador Bahrami et al. "Optimally Interacting Minds." *Science* 329, no. 5995 (2010): 1081–1085. (cited on page 118)

[16] F. R. S. Bayes. "An Essay towards Solving a Problem in the Doctrine of Chances." *Biometrika* 45, no. 3–4 (1958): 296–315. (cited on page 26)

[17] Suzanna Becker and Geoffrey E. Hinton. "Self-Organizing Neural Network that Discovers Surfaces in Random-Dot Stereograms." *Nature* 355, no. 6356 (1992): 161–163. (cited on page 307)

[18] Vikranth Rao Bejjanki, Meghan Clayards, David C. Knill, and Richard N. Aslin. "Cue Integration in Categorical Tasks: Insights from Audio-Visual Speech Perception." *PloS One* 6, no. 5 (2011): e19812. (cited on pages 118, 119)

[19] Christopher M. Bishop. "Pattern Recognition." *Machine Learning* 128, no. 9 (2006) . (cited on page 306)

[20] Ned Block. "If Perception Is Probabilistic, Why Does It Not Seem Probabilistic?" *Philosophical Transactions of the Royal Society B: Biological Sciences* 373, no. 1755 (2018): 20170341. (cited on page 76)

[21] Kathryn Bonnen, Johannes Burge, Jacob Yates, Jonathan Pillow, and Lawrence K. Cormack. "Continuous Psychophysics: Target-Tracking to Measure Visual Sensitivity." *Journal of Vision* 15, no. 3 (2015): 14. (cited on page 254)

[22] Jeffrey S. Bowers and Colin J. Davis. "Bayesian Just-so Stories in Psychology and Neuroscience." *Psychological Bulletin* 138, no. 3 (2012): 389–414. (cited on pages 302, 308)

[23] David H. Brainard and William T. Freeman. "Bayesian Color Constancy." *Journal of the Optical Society of America* 14, no. 7 (1997): 1393–1411. (cited on page 202)

[24] Anthony E. Brockwell, Alex L. Rojas, and Robert E. Kass. "Recursive Bayesian Decoding of Motor Cortical Signals by Particle Filtering." *Journal of Neurophysiology* 91, no. 4 (2004): 1899–1907. (cited on page 296)

[25] Anne-Marie Brouwer and David C. Knill. "The Role of Memory in Visually Guided Reaching." *Journal of Vision* 7, no. 5 (2007): 6.1–12. (cited on page 119)

[26] Peter Brugger and Susanne Brugger. "The Easter Bunny in October: Is It Disguised as a Duck?" *Perceptual and Motor Skills* 76, no. 2 (1993): 577–578. (cited on pages 26, 28)

[27] Heinrich H. Bulthoff. "Bayesian Decision Theory and Psychophysics". *Perception as Bayesian Inference*, edited by David C. Knill and Whitman Richards, 123–162. Cambridge: Cambridge University Press, 1996. (cited on page 119)

[28] Daniel Burdon. "Pigs Float down the Dawson." *Morning Bulletin*, February 9, 2011. (cited on page 21)

[29] Sherrington C. S. *The Integrative Action of the Nervous System*. C. Scribner and Sons: New York, 1906. (cited on page 284)

[30] Joshua Calder-Travis and Wei Ji Ma. "Explaining the Effects of Distractor Statistics in Visual Search." *Journal of Vision* 20, no. 13 (2020): 11. (cited on page 239)

[31] Claire Chambers, Taegh Sokhey, Deborah Gaebler-Spira, and Konrad Paul Kording. "The Development of Bayesian Integration in Sensorimotor Estimation." *Journal of Vision* 18, no. 12 (2018): 8. (cited on page 307)

[32] Gar Ming Chan. "Bayes' Theorem, COVID19, and Screening Tests." *American Journal of Emergency Medicine* 38, no. 10 (2020): 2011–2013. (cited on page 45)

[33] Nick Chater, Noah Goodman, Thomas L. Griffiths, Charles Kemp, Mike Oaksford, and Joshua B. Tenenbaum. "The Imaginary Fundamentalists: The Unshocking Truth about Bayesian Cognitive Science." *Behavioral and Brain Sciences* 34, no. 4 (2011): 194–196. (cited on page 308)

[34] Nick Chater and Mike Oaksford, eds. *The Probabilistic Mind: Prospects for Bayesian Cognitive Science*. New York: Oxford University Press, 2008. (cited on pages 4, 303, 308)

[35] Stephanie Y. Chen, Brian H. Ross, and Gregory L. Murphy. "Implicit and Explicit Processes in Category-Based Induction: Is Induction Best When We Don't Think?" *Journal of Experimental Psychology: General* 143, no. 1 (2014): 227–246. (cited on pages 303, 308)

[36] Youguo Chen, Bangwu Zhang, and Konrad P. Körding. "Speed Constancy or Only Slowness: What Drives the Kappa Effect." *PloS One* 11, no. 4 (2016): e0154013. (cited on page 75)

[37] Patricia W. Cheng. "From Covariation to Causation: A Causal Power Theory." *Psychological Review* 104, no. 2 (1997): 367–405. (cited on page 142)

[38] Anna Coenen, Bob Rehder, and Todd M. Gureckis. "Strategies to Intervene on Causal Systems Are Adaptively Selected." *Cognitive Psychology* 79 (2015): 102–133. (cited on page 142)

[39] Gloria Cooper. *Red Tape Holds up New Bridge, and More Flubs from the Nation's Press*. New York: TarcherPerigee, 1987. (cited on pages 23, 26)

[40] Carl F. Craver. *Explaining the Brain: Mechanisms and the Mosaic Unity of Neuroscience*. Oxford: Clarendon Press, 2007. (cited on page 303)

[41] Lorraine Daston. "How Probabilities Came to Be Objective and Subjective." *Historia Mathematica* 21, no. 3 (1994): 330–344. (cited on page 316)

[42] Sophie Deneve. "Bayesian Spiking Neurons I: Inference." *Neural Computation* 20, no. 1 (2008): 91–117. (cited on page 295)

[43] Rachel N. Denison. "Precision, Not Confidence, Describes the Uncertainty of Perceptual Experience: Comment on John Morrison's 'Perceptual Confidence.'" *Analytic Philosophy* 58, no. 1 (2017): 58–70. (cited on page 76)

[44] Rachel N. Denison, William T. Adler, Marisa Carrasco, and Wei Ji Ma. "Humans Incorporate Attention-Dependent Uncertainty into Perceptual Decisions and Confidence." *Proceedings of the National Academy of Sciences* 115, no. 43 (2018): 11090–11095. (cited on page 184)

[45] Kenji Doya, Shin Ishii, Alexandre Pouget, and Rajesh P. N. Rao. *Bayesian Brain: Probabilistic Approaches to Neural Coding*. Cambridge, MA: MIT Press, 2007. (cited on page 1)

[46] Miguel P. Eckstein. "Visual Search: A Retrospective." *Journal of Vision* 11, no. 5 (2011): 14. (cited on page 239)

[47] Marc O. Ernst and Martin S. Banks. "Humans Integrate Visual and Haptic Information in a Statistically Optimal Fashion." *Nature* 415, no. 6870 (2002): 429–433. (cited on pages 117, 119)

[48] J. St. B. T. Evans, P. G. Brooks, and P. Pollard. "Prior Beliefs and Statistical Inference." *British Journal of Psychology* 76, no. 4 (1985): 469–477. (cited on pages 4, 303)

[49] Jacob Feldman. "Bayesian Contour Integration." *Perception and Psychophysics* 63 (2001): 1171–1182. (cited on pages 219, 223)

[50] Norman Fenton. "Improve Statistics in Court." *Nature* 479, no. 7371 (2011): 36–37. (cited on pages 38, 46)

[51] Christopher R. Fetsch, Alexandre Pouget, Gregory C. DeAngelis, and Dora E. Angelaki. "Neural Correlates of Reliability-Based Cue Weighting During Multisensory Integration." *Nature Neuroscience* 15, no. 1 (2012): 146–154. (cited on page 297)

[52] Brian J. Fischer and José Luis Peña. "Owl's Behavior and Neural Representation Predicted by Bayesian Inference." *Nature Neuroscience* 14, no. 8 (2011): 1061–1066. (cited on pages 75, 77)

[53] József Fiser, Pietro Berkes, Gergő Orbán, and Máté Lengyel. "Statistically Optimal Perception and Learning: From Behavior to Neural Representations." *Trends in Cognitive Sciences* 14, no. 3 (2010): 119–130. (cited on pages 283, 297)

[54] Jerry A. Fodor. *Psychological Explanation: An Introduction to the Philosophy of Psychology*. New York: Random House, 1968. (cited on pages 303, 308)

[55] Peter Földiák. "The 'Ideal Homunculus': Statistical Inference from Neural Population Responses." *Computation and Neural Systems*, edited by Frank H. Eeckman and James M. Bower, 55–60. New York: Springer, 1993. (cited on pages 287, 297)

[56] William T. Freeman. "The Generic Viewpoint Assumption in a Framework for Visual Perception." *Nature* 368, no. 6471 (1994): 542–545. (cited on page 52)

[57] Wilson S. Geisler and Jeffrey S. Perry. "Contour Statistics in Natural Images: Grouping across Occlusions." *Visual Neuroscience* 26, no. 1 (2009): 109–121. (cited on pages 46, 72, 219, 223)

[58] Wilson S. Geisler and Randy L. Diehl. "A Bayesian Approach to the Evolution of Perceptual and Cognitive Systems." *Cognitive Science* 27, no. 3 (2003): 379–402. (cited on page 26)

[59] Andrew Gelman, John B. Carlin, Hal S. Stern, David B. Dunson, Aki Vehtari, and Donald B. Rubin. *Bayesian Data Analysis*. New York: Chapman and Hall/CRC, 1995. (cited on pages 4, 345)

[60] George A. Gescheider. *Psychophysics: The Fundamentals*. New York: Psychology Press, 2013. (cited on page 163)

[61] Laura S. Geurts, James R. H. Cooke, Ruben S. van Bergen, and Janneke F. M. Jehee. "Subjective Confidence Reflects Representation of Bayesian Probability in Cortex." *Nature Human Behaviour* 6 (2022): 294–305. (cited on pages 296, 297)

[62] Gerd Gigerenzer. "On Narrow Norms and Vague Heuristics: A Reply to Kahneman and Tversky." *Psychological Review* 103, no. 3 (1996): 592–596. (cited on page 308)

[63] Ahna R. Girshick, Michael S. Landy, and Eero P. Simoncelli. "Cardinal Rules: Visual Orientation Perception Reflects Knowledge of Environmental Statistics." *Nature Neuroscience* 14, no. 7 (2011): 926–932. (cited on pages 72, 76)

[64] Joshua I. Glaser, Ari S. Benjamin, Roozbeh Farhoodi, and Konrad P. Kording. "The Roles of Supervised Machine Learning in Systems Neuroscience." *Progress in Neurobiology* 175 (2019): 126–137. (cited on page 354)

[65] Daniel Goldreich. "A Bayesian Perceptual Model Replicates the Cutaneous Rabbit and Other Tactile Spatiotemporal Illusions." *PloS One* 2, no. 3 (2007): e333. (cited on page 75)

[66] Daniel Goldreich and Jonathan Tong. "Prediction, Postdiction, and Perceptual Length Contraction: A Bayesian Low-Speed Prior Captures the Cutaneous Rabbit and Related Illusions." *Frontiers in Psychology* 4 (2013): 221. (cited on pages 75, 77, 254)

[67] Daniel Goldreich and Mary A. Peterson. "A Bayesian Observer Replicates Convexity Context Effects in Figure-ground Perception." *Seeing and Perceiving* 25, no. 3–4 (2012): 365–395. (cited on page 223)

[68] Alison Gopnik, Clark Glymour, David M. Sobel, Laura E. Schulz, Tamar Kushnir, and David Danks. "A Theory of Causal Learning in Children: Causal Maps and Bayes Nets." *Psychological Review* 111, no. 1 (2004): 3–32. (cited on pages 4, 142, 303)

[69] Monica Gori, Michela Del Viva, Giulio Sandini, and David C. Burr. "Young Children Do Not Integrate Visual and Haptic Form Information." *Current Biology* 18, no. 9 (2008): 694–698. (cited on pages 307, 308)

[70] David M. Green and John A. Swets. *Signal Detection Theory and Psychophysics*. Vol. 1. New York: Wiley, 1966. (cited on page 163)

[71] Thomas L. Griffiths and Joshua B. Tenenbaum. "Optimal Predictions in Everyday Cognition." *Psychological Science* 17, no. 9 (2006): 767–773. (cited on pages 46, 303)

[72] Thomas L. Griffiths and Joshua B. Tenenbaum. "Theory-Based Causal Induction." *Psychological Review* 116, no. 4 (2009): 661–716. (cited on page 184)

[73] Ralf M. Haefner, Pietro Berkes, and József Fiser. "Perceptual Decision-Making as Probabilistic Inference by Neural Sampling." *Neuron* 90, no. 3 (2016): 649–660. (cited on page 283)

[74] Gary Hatfield. "Perception as unconscious inference." *Perception and the physical world: Psychological and philosophical issues in perception*, Citeseer. 2002. (cited on page 26)

[75] Michael J. Hautus, Neil A. Macmillan, and C. Douglas Creelman. "Detection Theory: A User's Guide." (2021) . (cited on page 164)

[76] Maija Honig, Wei Ji Ma, and Daryl Fougnie. "Humans Incorporate Trial-to-Trial Working Memory Uncertainty into Rewarded Decisions." *Proceedings of the National Academy of Sciences* 117, no. 15 (2020): 8391–8397. (cited on pages 272, 275)

[77] Neil M. T. Houlsby, Ferenc Huszár, Mohammad M. Ghassemi, Gergő Orbán, Daniel M. Wolpert, and Máté Lengyel. "Cognitive Tomography Reveals Complex, Task-Independent Mental Representations." *Current Biology* 23, no. 21 (2013): 2169–2175. (cited on page 305)

[78] Patrik O. Hoyer and Aapo Hyvärinen. "Interpreting Neural Response Variability as Monte Carlo Sampling of the Posterior". *Advances in Neural Information Processing Systems 15*, edited by Suzanna Becker, Sebastian Thrun and Klaus Obermayer, 293–300. Cambridge, MA: MIT Press, 2002. (cited on page 283)

[79] R. M. Isaacson and C. A. Was. "Building a Metacognitive Curriculum: An Educational Psychology to Teach Metacognition." *National Teaching & Learning Forum*, Vol. 19, no. 5. 2010: 1–4. (cited on page 279)

[80] Robert A. Jacobs. "Optimal Integration of Texture and Motion Cues to Depth." *Vision Research* 39, no. 21 (1999): 3621–3629. (cited on page 119)

[81] William James. *The Principles of Psychology*. New York: Henry Holt, 1890. (cited on page 215)

[82] Edwin T. Jaynes. *Probability Theory: The Logic of Science*. Cambridge: Cambridge University Press, 2003. (cited on page 98)

[83] Mehrdad Jazayeri and J. Anthony Movshon. "Optimal Representation of Sensory Information by Neural Populations." *Nature Neuroscience* 9, no. 5 (2006): 690–696. (cited on pages 287, 297)

[84] Mehrdad Jazayeri and Michael N. Shadlen. "Temporal Context Calibrates Interval Timing." *Nature Neuroscience* 13, no. 8 (2010): 1020–1026. (cited on pages 75, 77)

[85] Carolyn Y. Johnson, Yasmeen Abutaleb, and Joel Achenbach. "CDC Study Shows Three-Fourths of People Infected in Massachusetts Coronavirus Outbreak Were Vaccinated but Few Required Hospitalization." *Washington Post*, July 30, 2021. (cited on page 46)

[86] Matt Jones and Bradley C. Love. "Bayesian Fundamentalism or Enlightenment? On the Explanatory Status and Theoretical Contributions of Bayesian Models of Cognition." *Behavioral and Brain Sciences* 34, no. 4 (2011): 169–188. (cited on pages 302, 308)

[87] Daniel Kahneman. *Thinking, Fast and Slow*. New York: Macmillan, 2011. (cited on pages 303, 308)

[88] Daniel Kahneman, Paul Slovic, and Amos Tversky, eds. *Judgment under Uncertainty: Heuristics and Biases*. Cambridge: Cambridge University Press, 1982. (cited on page 303)

[89] David Kappel, Stefan Habenschuss, Robert Legenstein, and Wolfgang Maass. "Network Plasticity as Bayesian Inference." *PLoS Computational Biology* 11, no. 11 (2015): e1004485. (cited on page 295)

[90] Charles Kemp, Andrew Perfors, and Joshua B. Tenenbaum. "Learning Overhypotheses with Hierarchical Bayesian Models." *Developmental Science* 10, no. 3 (2007): 307–321. (cited on pages 142, 308)

[91] Daniel Kersten, Pascal Mamassian, and Alan Yuille. "Object Perception as Bayesian Inference." *Annual Review of Psychology* 55 (2004): 271–304. (cited on pages 184, 202)

[92] Daniel Kersten and Alan Yuille. "Bayesian Models of Object Perception." *Current Opinion in Neurobiology* 13 (2003): 1–9. (cited on page 200)

[93] David C. Knill. "Mixture Models and the Probabilistic Structure of Depth Cues." *Vision Research* 43, no. 7 (2003): 831–854. (cited on page 202)

[94] David C. Knill and Whitman Richards. *Perception as Bayesian Inference*. Cambridge: Cambridge University Press, 1996. (cited on page 1)

[95] David C. Knill and Jeffrey A. Saunders. "Do Humans Optimally Integrate Stereo and Texture Information for Judgments of Surface Slant?" *Vision Research* 43, no. 24 (2003): 2539–2558. (cited on pages 117, 119)

[96] Jan J. Koenderink and Andrea J. Van Doorn. "The Internal Representation of Solid Shape with Respect to Vision." *Biological Cybernetics* 32, no. 4 (1979): 211–216. (cited on page 52)

[97] Konrad Kording. "Decision Theory: What 'Should' the Nervous System Do?" *Science* 318, no. 5850 (2007): 606–610. (cited on page 274)

[98] Konrad P. Kording, Ulrik Beierholm, Wei Ji Ma, Steven Quartz, Joshua B. Tenenbaum, and Ladan Shams. "Causal Inference in Multisensory Perception." *PLoS One* 2, no. 9 (2007): e943. (cited on pages 211, 214, 223, 267)

[99] Konrad P. Kording, Izumu Fukunaga, Ian S. Howard, James N. Ingram, and Daniel M. Wolpert. "A Neuroeconomics Approach to Inferring Utility Functions in Sensorimotor Control." *PLoS Biology* 2, no. 10 (2004): e330. (cited on pages 272, 274, 275)

[100] Konrad P. Kording, Joshua B. Tenenbaum, and Reza Shadmehr. "The Dynamics of Memory as a Consequence of Optimal Adaptation to a Changing Body." *Nature Neuroscience* 10, no. 6 (2007): 779–786. (cited on page 254)

[101] Konrad P. Kording and Daniel M. Wolpert. "Bayesian Integration in Sensorimotor Learning." *Nature* 427, no. 6971 (2004): 244–247. (cited on pages 75, 77)

[102] Konrad P. Kording and Daniel M. Wolpert. "The Loss Function of Sensorimotor Learning." *Proceedings of the National Academy of Sciences* 101, no. 26 (2004): 9839–9842. (cited on pages 272, 275)

[103] John Kruschke. *Doing Bayesian Data Analysis: A Tutorial with R, JAGS, and Stan*. Amsterdam: Elsevier, 2014. (cited on pages 4, 345)

[104] Rosa Lafer-Sousa, Katherine L. Hermann, and Bevil R. Conway. "Striking Individual Differences in Color Perception Uncovered by 'the Dress' Photograph." *Current Biology* 25, no. 13 (2015): R545–R546. (cited on page 202)

[105] Pierre-Simon Laplace. "Memoir on the Probability of the Causes of Events." *Statistical Science* 1, no. 3 (1986): 364–378. (cited on page 142)

[106] Pierre-Simon Laplace. *Philosophical Essay on Probabilities.* Edited and translated by Andrew I. Dale. New York: Springer Science and Business Media, 2012. (cited on pages 23, 26)

[107] Tai Sing Lee and David Mumford. "Hierarchical Bayesian Inference in the Visual Cortex." *Journal of the Optical Society of America* 20, no. 7 (2003): 1434–1448. (cited on page 295)

[108] Hsin-Hung Li, Thomas C. Sprague, Aspen H. Yoo, Wei Ji Ma, and Clayton E. Curtis. "Joint Representation of Working Memory and Uncertainty in Human Cortex." *Neuron* 109, no. 22 (2021): 3699–3712. (cited on pages 296, 297)

[109] Falk Lieder and Thomas L. Griffiths. "Resource-Rational Analysis: Understanding Human Cognition as the Optimal Use of Limited Computational Resources." *Behavioral and Brain Sciences* 43, no. e1 (2020): 1–60. (cited on pages 304, 308)

[110] Zili Liu, David C. Knill, and Daniel Kersten. "Object Classification for Human and Ideal Observers." *Vision Research* 35, no. 4 (1995): 549–568. (cited on page 184)

[111] Lucretius [Titus Lucretius Carus]. *The Nature of Things*. Translated by A. E. Stallings. London: Penguin, 2007. (cited on page 7)

[112] Wei Ji Ma. "Organizing Probabilistic Models of Perception." *Trends in Cognitive Sciences* 16, no. 10 (2012): 511–518. (cited on pages 305, 308)

[113] Wei Ji Ma, Jeffrey M. Beck, Peter E. Latham, and Alexandre Pouget. "Bayesian Inference with Probabilistic Population Codes." *Nature Neuroscience* 9, no. 11 (2006): 1432–1438. (cited on pages 287, 295, 297)

[114] Wei Ji Ma and Mehrdad Jazayeri. "Neural Coding of Uncertainty and Probability." *Annual Review of Neuroscience* 37 (2014): 205–220. (cited on pages 304, 308)

[115] Wei Ji Ma, Vidhya Navalpakkam, Jeffrey M. Beck, Ronald van den Berg, and Alexandre Pouget. "Behavior and Neural Basis of Near-Optimal Visual Search." *Nature Neuroscience* 14, no. 6 (2011): 783–790. (cited on page 240)

[116] Wei Ji Ma, Shan Shen, Gintare Dziugaite, and Ronald van den Berg. "Requiem for the Max Rule?" *Vision Research* 116 (2015): 179–193. (cited on page 240)

[117] Wei Ji Ma, Xiang Zhou, Lars A. Ross, John J. Foxe, and Lucas C. Parra. "Lip-Reading Aids Word Recognition Most in Moderate Noise: A Bayesian Explanation Using High-Dimensional Feature Space." *PloS One* 4, no. 3 (2009): e4638. (cited on page 118)

[118] David J. C. MacKay. *Information Theory, Inference and Learning Algorithms*. Cambridge: Cambridge University Press, 2003. (cited on pages 221, 223, 358)

[119] Tamas J. Madarasz, Lorenzo Diaz-Mataix, Omar Akhand, Edgar A. Ycu, Joseph E. LeDoux, and Joshua P. Johansen. "Evaluation of Ambiguous Associations in the Amygdala by Learning the Structure of the Environment." *Nature Neuroscience* 19, no. 7 (2016): 965–972. (cited on page 142)

[120] Joseph G. Makin, Matthew R. Fellows, and Philip N. Sabes. "Learning Multisensory Integration and Coordinate Transformation via Density Estimation." *PLoS Computational Biology* 9, no. 4 (2013): e1003035. (cited on pages 295, 307)

[121] Laurence T. Maloney and Pascal Mamassian. "Bayesian Decision Theory as a Model of Human Visual Perception: Testing Bayesian Transfer." *Visual Neuroscience* 26, no. 1 (2009): 147–155. (cited on pages 304, 308)

[122] Laurence T. Maloney, Julia Trommershäuser, and Michael S. Landy. "Questions without Words: A Comparison between Decision Making under Risk and Movement Planning under Risk." *Integrated Models of Cognitive Systems*, edited by Wayne D. Gray, 297–313. New York: Oxford University Press, 2007. (cited on page 275)

[123] Helga Mazyar, Ronald van den Berg, and Wei Ji Ma. "Does Precision Decrease with Set Size?" *Journal of Vision* 12, no. 6 (2012): 10. (cited on pages 239, 240)

[124] Sharon Bertsch McGrayne. *The Theory that Would Not Die*. New Haven, CT: Yale University Press, 2011. (cited on page 46)

[125] Harry McGurk and John MacDonald. "Hearing Lips and Seeing Voices." *Nature* 264, no. 5588 (1976): 746–748. (cited on pages 118, 119)

[126] Barbara Mellers, Ralph Hertwig, and Daniel Kahneman. "Do Frequency Representations Eliminate Conjunction Effects? An Exercise in Adversarial Collaboration." *Psychological Science* 12, no. 4 (2001): 269–275. (cited on page 309)

[127] John P. Miller, Gwen A. Jacobs, and Frédéric E. Theunissen. "Representation of Sensory Information in the Cricket Cercal Sensory System. I. Response Properties of the Primary Interneurons." *Journal of Neurophysiology* 66, no. 5 (1991): 1680–1689. (cited on page 284)

[128] John Morrison. "Perceptual Confidence." *Analytic Philosophy* 57, no. 1 (2016): 15–48. (cited on page 76)

[129] Shane T. Mueller and Christoph T. Weidemann. "Decision Noise: An Explanation for Observed Violations of Signal Detection Theory." *Psychonomic Bulletin and Review* 15, no. 3 (2008): 465–494. (cited on page 98)

[130] Gregory L. Murphy, Stephanie Y. Chen, and Brian H. Ross. "Reasoning with Uncertain Categories." *Thinking and Reasoning* 18, no. 1 (2012): 81–117. (cited on page 185)

[131] In Jae Myung. "Tutorial on Maximum Likelihood Estimation." *Journal of Mathematical Psychology* 47, no. 1 (2003): 90–100. (cited on page 359)

[132] Jiri Najemnik and Wilson S. Geisler. "Optimal Eye Movement Strategies in Visual Search." *Nature* 434, no. 7031 (2005): 387–391. (cited on pages 239, 240)

[133] Marko Nardini, Peter Jones, Rachael Bedford, and Oliver Braddick. "Development of Cue Integration in Human Navigation." *Current Biology* 18, no. 9 (2008): 689–693. (cited on page 307)

[134] Daniel J. Navarro and Amy F. Perfors. "Similarity, Feature Discovery, and the Size Principle." *Acta Psychologica* 133, no. 3 (2010): 256–268. (cited on pages 303, 309)

[135] Andrew Y Ng, Stuart J Russell, et al. "Algorithms for inverse reinforcement learning." *Icml*, Vol. 1. 2000: 2. (cited on page 274)

[136] Loren W. Nolte. "An Adaptive Realization of the Optimum Receiver for a Sporadically Recurrent Waveform in Noise (Corresp.)." *IEEE Transactions on Information Theory* 13, no. 2 (1967): 308–311. DOI: 10.1109/TIT.1967.1053996. (cited on page 239)

[137] Mohamed A. F. Noor, Robin S. Parnell, and Bruce S. Grant. "A Reversible Color Polyphenism in American Peppered Moth (*Biston betularia cognataria*) Caterpillars." *PloS One* 3, no. 9 (2008): e3142. (cited on page 26)

[138] Elyse H. Norton, Luigi Acerbi, Wei Ji Ma, and Michael S. Landy. "Human Online Adaptation to Changes in Prior Probability." *PLoS Computational Biology* 15, no. 7 (2019): e1006681. (cited on pages 253, 254)

[139] Pamela Licalzi O'Connell. "Sweet Slips of the Ear: Mondegreens." *New York Times*, April 9, 1998 . (cited on page 22)

[140] A. Emin Orhan and Wei Ji Ma. "Efficient Probabilistic Inference in Generic Neural Networks Trained with Non-Probabilistic Feedback." *Nature Communications* 8, no. 1 (2017): 1–14. (cited on pages 295, 297, 307)

[141] John Palmer, Preeti Verghese, and Misha Pavel. "The Psychophysics of Visual Search." *Vision Research* 40, no. 10–12 (2000): 1227–1268. (cited on pages 239, 240)

[142] Ryan M. Peters, Phillip Staibano, and Daniel Goldreich. "Tactile Orientation Perception: An Ideal Observer Analysis of Human Psychophysical Performance in Relation to Macaque Area 3b Receptive Fields." *Journal of Neurophysiology* 114, no. 6 (2015): 3076–3096. (cited on page 298)

[143] W. Wesley Peterson, Theodore G. Birdsall, and William C. Fox. "The Theory of Signal Detectability." *Transactions of the IRE Professional Group on Information Theory* 4 (1954): 171–212. (cited on pages 164, 239)

[144] Karin Petrini, Alicia Remark, Louise Smith, and Marko Nardini. "When Vision Is Not an Option: Children's Integration of Auditory and Haptic Information Is Suboptimal." *Developmental Science* 17, no. 3 (2014): 376–387. (cited on page 307)

[145] Jonathan W. Pillow, Jonathon Shlens, Liam Paninski, Alexander Sher, Alan M. Litke, E. J. Chichilnisky, and Eero P. Simoncelli. "Spatio-Temporal Correlations and Visual Signalling in a Complete Neuronal Population." *Nature* 454, no. 7207 (2008): 995–999. (cited on page 298)

[146] Alexandre Pouget, Peter Dayan, and Richard Zemel. "Information Processing with Population Codes." *Nature Reviews Neuroscience* 1, no. 2 (2000): 125–132. (cited on pages 287, 298)

[147] J. R. Pruett Jr., R. J. Sinclair, and H. Burton. "Response Patterns in Second Somatosensory Cortex (SII) of Awake Monkeys to Passively Applied Tactile Gratings." *Journal of Neurophysiology* 84, no. 2 (2000): 780–797. (cited on page 284)

[148] Dale Purves, William T. Wojtach, and R. Beau Lotto. "Understanding Vision in Wholly Empirical Terms." *Proceedings of the National Academy of Sciences* 108, Supplement 3 (2011): 15588–15595. (cited on page 307)

[149] Ahmad T. Qamar, R. James Cotton, Ryan G. George, and Wei Ji Ma. "Trial-to-Trial, Uncertainty-Based Adjustment of Decision Boundaries in Visual Categorization." *Proceedings of the National Academy of Sciences* 110, no. 50 (2013): 20332–20337. (cited on page 185)

[150] Dobromir Rahnev and Rachel N. Denison. "Suboptimality in Perceptual Decision Making." *Behavioral and Brain Sciences* 41 (2018): e223. (cited on page 309)

[151] Rajesh P. N. Rao. "Bayesian Computation in Recurrent Neural Circuits." *Neural Computation* 16, no. 1 (2004): 1–38. (cited on page 295)

[152] Lionel Rigoux, Klaas E. Stephan, Karl J. Friston, and Jean Daunizeau. "Bayesian Model Selection for Group Studies Revisited." *Neuroimage* 84 (2014): 971–985. (cited on page 359)

[153] David L. Roberts and Andrew R. Solow. "Flightless Birds: When Did the Dodo Become Extinct?" *Nature* 426, no. 6964 (2003): 245.. (cited on page 46)

[154] Ruth Rosenholtz. "Visual Search for Orientation among Heterogeneous Distractors: Experimental Results and Implications for Signal-Detection Theory Models of Search." *Journal of Experimental Psychology: Human Perception and Performance* 27, no. 4 (2001): 985–999. (cited on pages 239, 240)

[155] Jason Rosenhouse. *The Monty Hall Problem: The Remarkable Story of Math's Most Contentious Brainteaser*. Oxford: Oxford University Press, 2009. (cited on pages 39, 46, 341)

[156] Jenny R. Saffran, Richard N. Aslin, and Elissa L. Newport. "Statistical Learning by 8-Month-Old Infants." *Science* 274, no. 5294 (1996): 1926–1928. (cited on page 142)

[157] Terence David Sanger. "Probability Density Estimation for the Interpretation of Neural Population Codes." *Journal of Neurophysiology* 76, no. 4 (1996): 2790–2793. (cited on pages 287, 298)

[158] Yoshiyuki Sato, Taro Toyoizumi, and Kazuyuki Aihara. "Bayesian Inference Explains Perception of Unity and Ventriloquism Aftereffect: Identification of Common Sources of Audiovisual Stimuli." *Neural Computation* 19, no. 12 (2007): 3335–3355. (cited on pages 211, 223)

[159] Gideon Schwarz. "Estimating the Dimension of a Model." *Annals of Statistics* 6, no. 2 (1978): 461–464. (cited on pages 351, 359)

[160] Robert Shapley, Michael Hawken, and Dario L. Ringach. "Dynamics of Orientation Selectivity in the Primary Visual Cortex and the Importance of Cortical Inhibition." *Neuron* 38, no. 5 (2003): 689–699. (cited on page 284)

[161] Shan Shen and Wei Ji Ma. "A Detailed Comparison of Optimality and Simplicity in Perceptual Decision Making." *Psychological Review* 123, no. 4 (2016): 452–480. (cited on page 240)

[162] Lei Shi, Thomas L. Griffiths, Naomi H. Feldman, and Adam N. Sanborn. "Exemplar Models as a Mechanism for Performing Bayesian Inference." *Psychonomic Bulletin and Review* 17, no. 4 (2010): 443–464. (cited on page 295)

[163] Chris R. Sims. "The Cost of Misremembering: Inferring the Loss Function in Visual Working Memory." *Journal of Vision* 15, no. 3 (2015) . (cited on page 275)

[164] Devinderjit Sivia and John Skilling. *Data Analysis: A Bayesian Tutorial*. Oxford: Oxford University Press, 2006. (cited on pages 4, 345)

[165] Russell Smith. "Milk Drinkers Turn to Powder and Other Pun-ishing Headlines." *Globe and Mail*, September 23, 2009 . (cited on page 27)

[166] Charles Spence. "Noise and Its Impact on the Perception of Food and Drink." *Flavour* 3, no. 1 (2014): 1–17. (cited on page 106)

[167] Stephen M. Stigler. "Who Discovered Bayes' Theorem?" *American Statistician* 37, no. 4a (1983): 290–296. (cited on page 27)

[168] Alan A. Stocker and Eero P. Simoncelli. "Noise Characteristics and Prior Expectations in Human Visual Speed Perception." *Nature Neuroscience* 9, no. 4 (2006): 578–585. (cited on pages 75, 77, 274)

[169] J. V. Stone, I. S. Kerrigan, and J. Porrill. "Where Is the Light? Bayesian Perceptual Priors for Lighting Direction." *Proceedings of the Royal Society B: Biological Sciences* 276, no. 1663 (2009): 1797–1804. (cited on page 27)

[170] James V. Stone. *Vision and Brain: How We Perceive the World*. Cambridge, MA: MIT Press, 2012. (cited on page 202)

[171] James V. Stone. *Bayes' Rule: A Tutorial Introduction to Bayesian Analysis*. n.p.: Sebtel, 2013. (cited on pages 1, 46)

[172] Lawrence D. Stone, Colleen M. Keller, Thomas M. Kratzke, and Johan P. Strumpfer. "Search Analysis for the Underwater Wreckage of Air France Flight 447." *14th International Conference on Information Fusion*, IEEE (2011), 1–8. (cited on page 275)

[173] Leland S. Stone and Peter Thompson. "Human Speed Perception Is Contrast Dependent." *Vision Research* 32, no. 8 (1992): 1535–1549. (cited on page 75)

[174] Joshua B. Tenenbaum. "Bayesian Modeling of Human Concept Learning". *Advances in Neural Information Processing Systems 11*, edited by Michael S. Kearns, Sara A. Solla and David A. Cohn, 59–68. Cambridge, MA: MIT Press, 1998. (cited on page 142)

[175] Joshua B. Tenenbaum, Charles Kemp, Thomas L. Griffiths, and Noah S. Goodman. "How to Grow a Mind: Statistics, Structure, and Abstraction." *Science* 331, no. 6022 (2011): 1279–1285. (cited on page 307)

[176] Emanuel Todorov. "Optimality Principles in Sensorimotor Control." *Nature Neuroscience* 7, no. 9 (2004): 907–915. (cited on pages 274, 275)

[177] Dave Tompkins. *How to Wreck a Nice Beach: The Vocoder from World War II to Hip-Hop, the Machine Speaks*. Melville House, 2011. (cited on pages 22, 27)

[178] Jonathan Tong, Vy Ngo, and Daniel Goldreich. "Tactile Length Contraction as Bayesian Inference." *Journal of Neurophysiology* 116, no. 2 (2016): 369–379. (cited on page 75)

[179] Michel Treisman. "Motion Sickness: An Evolutionary Hypothesis." *Science* 197, no. 4302 (1977): 493–495. (cited on page 29)

[180] Julia Trommershäuser, Sergei Gepshtein, Laurence T. Maloney, Michael S. Landy, and Martin S. Banks. "Optimal Compensation for Changes in Task-Relevant Movement Variability." *Journal of Neuroscience* 25, no. 31 (2005): 7169–7178. (cited on pages 272, 275)

[181] Julia Trommershäuser, Konrad P. Körding, and Michael S. Landy. *Sensory Cue Integration*. Oxford: Oxford University Press, 2011. (cited on pages 1, 119)

[182] Julia Trommershäuser, Laurence T. Maloney, and Michael S. Landy. "Statistical Decision Theory and the Selection of Rapid, Goal-Directed Movements." *Journal of the Optical Society of America A* 20, no. 7 (2003): 1419–1433. (cited on pages 272, 275, 303)

[183] Julia Trommershäuser, Laurence T. Maloney, and Michael S. Landy. "Decision Making, Movement Planning and Statistical Decision Theory." *Trends in Cognitive Sciences* 12, no. 8 (2008): 291–297. (cited on pages 271, 309)

[184] Robert J. van Beers, Anne C. Sittig, and Jan J. van der Gon Denier. "How Humans Combine Simultaneous Proprioceptive and Visual Position Information." *Experimental Brain Research* 111, no. 2 (1996): 253–261. (cited on pages 118, 119)

[185] Ruben S. van Bergen, Wei Ji Ma, Michael S. Pratte, and Janneke F. M. Jehee. "Sensory Uncertainty Decoded from Visual Cortex Predicts Behavior." *Nature Neuroscience* 18, no. 12 (2015): 1728–1730. (cited on pages 296, 298)

[186] Ronald van den Berg, Michael Vogel, Krešimir Josić, and Wei Ji Ma. "Optimal Inference of Sameness." *Proceedings of the National Academy of Sciences* 109, no. 8 (2012): 3178–3183. (cited on pages 215, 217, 223)

[187] Bas van Opheusden, Luigi Acerbi, and Wei Ji Ma. "Unbiased and Efficient Log-Likelihood Estimation with Inverse Binomial Sampling." *PLoS Computational Biology* 16, no. 12 (2020): e1008483. (cited on pages 104, 359)

[188] Iris Van Rooij. "The Tractable Cognition Thesis." *Cognitive science* 32, no. 6 (2008): 939–984. (cited on page 305)

[189] Meel Velliste, Sagi Perel, M. Chance Spalding, Andrew S. Whitford, and Andrew B. Schwartz. "Cortical Control of a Prosthetic Arm for Self-Feeding." *Nature* 453, no. 7198 (2008): 1098–1101. (cited on page 296)

[190] Hermann von Helmholtz, *Treatise on Physiological Optics. Vol. 3, The Perceptions of Vision*, edited by James P. C. Southall. Rochester, NY: Optical Society of America, 1925. (cited on pages 24, 26)

[191] Scott I. Vrieze. "Model Selection and Psychological Theory: A Discussion of the Differences between the Akaike Information Criterion (AIC) and the Bayesian Information Criterion (BIC)." *Psychological Methods* 17, no. 2 (2012): 228–243. (cited on page 359)

[192] Edward Vul, Noah Goodman, Thomas L. Griffiths, and Joshua B. Tenenbaum. "One and Done? Optimal Decisions from Very Few Samples." *Cognitive Science* 38, no. 4 (2014): 599–637. (cited on pages 306, 309)

[193] Edgar Y. Walker, R. James Cotton, Wei Ji Ma, and Andreas S. Tolias. "A Neural Basis of Probabilistic Computation in Visual Cortex." *Nature Neuroscience* 23, no. 1 (2020): 122–129. (cited on pages 295, 298)

[194] Mark T. Wallace, G. E. Roberson, W. David Hairston, B. E. Stein, J. W. Vaughan, and J. A. Schirillo. "Unifying Multisensory Signals across Time and Space." *Experimental Brain Research* 158, no. 2 (2004): 252–258. (cited on page 107)

[195] Pascal Wallisch. "Illumination Assumptions Account for Individual Differences in the Perceptual Interpretation of a Profoundly Ambiguous Stimulus in the Color Domain. 'The Dress.'" *Journal of Vision* 17, no. 4 (2017): 5. (cited on page 202)

[196] Paul A. Warren, Erich W. Graf, Rebecca A. Champion, and Laurence T. Maloney. "Visual Extrapolation under Risk: Human Observers Estimate and Compensate for Exogenous Uncertainty." *Proceedings of the Royal Society B: Biological Sciences* 279, no. 1736 (2012): 2171–2179. (cited on pages 272, 275)

[197] Kunlin Wei and Konrad P. Körding. "Uncertainty of Feedback and State Estimation Determines the Speed of Motor Adaptation." *Frontiers in Computational Neuroscience* 4 (2010): 11. (cited on page 254)

[198] Xue-Xin Wei and Alan A. Stocker. "A Bayesian Observer Model Constrained by Efficient Coding Can Explain 'Anti-Bayesian' Percepts." *Nature Neuroscience* 18, no. 10 (2015): 1509–1517. (cited on page 77)

[199] Yair Weiss, Eero P. Simoncelli, and Edward H. Adelson. "Motion Illusions as Optimal Percepts." *Nature Neuroscience* 5, no. 6 (2002): 598–604. (cited on pages 75, 77)

[200] Max Wertheimer, "Gestalt theory." *A Source Book of Gestalt Psychology*, edited by Willis D. Ellis, 1–11. London: Kegan Paul, Trench, Trubner, 1938. (cited on pages 42, 46, 223)

[201] Louise Whiteley and Maneesh Sahani. "Implicit Knowledge of Visual Uncertainty Guides Decisions with Asymmetric Outcomes." *Journal of Vision* 8, no. 3 (2008): 1–15. (cited on pages 271, 275, 305, 309)

[202] Frederick A. A. Kingdom and Nicolaas Prins. *Psychophysics: A Practical Introduction*. 2nd ed. London: Academic Press, 2016. (cited on pages 164, 357)

[203] Felix A. Wichmann and N. Jeremy Hill. "The Psychometric Function: I. Fitting, Sampling, and Goodness of Fit." *Perception and Psychophysics* 63, no. 8 (2001): 1293–1313. (cited on page 164)

[204] Thomas D. Wickens. *Elementary Signal Detection Theory*. Oxford: Oxford University Press, 2001. (cited on page 164)

[205] Robert C. Wilson and Anne G. E. Collins. "Ten Simple Rules for the Computational Modeling of Behavioral Data." *Elife* 8 (2019): e49547. (cited on page 359)

[206] Robert C. Wilson, Matthew R. Nassar, and Joshua I. Gold. "Bayesian Online Learning of the Hazard Rate in Change-Point Problems." *Neural Computation* 22, no. 9 (2010): 2452–2476. (cited on pages 253, 254)

[207] Daniel M. Wolpert. "Computational Approaches to Motor Control." *Trends in Cognitive Sciences* 1, no. 6 (1997): 209–216. (cited on page 254)

[208] Sylvia Wright. "The Death of Lady Mondegreen." *Harper's Magazine* 209, no. 1254 (1954): 48–51. (cited on pages 21, 27)

[209] Ting Xiang, Terry Lohrenz, and P. Read Montague. "Computational Substrates of Norms and Their Violations during Social Exchange." *Journal of Neuroscience* 33, no. 3 (2013): 1099–1108. (cited on page 142)

[210] Fei Xu and Joshua B. Tenenbaum. "Word Learning as Bayesian Inference." *Psychological Review* 114, no. 2 (2007): 245–272. (cited on pages 4, 142, 303, 309)

[211] Scott Cheng-Hsin Yang, Mate Lengyel, and Daniel M. Wolpert. "Active Sensing in the Categorization of Visual Patterns." *Elife* 5 (2016): e12215. (cited on pages 239, 240)

[212] Aspen H. Yoo, Zuzanna Klyszejko, Clayton E. Curtis, and Wei Ji Ma. "Strategic Allocation of Working Memory Resource." *Scientific Reports* 8, no. 1 (2018): 1–8. (cited on page 272)

[213] Richard S. Zemel, Peter Dayan, and Alexandre Pouget. "Probabilistic Interpretation of Population Codes." *Neural Computation* 10, no. 2 (1998): 403–430. (cited on page 298)

[214] Yanli Zhou, Luigi Acerbi, and Wei Ji Ma. "The Role of Sensory Uncertainty in Simple Contourition." *PLoS Computational Biology* 16, no. 11 (2020): e1006308. (cited on pages 219, 223)

Index

Note: Photos, Figures and Tables are indicated by italicized page numbers.